T0395433

VOLUME SEVEN HUNDRED AND THIRTEEN

METHODS IN ENZYMOLOGY

Apobec Enzymes

METHODS IN ENZYMOLOGY

Editors-in-Chief

ANNA MARIE PYLE

Departments of Molecular, Cellular and Developmental Biology and Department of Chemistry
Investigator, Howard Hughes Medical Institute
Yale University

DAVID W. CHRISTIANSON

Roy and Diana Vagelos Laboratories
Department of Chemistry
University of Pennsylvania
Philadelphia, PA

Founding Editors

SIDNEY P. COLOWICK and NATHAN O. KAPLAN

METHODS IN ENZYMOLOGY

Apobec Enzymes

Edited by

F. NINA PAPAVASILIOU
Division of Immune Diversity, German Cancer Research Centre (DKFZ), Research Topic Immunology, Infection and Cancer, Heidelberg, Germany

RICCARDO PECORI
Division of Immune Diversity, German Cancer Research Centre (DKFZ), Research Topic Immunology, Infection and Cancer, Heidelberg, Germany

Academic Press is an imprint of Elsevier
50 Hampshire Street, 5th Floor, Cambridge, MA 02139, United States
525 B Street, Suite 1650, San Diego, CA 92101, United States
125 London Wall, London, EC2Y 5AS, United Kingdom

First edition 2025

Copyright © 2025 Elsevier Inc. All rights are reserved, including those for text and data mining, AI training, and similar technologies.

For accessibility purposes, images in electronic versions of this book are accompanied by alt text descriptions provided by Elsevier. For more information, see https://www.elsevier.com/about/accessibility.

Publisher's note: Elsevier takes a neutral position with respect to territorial disputes or jurisdictional claims in its published content, including in maps and institutional affiliations.

No part of this publication may be reproduced or transmitted in any form or by any means, electronic or mechanical, including photocopying, recording, or any information storage and retrieval system, without permission in writing from the publisher. Details on how to seek permission, further information about the Publisher's permissions policies and our arrangements with organizations such as the Copyright Clearance Center and the Copyright Licensing Agency, can be found at our website: www.elsevier.com/permissions.

This book and the individual contributions contained in it are protected under copyright by the Publisher (other than as may be noted herein).

Notices

Knowledge and best practice in this field are constantly changing. As new research and experience broaden our understanding, changes in research methods, professional practices, or medical treatment may become necessary.

Practitioners and researchers must always rely on their own experience and knowledge in evaluating and using any information, methods, compounds, or experiments described herein. In using such information or methods they should be mindful of their own safety and the safety of others, including parties for whom they have a professional responsibility.

To the fullest extent of the law, neither the Publisher nor the authors, contributors, or editors, assume any liability for any injury and/or damage to persons or property as a matter of products liability, negligence or otherwise, or from any use or operation of any methods, products, instructions, or ideas contained in the material herein.

ISBN: 978-0-443-31786-6
ISSN: 0076-6879

For information on all Academic Press publications
visit our website at https://www.elsevier.com/books-and-journals

Publisher: Zoe Kruze
Editorial Project Manager: Saloni Vohra
Production Project Manager: James Selvam
Cover Designer: Bakyalakshmi S

Typeset by MPS Limited, India

Contents

Contributors

Madison B. Adolph
Department of Biochemistry and Molecular Biology, Saint Louis University School of Medicine, St. Louis, MO, United States

Sonali Bhakta
Area of Bioscience and Biotechnology, School of Materials Science, Japan Advanced Institute of Science and Technology, 1–1 Asahidai, Nomicity, Ishikawa 923–1292, Japan; Department of Anatomy and Histology, Bangladesh Agricultural University, Mymensingh 2202, Bangladesh

Alexander J. Brown
School of Molecular Biosciences, Washington State University, Pullman, WA, United States

Rémi Buisson
Department of Biological Chemistry, School of Medicine; Chao Family Comprehensive Cancer Center; Center for Virus Research; Department of Pharmaceutical Sciences, School of Pharmacy & Pharmaceutical Sciences, University of California Irvine, Irvine, California, United States

Linda Chelico
Department of Biochemistry, Microbiology, and Immunology, University of Saskatchewan, Saskatoon, SK, Canada

Xiaojiang S. Chen
Molecular and Computational Biology, Department of Biological Sciences and Chemistry, University of Southern California, Los Angeles, CA, United States

Cameron Cordero
Department of Microbiology and Molecular Genetics, University of Vermont Cancer Center, University of Vermont, Burlington, VT, United States

Margo Coxon
School of Molecular Biosciences, Washington State University, Pullman, WA, United States

Rachel A. DeWeerd
Department of Pediatrics; Center for Genome Integrity, Siteman Cancer Center, Washington University School of Medicine, St. Louis, MO, United States

Michelle Eidelman
Mina and Everard Goodman Faculty of Life Sciences; The Institute of Nanotechnology and Advanced Materials, Bar-Ilan University, Ramat Gan, Israel

Eli Eisenberg
Raymond and Beverly Sackler School of Physics and Astronomy, Tel Aviv University, Tel Aviv, Israel

Yuqing Feng
Department of Biology, York University, Toronto, ON, Canada

Benjamin Fixman
Molecular and Computational Biology, Department of Biological Sciences and Chemistry, University of Southern California, Los Angeles, CA, United States

Adriano Fonzino
Department of Biosciences, Biotechnology and Environment, University of Bari Aldo Moro, Bari BA, Italy

Dmitry A. Gordenin
Genome Integrity & Structural Biology Laboratory, National Institute of Environmental Health Sciences, Durham, NC, United States

Priyanka Govindarajan
Simon Fraser University, Burnaby, BC, Canada

Abby M. Green
Department of Pediatrics; Center for Genome Integrity, Siteman Cancer Center, Washington University School of Medicine, St. Louis, MO, United States

Min Hao
State Key Laboratory of Natural and Biomimetic Drugs, Chemical Biology Center , Department of Molecular and Cellular Pharmacology, Pharmaceutical Sciences, Peking University, Beijing, P.R. China

Victoria Harcy
School of Molecular Biosciences, Washington State University, Pullman, WA, United States

Zachary W. Kockler
Genome Integrity & Structural Biology Laboratory, National Institute of Environmental Health Sciences, Durham, NC, United States

Mani Larijani
Simon Fraser University, Burnaby, BC, Canada

Erez Y. Levanon
Mina and Everard Goodman Faculty of Life Sciences; The Institute of Nanotechnology and Advanced Materials, Bar-Ilan University, Ramat Gan, Israel

Qishan Liang
Department of Cellular and Molecular Medicine; Sanford Stem Cell Institute and Stem Cell Program; Institute for Genomic Medicine; Center for RNA Technologies and Therapeutics, University of California San Diego; Sanford Laboratories for Innovative Medicines, La Jolla, CA, United States

Tao Liu
State Key Laboratory of Natural and Biomimetic Drugs, Chemical Biology Center , Department of Molecular and Cellular Pharmacology, Pharmaceutical Sciences, Peking University, Beijing, P.R. China

Pietro Luca Mazzacuva
Institute of Biomembranes, Bioenergetics and Molecular Biotechnology, National Research Council, Bari; Department of Engineering, University Campus Bio-Medico of Rome, RM, Italy

Tony M. Mertz
Department of Microbiology and Molecular Genetics, University of Vermont Cancer Center, University of Vermont, Burlington, VT, United States

Eszter Németh
Institute of Molecular Life Sciences, HUN-REN Research Centre for Natural Sciences, Budapest, Hungary

Graziano Pesole
Department of Biosciences, Biotechnology and Environment, University of Bari Aldo Moro, Bari BA; Institute of Biomembranes, Bioenergetics and Molecular Biotechnology, National Research Council, Bari, Italy

Ernesto Picardi
Department of Biosciences, Biotechnology and Environment, University of Bari Aldo Moro, Bari BA; Institute of Biomembranes, Bioenergetics and Molecular Biotechnology, National Research Council, Bari, Italy

Atri K. Raval
Department of Microbiology and Molecular Genetics, University of Vermont Cancer Center, University of Vermont, Burlington, VT, United States

Steven A. Roberts
Department of Microbiology and Molecular Genetics, University of Vermont Cancer Center, University of Vermont, Burlington, VT, United States

Ambrocio Sanchez
Department of Biological Chemistry, School of Medicine; Chao Family Comprehensive Cancer Center; Center for Virus Research, University of California Irvine, Irvine, California, United States

Dávid Szüts
Institute of Molecular Life Sciences, HUN-REN Research Centre for Natural Sciences, Budapest, Hungary

Toshifumi Tsukahara
Area of Bioscience and Biotechnology, School of Materials Science, Japan Advanced Institute of Science and Technology, 1–1 Asahidai, Nomicity, Ishikawa 923–1292; GeCoRT Co., Ltd., Kanagawa, 220–0011, Japan

Shanshan Wang
Molecular and Computational Biology, Department of Biological Sciences and Chemistry, University of Southern California, Los Angeles, CA, United States

Shuhao Xu
Department of Cellular and Molecular Medicine; Sanford Stem Cell Institute and Stem Cell Program; Institute for Genomic Medicine; Center for RNA Technologies and Therapeutics, University of California San Diego, La Jolla, CA, United States

Gene W. Yeo
Department of Cellular and Molecular Medicine; Sanford Stem Cell Institute and Stem Cell Program; Institute for Genomic Medicine; Center for RNA Technologies and Therapeutics, University of California San Diego; Sanford Laboratories for Innovative Medicines; Sanford Laboratories for Innovative Medicines, La Jolla, CA, United States

Tao Yu
Department of Cellular and Molecular Medicine; Sanford Stem Cell Institute and Stem Cell Program; Institute for Genomic Medicine; Center for RNA Technologies and Therapeutics, University of California San Diego; Sanford Laboratories for Innovative Medicines, La Jolla, CA, United States

Ying Zeng
Simon Fraser University, Burnaby, BC, Canada

Preface

Apolipoprotein B mRNA-editing enzyme catalytic polypeptide-like (APOBEC) and activation-induced cytidine deaminase (AID) enzymes play critical roles in various biological processes, including adaptive immunity, viral restriction, and genome stability. Over the past two decades, our understanding of these enzymes has expanded significantly, revealing both their physiological importance and their potential as therapeutic targets.

Additionally, while these deaminases have historically been classified as either DNA mutators or RNA editors, some family members are capable of deaminating both DNA and RNA, making it essential to dissect these two similar but distinct activities. Furthermore, taking advantage of their highly efficient and diverse editing activity, many of these enzymes have been employed for targeted genome (or transcriptome) editing technologies or as tools to track RNA-binding protein binding sites or label endogenous proteins. With this growing interest comes the need for robust and reproducible methodologies to study APOBEC and AID enzymes across diverse experimental settings.

This book serves as a comprehensive collection of scientific protocols dedicated to the study of AID/APOBEC enzymes. The chapters included cover a broad spectrum of methodologies, from assays to characterize DNA or RNA editing activity to techniques for monitoring editing events in real-time, identifying inhibitors, and harnessing deaminases for programmable base editing for different purposes.

The compilation of these protocols has been a collaborative effort, and we are deeply grateful to the contributing authors who have generously shared their expertise and time. Their meticulous work ensures that these protocols are not only technically rigorous but also accessible to a broad range of researchers. We believe that this volume will help the AID/APOBEC community to standardize key methodologies and lower the barrier to entry for new investigators seeking to explore this fascinating field.

While it is true that failure and troubleshooting are inseparable aspects of science, our hope is that this book will serve as a valuable resource for both seasoned researchers and newcomers alike. Scientific progress thrives on the sharing of knowledge and the refinement of techniques, and it is our belief that making these methodologies widely available will accelerate discoveries in this dynamic area of research.

Nina F. Papavasiliou
Riccardo Pecori
Heidelberg, Germany

CHAPTER ONE

Fluorescent shift assay for APOBEC-mediated RNA editing

Shanshan Wang, Benjamin Fixman, and Xiaojiang S. Chen*
Molecular and Computational Biology, Department of Biological Sciences and Chemistry, University of Southern California, Los Angeles, CA, United States
*Corresponding author. e-mail address: xiaojiac@usc.edu

Contents

Abstract

Cytidine (C) to Uridine (U) RNA editing is a post-transcriptional modification that is involved in diverse biological processes. The APOBEC deaminase family acts in various cellular processes mostly through inducing C-to-U mutation in single-stranded RNA (or DNA). However, comparing the activity of different RNA editing enzymes to one another is difficult due to the limited number of systems that can provide direct and efficient readout. In this report, a system in which RNA editing directly prompts a change in the subcellular localization of a modified eGFP structure is described in detail. This approach allows us to compare relative fluorescence intensity based on the RNA editing level. When observed through a fluorescence detection system, like a scanning confocal microscope, the cellular nucleus can be readily identified using a DNA-binding stain, such as DAPI or Hoechst, so that the accurate calculation of the ratio of nuclear to cytosolic eGFP intensity can be applied for an individual cell. This method provides a useful and flexible tool to examine and quantify RNA editing activity within cells, and it is not only limited to APOBEC proteins, but can also be applied more generally to other RNA editing enzymatic assays.

Methods in Enzymology, Volume 713
ISSN 0076-6879, https://doi.org/10.1016/bs.mie.2024.12.002
Copyright © 2025 Elsevier Inc. All rights are reserved, including those for text and data mining, AI training, and similar technologies.

1. Introduction

The APOBEC deaminase family plays a crucial role in a diverse array of biological functions by catalyzing the conversion of Cytidine (C) to Uridine (U) on RNA and/or DNA (Chen, Eggerman, & Patterson, 2007; Salter, Bennett, & Smith, 2016; Saraconi, Severi, Sala, Mattiuz, & Conticello, 2014). The first instance of endogenous C-to-U RNA editing was documented in the ApoB mRNA at position 6666, leading to the creation of an early stop codon and the production of two distinct ApoB protein isoforms: ApoB100 and ApoB48, which are involved in lipid and cholesterol metabolism (Blanc, Xie, Luo, Kennedy, & Davidson, 2012; Chen et al., 1987; Driscoll, Wynne, Wallis, & Scott, 1989; Powell et al., 1987). This RNA editing event is observed in the small intestine of humans and in both the small intestine and liver of mice (Teng, Black, & Davidson, 1990). The protein APOBEC1 (A1) is responsible for this ApoB RNA editing activity, and the rest of the APOBEC family protein members were named after this founding member (Lerner, Papavasiliou, & Pecori, 2018). A1 requires a cofactor to form a so-called "editosome" complex to show detectible RNA editing activity in physiological contexts. Multiple cofactors have been identified to function together with A1 to catalyze RNA editing, including A1CF, RBM47, and RBM46 (Blanc et al., 2019; Fossat et al., 2014; Henderson, Blanc, & Davidson, 2001; Hirano et al., 1996; Lellek et al., 2000; Nakamuta, Taniguchi, Ishida, Kobayashi, & Chan, 1998; Teng et al., 1997; Wang et al., 2023).

The human APOBEC proteins include a total of 11 members with different biological roles and various physiological functions (Salter et al., 2016). Originally, only A1 was identified to have single-stranded RNA editing activity (Prohaska, Bennett, Salter, & Smith, 2014). Recently, some studies have reported APOBEC3A (A3A) RNA editing in specific cell types (Kim, Shi, Kelley, & Chen, 2023; Niavarani et al., 2015; Sharma et al., 2015; Sharma, Patnaik, Kemer, & Baysal, 2017), and APOBEC3G (A3G)-mediated RNA activities have also been reported and described in HEK293T cells (Sharma, Patnaik, Taggart, & Baysal, 2016). A3G-mediated RNA editing is also reported to function in response to mitochondrial hypoxic stress in natural killer cells (Sharma et al., 2019). More recently, APOBECs have been found to show RNA editing on RNA genomes of viral pathogens (Kim et al., 2022; Nakata et al., 2023). These findings suggest that many members of the APOBEC family could be involved in RNA editing activity; therefore, additional methods for examining and quantifying RNA editing activities are needed.

Examining RNA editing is difficult when comparing similar editing activities on different RNA substrates by the same enzyme or on the same RNA substrate by different editing enzymes due to the limited number of systems that can effectively provide the RNA editing readout for quantification. This is especially problematic when RNA editing levels are provided by different enzymes on different polynucleotide substrates. A few strategies for characterizing adenosine deamination have been previously described (Gommans, McCane, Nacarelli, & Maas, 2010; Reenan, 2005; Wang, Park, & Beal, 2018), some which have been applied for the characterization of cytidine deamination (Driscoll et al. 1989; Kankowski et al., 2018; Severi & Conticello, 2015), although these can sometimes be limited due to time-consuming techniques, low sensitivity, or imprecise quantification. Building upon these prior techniques resulted in our creation of a system where RNA editing directly prompts a change in the subcellular positioning of a modified eGFP structure. Here, editing of the RNA transcript, results in a shift from nuclear-localized to cytosolic-localized eGFP (Wolfe, Arnold, & Chen, 2019). This approach for comparing relative fluorescence intensity is favored because it should not be affected by different fluorophore expression levels or excitation strength. When observed through a fluorescence imaging system like a scanning confocal microscope, the cellular nucleus can be readily identified using a DNA-binding stain, so that the accurate calculation of the nuclear to cytosolic eGFP ratio, and therefore the C > U RNA editing activity, can be applied for an individual cell (Di Ventura & Kuhlman, 2016; Henderson & Eleftheriou, 2000; Lange et al., 2007). Here, we present a system in which RNA editing triggers a direct shift in the subcellular localization of a modified eGFP between the nucleus and the cytosol. This change enables the comparison of relative fluorescence intensity distribution between the nucleus and cytosol corresponding to varying levels of RNA editing.

2. Key materials

- In-Fusion Snap Assembly Master Mix (TaKaRa, 638948)
- PrimeStar MAX (TaKaRa, R047A)
- HEK293T cell (ATCC)
- pcDNA 3.1 (+) vector
- GeneJET Plasmid MiniPrep kit (ThermoFisher, K0702)

- DMEM (Dulbecco's Modified Eagle's Medium) (Corning, 10-013-CV)
- Fetal Bovine Serum (Sigma-Aldrich, F2442)
- Penicillin-Streptomycin (10,000 U/mL) (Gibco, 15140122)
- Trypsin–EDTA (0.25) (Gibco, 25200072)
- DPBS (Gibco, 14190144)
- Poly-D-lysine (Sigma, A-003-M)
- 8-well glass slides (CellVis, C8-1.5H-N)
- X-tremeGENE 9 transfection reagent (Sigma, XTG9-RO)
- OPTI-mem reduced serum media (ThermoFisher, 31985070)
- Zeiss LSM-700 inverted confocal microscope
- Hoechst 33342 nuclear stain (Invitrogen, H1399)
- Imaging buffer: 140 mM NaCl, 2.5 mM KCl, 1.8 mM $CaCl_2$, 1.0 mM MgCl2, 20 mM Hepes (pH 7.4), 5 mM glucose.

3. Protocol

3.1 Construct design and generation

1. The proposed method contains separate reporter and editor constructs as the core components (Fig. 1A); plasmids encoding these two components will be co-transfected. Here, the assay is designed for A1-mediated RNA editing as previously described (Wolfe et al., 2019). Standard cloning techniques can be used to assemble these constructs.
2. To design the editor construct, A1 and an appropriate co-factor, such as A1CF, are co-expressed in a single reading frame. A fluorescent tag such as mCherry (Shaner et al., 2004) is used to validate expression of the two proteins in a particular cell. These components should be inserted as a single reading frame, separated by self-cutting 2A-peptide sequences (Liu et al., 2017; Ryan, King, & Thomas, 1991) to ensure equal expression of the individual proteins in the same cell.
3. For the reporter construct, the region targeted for editing is placed between eGFP and the MAPKK nuclear-export sequence. A1 and A1CF combine to edit a single cytosine within a minimal 27-base target sequence of the ApoB mRNA (see **Note 1**) (Maris, Masse, Chester, Navaratnam, & Allain, 2005). The length of the RNA tested can be adjusted as needed, typically between 20-200 nt, but can be up to 2 kb. A key consideration is whether the local secondary structure and global tertiary structure of the inserted RNA affect the accessibility of the editing site. Deamination of this cytosine results in the formation of an

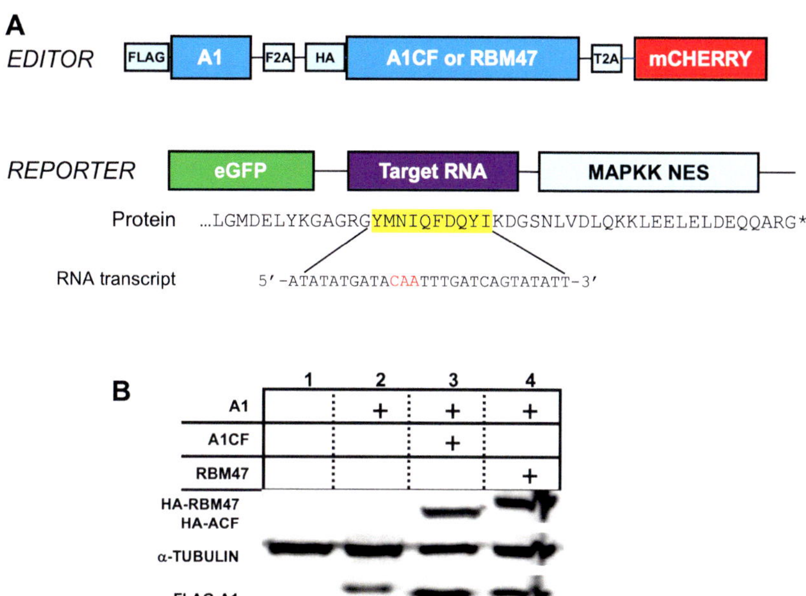

Fig. 1 Design of an assay system for RNA editing through C-to-U deamination by the A1 editosome. (A) Cartoon depiction of the editor and reporter constructs. The editor construct expresses FLAG-A1, an HA-tagged cofactor (HA-A1CF or HA-RBM47), and mCherry as individual proteins, cleaved from a single open reading frame via self-cleaving F2A peptide. For the reporter construct, eGFP is fused to a MAPKK nuclear export sequence (NES) at the C-terminus with a target RNA substrate transcript sequence inserted between eGFP and NES. Shown here is the minimal 27 base ApoB target RNA transcript sequence known to be edited by A1. The length of the RNA tested can be adjusted as needed (see explanation in the protocol). Editing of the target RNA transcript (deamination of the target C) from the reporter by the co-expressed editor A1/cofactor proteins results in an early stop codon before the NES, leading to retention of the NES-less eGFP in the nucleus. mCherry and eGFP were always included so that only cells + for both fluorophores were counted as an observed cell. (B) Characteristic western blot from three independent co-transfections of the eGFP reporter with either mCherry alone (1), FLAG-A1 (2), or FLAG-A1 with either HA-A1CF (3) or HA-RBM47 (4), showing the successful expression of all proteins after self-cleavage in cells. α-Tubulin is the internal control for protein load.

early stop codon, eliminating the translation of the nuclear-export sequence present downstream the target region (see **Note 2**).

4. Individual components can be ordered as double-stranded gene blocks using codon-optimization for human expression. Ensure that different components are ordered with epitopes for quantifying expression via Western blot (an example is shown in Fig. 1A) (see **Note 3**).

5. Insert the components into an expression vector such as a pcDNA3.1(+) vector driven by a strong promoter such as the CMV promoter (see **Note 4**). These two constructs will then be co-expressed. Western blot can be used to verify protein expression (Fig. 1B).

3.2 Cell culture and transfection

1. Human embryonic kidney (HEK293T) cells are cultured in a T75 flask with Dulbecco's Modified Eagle Medium (DMEM) at 37 °C and 5 % CO_2. All cultures are supplemented with 10 % fetal bovine serum (FBS), streptomycin (100 µg/mL) and penicillin (100 U/mL).
2. Once cell cultures reach approximately 80-90 % confluency, take out the medium and wash the cell with 3 mL to 5 mL DPBS, then add 1 mL of 0.25 % trypsin-EDTA.
3. Put the T75 into an incubator for 2 min (See **Note 5**)
4. After 2 min incubation, resuspend the cell with 10 mL DMEM media and put it into a 15 mL centrifuge tube.
5. Centrifuge the cell at 500 g for 5 min
6. Aspirate the supernatant, resuspend the cell pellet with 12 mL DMEM media. Take 1 mL cell suspension, mix it with 11 mL of fresh media, and transfer the mixture to a new a T75 flask. Adjust the dilution factor as needed.

3.3 Transfection

1. Once the HEK293T cell reaches ~90 % confluency, resuspend the cell with DMEM media via trypsin digestion mentioned in the cell culture maintenance part.
2. Dilute the cells to an approximate concentration of 250,000 to 300,000 cells/mL—a relatively low value, in order to ensure a clean monolayer for visualization. (See **Note 6**)
3. Cells (250 µL) were added to each well of the 8-well glass slides (CellVis) previously coated with 0.1 mg/mL poly-D-lysine (Sigma).
4. Poly-D-lysine coat: stock poly-D-lysine is diluted in DPBS (Gibco) to a final concentration 0.1 mg/mL. 250 µL of 0.1 mg/mL poly-D-lysine is added to each well of the 8-well glass slides (CellVis). After 10 min incubation, poly-D-lysine is removed, and washed three times using DPBS. The 8-well glass slides are now ready for use. (see **Note 7**)
5. Within 24 h after seeding the cells, the cells are transfected with X-tremeGENE 9 transfection reagent (Sigma): 50 µL master mixes are made by combining 1 µL of a reporter construct at 50 ng/µL and 5 µL of

an editor construct at 100 ng/μL with 44 μL of OPTImem reduced serum media (ThermoFisher), adding 1.5 μL of reagent. (see **Note 8**)

6. Incubate the master mix at room temperature for 30 min
7. Fifteen microliters of master mix are added dropwise to each specific well, and expression is allowed to occur for 48 h at 37 °C, 5 % CO_2.

3.4 Fluorescence confocal microscopy

This method uses live cell microscopy on a Zeiss LSM-700 inverted confocal microscope. We found that the ideal visualization for analysis is through a 40 × water-immersion objective to maximize the number of easily countable cells per image.

1. Before visualization, aspirate the DMEM medium, and wash cells once with 250 μL of DPBS.
2. Add 200 μL 5-μg/mL solution of Hoechst 33342 nuclear stain diluted in DPBS to each well, and incubate for 15 min
3. After 15 min incubation, aspirate the stain and rinse the cell with DPBS two times.
4. After aspirating the DPBS, add 250 μL imaging buffer [140 mM NaCl, 2.5 mM KCl, 1.8 mM $CaCl_2$, 1.0 mM $MgCl_2$, 20 mM Hepes (pH 7.4), 5 mM glucose]. The cells are ready for imaging now.
5. The imaging process is done at a higher laser intensity (generally around 15–20 %) and lower gain (approximately 500–600 units) to maximize the observed signal-to-noise.
6. The excitation wavelengths for Hoechst 33342, eGFP, and mCherry are 405, 488, and 555 nm, respectively.
7. Set the emission band-pass filters to 400–480, 490– 555, and 555–700 nm, respectively.
8. Images are captured as multichannel 16-bit grayscale intensity images 1012 × 1012 pixels across, using two-pass line averaging for smoothing and a pixel dwell time of 0.80 μs.
9. For each well, approximately three to five images are captured, allowing for measurement of around 20–30 cells. The initial fluorescent image of the editor and reporter is summarized in Fig. 2A.

3.5 Analysis of fluorescence images

All image analyses are done using the LSM Toolbox plugin built into the FIJI distribution of ImageJ2 (Rueden et al., 2017; Schindelin et al., 2012).

1. Open the output LSM image file in single-channel color mode and use the Hoechst stain channel to assess the location of the nucleus for each cell. Use

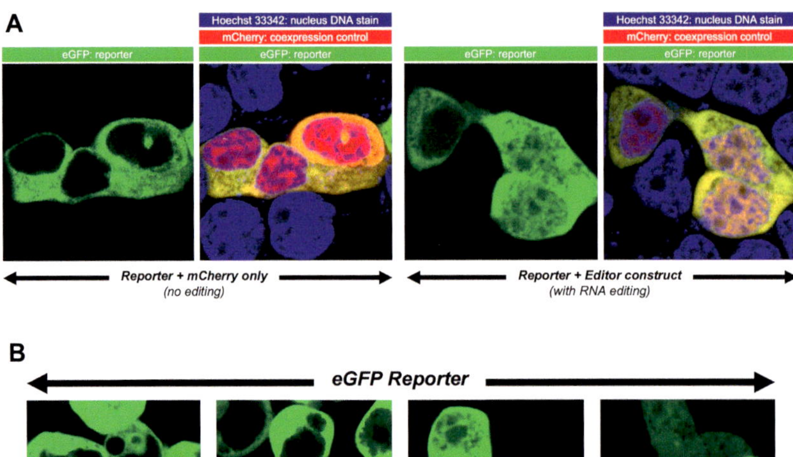

Fig. 2 Visualization of the RNA editing via fluorescent shift assay. (A) Initial test results of the assay system, showing the live-cell eGFP fluorescence images for either an unedited ApoB reporter construct in the absence of the editor (left) or a condition where editing has occurred in the presence of the editor (right). Presence of both eGFP and mCherry fluorescence in the same cells represents co-expression of proteins from both editor and report constructs co-expression. The nuclear periphery is delineated by Hoechst 33342 staining to enable precise selection of nuclear and cytoplasmic regions for fluorescence quantification in cells where the fluorescence intensity values are similar between these compartments. RNA editing shows a clear shift in the nuclear localization of eGFP fluorescence, which can then be quantified through ImageJ or comparable software. (B) Live-cell fluorescence images showing that RNA editing can be visualized directly by a clear shift in subcellular localization; RNA editing only occurred when A1 is co-expressed with A1CF or RBM47 cofactors. Shown are representative eGFP fluorescence images of the APOB reporter co-transfected with either mCherry alone, A1 alone, A1 + A1CF, or A1 + RBM47. The Hoechst 33342 and mCherry signals have been excluded for clarity.

the mCherry to confirm that a cell has both the eGFP reporter and mCherry-containing editor construct. RNA editing activity on ApoB RNA can be visualized directly by a clear shift in subcellular localization (Fig. 2).

2. Use freehand selection to outline the nuclear and the cytoplasmic regions within the eGFP channel and record the average intensity of each region.
3. At least 21 cells, but ideally 30, are needed to do the calculation with very little variation in the sample size.

4. Calculate the ratio of average nuclear to cytosolic intensity for each cell. These values are assessed in Graphpad Prism 7.0d using a one-way ANOVA analysis. If, instead, only one comparison is being made use a Student's *t*-test
5. The data are assumed to have a Gaussian distribution and a Tukey post-test is done to compare the means within experiments.
6. The calculated P values from the individual significance tests are accordingly adjusted for the inherent multiplicities, and conclusions are drawn based on the observed levels of significance, with all data displayed as box-and-whisker plots showing the median and upper/lower quartiles within the box and whiskers extended to the minimum and maximum for each sample. The initial assessment result of the editor and reporter system is summarized in Fig. 3A.

4. Discussion and conclusion

The fluorescent shift assay was initially developed and utilized for comparison of RNA editing activity of A1-A1CF and A1-RBM47 complexes in HEK293T cells, and the editing results are summarized in Fig. 3B (Wolfe et al., 2019). This detailed protocol utilized RNA editing examination by A1-A1CF and A1-RBM47 as an example. Moreover, due to its high flexibility, this fluorescent shift assay was successfully applied to other APOBEC proteins and different RNA substrates (Kim et al., 2022; Kim et al., 2023; Wang et al., 2023). To adapt this fluorescent assay, modifications can be made to the construct design and generation process. For instance, in the study examining A3A-mediated RNA editing activities (Kim et al., 2023), the A1-A1CF gene fragment in the editor vector was replaced with A3A using standard cloning techniques. Additionally, the target sequence of ApoB mRNA was substituted by the sequence of an A3A substrate. This assay can potentially be adapted for other RNA deaminases and targets, provided that the appropriate editor and reporter vectors are constructed. For instance, to examine the ability of an RNA deaminase to edit viral genomic RNA or mRNA, the transcript sequence(s) of interest can be inserted into the reporter vector, and relative editing can be measured. The only requirement is that a Glutamine (CAA, CAG) or Arginine (CGA) residue be present so that upon cytosine deamination, a stop codon is introduced, and the NES is not translated. It is important to note that if the CAA/CAG/CGA

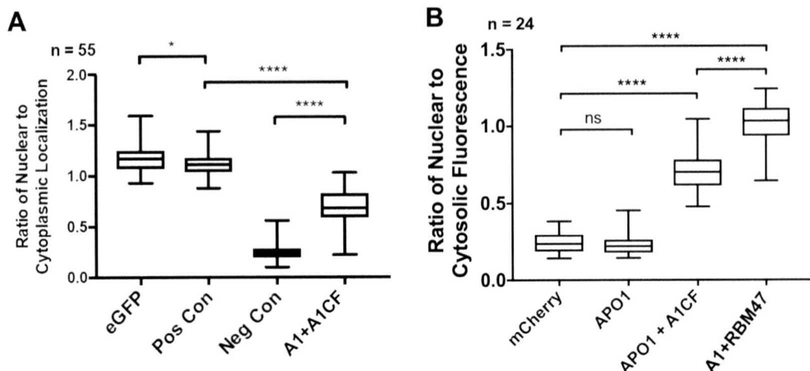

Fig. 3 Characterization of the editing activity on APOB RNA by A1 paired with either cofactor A1CF or RBM47. (A) An initial assessment of the reporter and editor system to confirm the relative sensitivity of the assay. The fluorescence localization ratios of wild-type eGFP were compared to the reporter co-expressed with mCherry alone but with the target cytosine mutated to thymine (positive control, or Pos Con), co-expressed with mCherry alone (negative control, or Neg Con), and coexpressed with the editor construct (A1 +A1CF). These respectively acted as a positive control to show a maximum baseline for the observed localization ratio, a negative control to give a minimum baseline for the localization ratio, and a test condition to show the editor results in a localization change that is statistically different from both controls. The positive control was also found to be significantly different from eGFP alone (multiplicity-adjusted P = 0.0308) but this was not considered a detrimental characteristic of the tag. (B) Quantification of the fluorescent localization changes (expressed as changes in the ratio of nuclear eGFP fluorescence divided by cytosolic), with indication of statistical significance with the co-expression of A1 with either A1CF or RBM47. Displayed is a box-and-whisker plot comparing the mean values of the ratio of nuclear to cytosolic fluorescence for the same transfections described in panel A; analysis is via ANOVA with Tukey post-test for significance, with reported significance levels shown from multiplicity-adjusted P values (n.s. = not significant when P N 0.05 and ****P b 0.0001, n = number of cells used in analysis).

trinucleotide is out of frame, one or two nucleotides can be introduced at the 5′ end of the transcript, without affecting the remainder of the target RNA sequence or structure.

Bringing further attention to RNA editing, numerous groups are racing to design programmable RNA editors for therapeutic use (Booth et al., 2023; Song, Zhuang, & Yi, 2024). In addition to monitoring RNA editing activities of endogenous proteins, this fluorescent shift assay can quickly be used to estimate the specific on- and off-target editing rate of novel RNA base-editors. Moreso, it is conceivable that, as a mechanism to measure ongoing RNA editing activity of different endogenous proteins, multiple reporter vectors, with different fluorescent proteins and target sequences optimized for different endogenous editors, could be transfected into various cell types too allow one to distinguish the operative effect of each endogenous RNA editor. All in all, this fluorescent shift assay provides an efficient and reliable way to evaluate and quantify RNA editing activity on certain RNA substrates by various C > U RNA editors in a cell-based experimental setting.

5. Notes

1. The ApoB mRNA editing site is known to contain a hairpin structure that is necessary for editing to occur. Design of other editing constructs should use as much of the transcript as is necessary to include predicted secondary structure.
2. Different RNA sequences can be inserted for testing, but creation of a stop codon is necessary for this assay.
3. As illustrated in Fig. 1, A1 is tagged with an N-terminal FLAG tag, while A1CF/RBM47 is tagged with an N-terminal HA tag to enable expression verification via Western blot analysis.
4. We observed difficulty cloning the editor construct within E. coli, possibly due to leaky expression by the CMV promoter being toxic. If observed, use a known intron containing a stop codon or frame shift to prevent expression in bacterial systems.
5. Incubation time can be different based on different cell lines, but it is better to use a short trypsin digestion time to keep the cells in good condition.
6. The initial cell amount seeded should be optimized according to the specific transfection kit. Different transfection reagents have different toxicity to various cell types, and the goal is to have a monolayer of cells when visualized.
7. Three times wash is important because the remaining poly-D-lysine can cause some level of toxicity to the cells.
8. Different transfection reagents can be used by following the respective manufacturer protocol

Acknowledgements

This work was supported by NIH grant R01AI150524 to X.S.C.

References

Blanc, V., Xie, Y., Kennedy, S., Riordan, J. D., Rubin, D. C., Madison, B. B., ... Davidson, N. O. (2019). Apobec1 complementation factor (A1CF) and RBM47 interact in tissue-specific regulation of C-to-U RNA editing in mouse intestine and liver. *RNA (New York, N. Y.), 25*(1), 70–81. https://doi.org/10.1261/rna.068395.118.

Blanc, V., Xie, Y., Luo, J., Kennedy, S., & Davidson, N. O. (2012). Intestine-specific expression of Apobec-1 rescues apolipoprotein B RNA editing and alters chylomicron production in Apobec1 −/− mice. *Journal of Lipid Research, 53*(12), 2643–2655. https://doi.org/10.1194/jlr.M030494.

Booth, B. J., Nourreddine, S., Katrekar, D., Savva, Y., Bose, D., Long, T. J., ... Mali, P. (2023). RNA editing: Expanding the potential of RNA therapeutics. *Molecular Therapy: The Journal of the American Society of Gene Therapy, 31*(6), 1533–1549. https://doi.org/10.1016/j.ymthe.2023.01.005.

Chen, S. H., Habib, G., Yang, C. Y., Gu, Z. W., Lee, B. R., Weng, S. A., ... Rosseneu, M. (1987). Apolipoprotein B-48 is the product of a messenger RNA with an organ-specific in-frame stop codon. *Science (New York, N. Y.), 238*(4825), 363–366. https://doi.org/10.1126/science.3659919.

Chen, Z., Eggerman, T. L., & Patterson, A. P. (2007). ApoB mRNA editing is mediated by a coordinated modulation of multiple apoB mRNA editing enzyme components. *American Journal of Physiology. Gastrointestinal and Liver Physiology, 292*(1), G53–G65. https://doi.org/10.1152/ajpgi.00118.2006.

Di Ventura, B., & Kuhlman, B. (2016). Go in! Go out! Inducible control of nuclear localization. *Current Opinion in Chemical Biology, 34*, 62–71.

Driscoll, D. M., Wynne, J. K., Wallis, S. C., & Scott, J. (1989). An in vitro system for the editing of apolipoprotein B mRNA. *Cell, 58*(3), 519–525. https://doi.org/10.1016/0092-8674(89)90432-7.

Fossat, N., Tourle, K., Radziewic, T., Barratt, K., Liebhold, D., Studdert, J. B., ... Tam, P. P. (2014). C-to-U RNA editing mediated by APOBEC1 requires RNA-binding protein RBM47. *EMBO Reports, 15*(8), 903–910. https://doi.org/10.15252/embr.201438450.

Gommans, W. M., McCane, J., Nacarelli, G. S., & Maas, S. (2010). A mammalian reporter system for fast and quantitative detection of intracellular A-to-I RNA editing levels. *Analytical Biochemistry, 399*(2), 230–236. https://doi.org/10.1016/j.ab.2009.12.037.

Henderson, B. R., & Eleftheriou, A. (2000). A comparison of the activity, sequence specificity, and CRM1-dependence of different nuclear export signals. *Experimental Cell Research, 256*(1), 213–224. https://doi.org/10.1006/excr.2000.4825.

Henderson, J. O., Blanc, V., & Davidson, N. O. (2001). Isolation, characterization and developmental regulation of the human apobec-1 complementation factor (ACF) gene. *Biochimica et Biophysica Acta, 1522*(1), 22–30. https://doi.org/10.1016/s0167-4781(01)00295-0.

Hirano, K., Young, S. G., Farese, R. V., Jr., Ng, J., Sande, E., Warburton, C., ... Davidson, N. O. (1996). Targeted disruption of the mouse apobec-1 gene abolishes apolipoprotein B mRNA editing and eliminates apolipoprotein B48. *The Journal of Biological Chemistry, 271*(17), 9887–9890. https://doi.org/10.1074/jbc.271.17.9887.

Kankowski, S., Förstera, B., Winkelmann, A., Knauff, P., Wanker, E. E., You, X. A., ... Meier, J. C. (2018). A novel RNA editing sensor tool and a specific agonist determine neuronal protein expression of RNA-edited glycine receptors and identify a genomic APOBEC1 dimorphism as a new genetic risk factor of epilepsy. *Frontiers in Molecular Neuroscience, 10*, 439.

Kim, K., Calabrese, P., Wang, S., Qin, C., Rao, Y., Feng, P., & Chen, X. S. (2022). The roles of APOBEC-mediated RNA editing in SARS-CoV-2 mutations, replication and fitness. *Research Square*. https://doi.org/10.21203/rs.3.rs-1524060/v1.

Kim, K., Shi, A. B., Kelley, K., & Chen, X. S. (2023). Unraveling the enzyme-substrate properties for APOBEC3A-mediated RNA editing. *Journal of Molecular Biology, 435*(17), 168198. https://doi.org/10.1016/j.jmb.2023.168198.

Lange, A., Mills, R. E., Lange, C. J., Stewart, M., Devine, S. E., & Corbett, A. H. (2007). Classical nuclear localization signals: Definition, function, and interaction with importin alpha. *The Journal of Biological Chemistry, 282*(8), 5101–5105. https://doi.org/10.1074/jbc.R600026200.

Lellek, H., Kirsten, R., Diehl, I., Apostel, F., Buck, F., & Greeve, J. (2000). Purification and molecular cloning of a novel essential component of the apolipoprotein B mRNA editing enzyme-complex. *The Journal of Biological Chemistry, 275*(26), 19848–19856. https://doi.org/10.1074/jbc.M001786200.

Lerner, T., Papavasiliou, F. N., & Pecori, R. (2018). RNA editors, cofactors, and mRNA targets: An overview of the C-to-U RNA editing machinery and its implication in human disease. *Genes (Basel), 10*(1), https://doi.org/10.3390/genes10010013.

Liu, Z., Chen, O., Wall, J. B. J., Zheng, M., Zhou, Y., Wang, L., ... Liu, J. (2017). Systematic comparison of 2A peptides for cloning multi-genes in a polycistronic vector. *Scientific Reports, 7*(1), 2193. https://doi.org/10.1038/s41598-017-02460-2.

Maris, C., Masse, J., Chester, A., Navaratnam, N., & Allain, F. H. (2005). NMR structure of the apoB mRNA stem-loop and its interaction with the C-to-U editing APOBEC1 complementary factor. *RNA (New York, N. Y.), 11*(2), 173–186. https://doi.org/10.1261/rna.7190705.

Nakamuta, M., Taniguchi, S., Ishida, B. Y., Kobayashi, K., & Chan, L. (1998). Phenotype interaction of apobec-1 and CETP, LDLR, and ApoE gene expression in mice. *Arteriosclerosis, Thrombosis, and Vascular Biology, 18*(5), 747–755. https://doi.org/10.1161/01.ATV.18.5.747.

Nakata, Y., Ode, H., Kubota, M., Kasahara, T., Matsuoka, K., Sugimoto, A., ... Iwatani, Y. (2023). Cellular APOBEC3A deaminase drives mutations in the SARS-CoV-2 genome. *Nucleic Acids Research, 51*(2), 783–795. https://doi.org/10.1093/nar/gkac1238.

Niavarani, A., Currie, E., Reyal, Y., Anjos-Afonso, F., Horswell, S., Griessinger, E., ... Bonnet, D. (2015). APOBEC3A is implicated in a novel class of G-to-A mRNA editing in WT1 transcripts. *PLoS One, 10*(3), e0120089.

Powell, L. M., Wallis, S. C., Pease, R. J., Edwards, Y. H., Knott, T. J., & Scott, J. (1987). A novel form of tissue-specific RNA processing produces apolipoprotein-B48 in intestine. *Cell, 50*(6), 831–840. https://doi.org/10.1016/0092-8674(87)90510-1.

Prohaska, K. M., Bennett, R. P., Salter, J. D., & Smith, H. C. (2014). The multifaceted roles of RNA binding in APOBEC cytidine deaminase functions. *Wiley Interdisciplinary Reviews RNA, 5*(4), 493–508. https://doi.org/10.1002/wrna.1226.

Reenan, R. A. (2005). Molecular determinants and guided evolution of species-specific RNA editing. *Nature, 434*(7031), 409–413. https://doi.org/10.1038/nature03364.

Rueden, C. T., Schindelin, J., Hiner, M. C., DeZonia, B. E., Walter, A. E., Arena, E. T., & Eliceiri, K. W. (2017). ImageJ2: ImageJ for the next generation of scientific image data. *BMC Bioinformatics, 18*(1), 529. https://doi.org/10.1186/s12859-017-1934-z.

Ryan, M. D., King, A. M., & Thomas, G. P. (1991). Cleavage of foot-and-mouth disease virus polyprotein is mediated by residues located within a 19 amino acid sequence. *Journal of General Virology, 72*(11), 2727–2732.

Salter, J. D., Bennett, R. P., & Smith, H. C. (2016). The APOBEC protein family: United by structure, divergent in function. *Trends in Biochemical Sciences, 41*(7), 578–594. https://doi.org/10.1016/j.tibs.2016.05.001.

Saraconi, G., Severi, F., Sala, C., Mattiuz, G., & Conticello, S. G. (2014). The RNA editing enzyme APOBEC1 induces somatic mutations and a compatible mutational signature is present in esophageal adenocarcinomas. *Genome Biology, 15*(7), 1–10.

Schindelin, J., Arganda-Carreras, I., Frise, E., Kaynig, V., Longair, M., Pietzsch, T., ... Cardona, A. (2012). Fiji: An open-source platform for biological-image analysis. *Nature Methods, 9*(7), 676–682. https://doi.org/10.1038/nmeth.2019.

Severi, F., & Conticello, S. G. (2015). Flow-cytometric visualization of C > U mRNA editing reveals the dynamics of the process in live cells. *RNA Biology, 12*(4), 389–397.

Shaner, N. C., Campbell, R. E., Steinbach, P. A., Giepmans, B. N., Palmer, A. E., & Tsien, R. Y. (2004). Improved monomeric red, orange and yellow fluorescent proteins derived from Discosoma sp. red fluorescent protein. *Nature Biotechnology, 22*(12), 1567–1572. https://doi.org/10.1038/nbt1037.

Sharma, S., Patnaik, S. K., Kemer, Z., & Baysal, B. E. (2017). Transient overexpression of exogenous APOBEC3A causes C-to-U RNA editing of thousands of genes. *RNA Biology, 14*(5), 603–610. https://doi.org/10.1080/15476286.2016.1184387.

Sharma, S., Patnaik, S. K., Taggart, R. T., & Baysal, B. E. (2016). The double-domain cytidine deaminase APOBEC3G is a cellular site-specific RNA editing enzyme. *Scientific Reports, 6*, 39100. https://doi.org/10.1038/srep39100.

Sharma, S., Patnaik, S. K., Taggart, R. T., Kannisto, E. D., Enriquez, S. M., Gollnick, P., & Baysal, B. E. (2015). APOBEC3A cytidine deaminase induces RNA editing in monocytes and macrophages. *Nature Communications, 6*, 6881. https://doi.org/10.1038/ncomms7881.

Sharma, S., Wang, J., Alqassim, E., Portwood, S., Cortes Gomez, E., Maguire, O., ... Baysal, B. E. (2019). Mitochondrial hypoxic stress induces widespread RNA editing by APOBEC3G in natural killer cells. *Genome Biology, 20*(1), 37. https://doi.org/10.1186/s13059-019-1651-1.

Song, J., Zhuang, Y., & Yi, C. (2024). Programmable RNA base editing via targeted modifications. *Nature Chemical Biology, 20*(3), 277–290.

Teng, B., Black, D. D., & Davidson, N. O. (1990). Apolipoprotein B messenger RNA editing is developmentally regulated in pig small intestine: Nucleotide comparison of apolipoprotein B editing regions in five species. *Biochemical and Biophysical Research Communications, 173*(1), 74–80. https://doi.org/10.1016/s0006-291x(05)81023-x.

Teng, B., Ishida, B., Forte, T. M., Blumenthal, S., Song, L. Z., Gotto, A. M., Jr., & Chan, L. (1997). Effective lowering of plasma, LDL, and esterified cholesterol in LDL receptor-knockout mice by adenovirus-mediated gene delivery of ApoB mRNA editing enzyme (Apobec1). *Arteriosclerosis, Thrombosis, and Vascular Biology, 17*(5), 889–897. https://doi.org/10.1161/01.atv.17.5.889.

Wang, S., Kim, K., Gelvez, N., Chung, C., Gout, J. F., Fixman, B., ... Chen, X. S. (2023). Identification of RBM46 as a novel APOBEC1 cofactor for C-to-U RNA-editing activity. *Journal of Molecular Biology, 435*(24), https://doi.org/10.1016/j.jmb.2023.168333.

Wang, Y., Park, S., & Beal, P. A. (2018). Selective recognition of RNA substrates by ADAR deaminase domains. *Biochemistry, 57*(10), 1640–1651. https://doi.org/10.1021/acs.biochem.7b01100.

Wolfe, A. D., Arnold, D. B., & Chen, X. S. (2019). Comparison of RNA editing activity of APOBEC1-A1CF and APOBEC1-RBM47 complexes reconstituted in HEK293T cells. *Journal of Molecular Biology, 431*(7), 1506–1517. https://doi.org/10.1016/j.jmb.2019.02.025.

CHAPTER TWO

Low-error RNA sequencing techniques for detecting RNA editing by APOBECs: Circular RNAseq assay and safe-sequencing system (SSS)

Shanshan Wang, Benjamin Fixman, and Xiaojiang S. Chen*
Molecular and Computational Biology, Department of Biological Sciences and Chemistry, University of Southern California, Los Angeles, CA, United States
*Corresponding author. e-mail address: xiaojiac@usc.edu

Contents

Methods in Enzymology, Volume 713
ISSN 0076-6879, https://doi.org/10.1016/bs.mie.2024.12.004
Copyright © 2025 Elsevier Inc. All rights are reserved, including those for text and data mining, AI training, and similar technologies.

Abstract

Cytidine-to-Uridine (C-to-U) RNA editing is a post-transcriptional modification essential for various biological processes. APOBEC deaminases mediate C-to-U editing which play critical role in cellular function and regulation. Advances in next-generation sequencing (NGS) technologies and analytical tools have provided powerful means to assess RNA editing activities and their physiological implications. However, inherent errors in NGS workflows—including reverse transcription, PCR amplification, and sequencing—complicate the detection of actual editing events. With error rates ranging from 10^{-2} to 10^{-3} per nucleotide, these technical artifacts can obscure APOBEC-mediated editing events occurring at similar frequencies. To address these challenges, in this chapter, we describe two established and optimized RNA sequencing strategies explicitly designed to detect low-frequency RNA editing events accurately while distinguishing them from NGS-associated errors. These methods are termed "circular RNA Sequencing Assay" and "Safe-Sequencing System (SSS)" and enable the reliable identification of RNA editing events (and also somatic mutations) at or below typical error thresholds.

1. Introduction

Over nearly two decades, RNA sequencing has significantly advanced, becoming a powerful tool for studying gene expression, RNA editing, and transcriptional errors (Stark et al., 2019). This technology enables the analysis of RNA editing activity by enzymes such as APOBEC proteins and their physiological functions. However, these studies are often affected by technical errors introduced during library preparation, including reverse transcription, PCR amplification, and errors during the sequencing process itself (Gout et al., 2013).

In addition, the transcriptome, encompassing the complete set of transcripts in a cell, is fundamental for deciphering the functional elements of the genome. While accurate transcription is vital, the process is inherently prone to errors. Transcription mutagenesis, though transient due to the short half-life of RNA molecules (typically on the scale of hours), can have significant consequences for cellular function and human health (Eisen et al., 2020). The rapid turnover of RNA transcripts, thus, complicates the detection of rare mutations arising from transcription errors. Additionally, the accurate quantification of such rare mutations using high-throughput NGS is hindered by technical error rates of approximately 10^{-2} to 10^{-3} per nucleotide, stemming from errors during library preparation and sequencing processes (Eboreime et al., 2016).

To overcome these challenges, two optimized sequencing strategies, the circular RNA sequencing assay and the bar-coded safe-sequencing system (SSS), are described in detail below. These methods enable the differentiation of true low-frequency somatic mutations or RNA editing events from technical errors introduced during library construction, providing robust tools for studying RNA editing and transcriptional fidelity.

2. Circular-sequencing assay

Reverse transcriptase commits approximately one error every ~20,000 bases, while RNA polymerases (RNAPs) are expected to make only one error every 300,000 bases (Gout et al., 2017; Gout et al., 2013). Because the error rate of reverse transcription alone dwarfs the error rate of RNA polymerases inside the cell, it is virtually impossible to distinguish true base changes from technical errors caused by reverse transcription during library preparation in traditional RNA-Seq data (Fig. 1A). To solve this problem, an optimized version of the Circle-sequencing assay was developed by the Vermulst lab at the University of Southern California (Acevedo & Andino, 2014; Gout et al., 2017). This assay allows the user to detect transcription errors and other rare variants in RNA throughout the transcriptome (Gout et al., 2017). Therefore, this assay can be applied to detect low-frequency C-to-U RNA edits by APOBECs. Novel RNA targets by APOBEC1 (A1) and its newly discovered cofactor-RBM46 have been identified and reported using Circle-sequencing assay (Wang et al., 2023).

Compared to standard RNA sequencing library preparation, the key step in the Circle-Sequencing assay is RNA circularization. After RNA fragmentation, the target RNA is circularized and then reverse transcribed in a rolling-circle fashion, resulting in linear cDNA molecules containing multiple tandem repeats of the same RNA template. If a technical error or a C-to-U edit is present in the original RNA template, it will appear in every single repeat within the cDNA molecule (Fig. 1A. Conversely, technical errors introduced during reverse transcription, PCR amplification, or sequencing are random and typically occur in only one or a few repeats in different positions. This distinction allows random errors introduced during library construction to be separated from technical errors or RNA editing events (Fig. 1B).

The detailed protocol of a circle-sequencing assay is summarized in the paper published by Vermulst lab (Fritsch et al., 2018). Here we will summarize the method that is applied for detecting novel RNA editing targets by APOBEC1 and RBM46 (Wang et al., 2023).

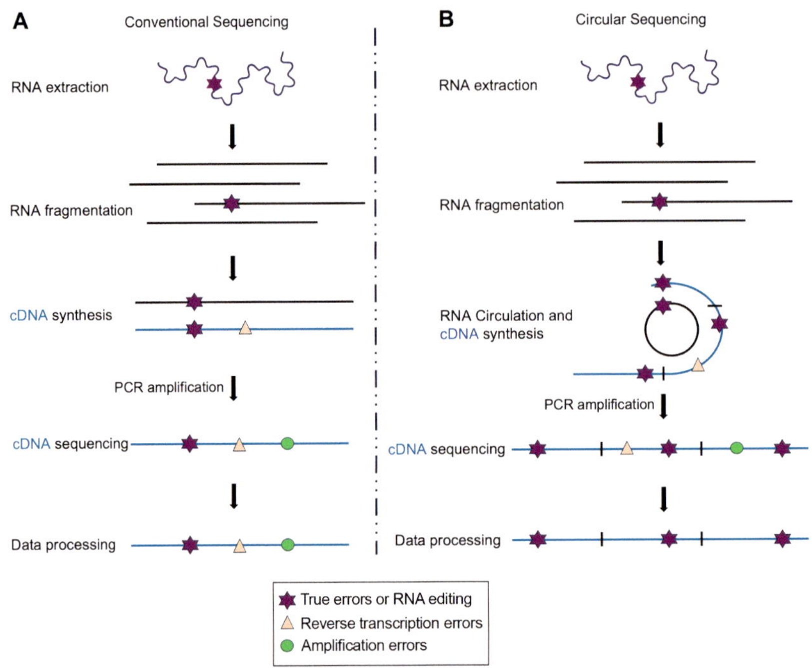

Fig. 1 Schematic representation of RNA-seq versus Circular RNAseq. (A) Traditional RNA-seq experiments isolate RNA from a sample of interest, fragment the RNA, and reverse transcribe it before the final library preparation and sequencing. These preparation procedures introduce numerous technical artifacts into the library in the form of reverse transcription errors, PCR amplification errors, and sequencing errors. (B) This optimized Circular RNAseq Assay allows for the correction of these technical artifacts by circularizing the fragmented RNA molecules before reverse transcription, which allows them to be reverse transcribed in a rolling circle fashion to produce linear cDNA molecules that contain several copies of the original RNA template in tandem repeat. These tandem repeats can then be used to distinguish true variants (transcriptional errors or changes due to RNA editing) from artifacts, as true errors or RNA editing will be present in all repeats at the same location, whereas artifacts such as reverse transcription errors and PCR amplification errors are only present in one or two repeats of any given cDNA molecule.

3. Materials and reagents

3.1 Cell culture reagent

- DMEM (Dulbecco's Modified Eagle's Medium) (Corning, 10-013-CV)
- Fetal Bovine Serum (Sigma-Aldrich, F2442)
- Penicillin-Streptomycin (10,000 U/mL) (Gibco, 15140122)

- Trypsin–EDTA (0.25) (Gibco, 25200072)
- DPBS (Gibco, 14190144)
- 10 cm plates (VWR, 10062-890)
- Lipofectamine 3000 Transfection Reagent (Thermo Fisher, L3000015)

3.2 Library preparation and sequencing reagents

- RiboPure™ RNA Purification Kit (Invitrogen, AM1924)
- NEBNext RNase III RNA Fragmentation Module (NEB, E6146S)
- Oligo Clean & Concentrator kit (Zymo Research, D4061)
- NEBNext Ultra RNA Library Prep Kit for Illumina (E7530L)
- NEBNext Multiplex Oligos for Illumina (E7335S, E7500S)
- NEBNext® Ultra™ II Non-Directional RNA Second Strand Synthesis Module (NEB, E6111S)
- AMPure XP beads (Beckman Coulter, A63880)
- MiSeq system (illumina)

4. Methods

4.1 The cell-based RNA editing system

Before applying the circular RNAseq assay, Sanger sequencing can be employed to quickly check the editing activities. The cell-based RNA editing system is adapted from a previously reported study (Wolfe et al., 2019) and described in the Chapter 1. Editor vectors, containing A1 and cofactors (A1CF, RBM47, RBM46, SYNCRIP), are constructed as one open reading frame (ORF) with the self-cleaving peptide T2A inserted between A1 and the cofactor, resulting in individual A1 and cofactor proteins in a 1:1 ratio (Liu et al., 2017; Wolfe et al., 2019). Reporter vectors contain an eGFP gene along with a DNA fragment insert encoding a target RNA. Once transfected inside the nucleus, the DNA is transcribed into a large RNA precursor. Following splicing, a mature mRNA is produced that encodes eGFP and includes the RNA substrate insert with the target C site for potential editing (Fig. 2).

4.2 Cell culture maintenance

1. Human embryonic kidney (HEK293T) cells are cultured in a T75 flask with Dulbecco's Modified Eagle Medium (DMEM) at 37 °C and 5 % CO_2. All cultures are supplemented with 10 % fetal bovine serum (FBS), streptomycin (100 μg/mL) and penicillin (100 U/mL).

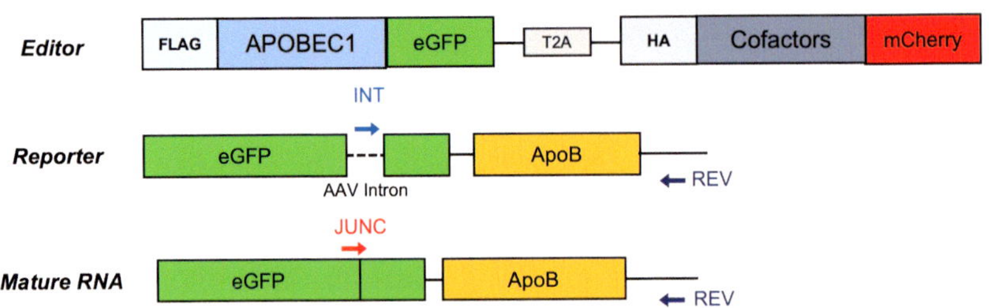

Fig. 2 Cartoon representation of reporter and editor constructs utilized in a cell-based RNA editing assay. The editor construct comprises FLAG-A1 and cofactors (HA-A1CF, HA-RBM46, HA-RBM47, or HA-SYNCRIP), alongside mCherry, expressed as a singular, cleaved protein from a single open reading frame via self-cleaving T2A peptides. The reporter construct includes an AAV intron (indicated by a blue arrow) inserted within the coding region of the eGFP fusion protein. This design prevents sequencing of the plasmid DNA harboring the AAV intron, ensuring only the sequencing of the mature spliced RNA using the JUNC primer (indicated by a red arrow). Additionally, a target RNA substrate transcript sequence (such as ApoB) is inserted. Displayed here is the minimal 33-base ApoB target RNA transcript, known to be edited by A1 in the presence of a cofactor.

2. Once cell cultures reach approximately 80–90 % confluency, take out the medium and wash the cell with 3 to 5 mL DPBS, then add 1 mL of 0.25 % trypsin-EDTA.
3. Put the T75 into an incubator for 2 min
4. Resuspend the cells in 10 mL DMEM media and transfer the volume into a 15 mL centrifuge tube.
5. Centrifuge the cell at 500 g for 5 min
6. Aspirate the supernatant, resuspend the cell pellet in 12 mL DMEM. Take 1 mL cell suspension, mix it with 11 mL of fresh media and transfer the mixture to a new T75 flask. Adjust the dilution factor as needed.

4.3 Transfection

1. Once the HEK293T cell reaches ~90 % confluency, resuspend the cell in DMEM via trypsin digestion as described in the cell culture maintenance immediately.
2. Dilute the cells before seeding them on 8-well glass slides (the next day of transfection, the cell should reach ~70 % confluency).
3. Cells (250 μL) are added to each well of an 8-well glass slides (CellVis).
4. Within 24 h after seeding, cells are transfected with X-tremeGENE 9 transfection reagent (Sigma) as follows: 50 μL master mixes are made by

combining 1 μL of a reporter construct at 50 ng/μL and 5 μL of an editor construct at 100 ng/μL with 44 μL of Opti-MEM reduced serum media (ThermoFisher), adding 1.5 μL of transfection reagent.

5. Incubate the master mix at room temperature for 30 min.
6. 25 μL of each master mix are added dropwise to a particular well, and expression is allowed to occur for 48 h at 37 °C, 5% CO_2.

4.4 RNA extraction and Sanger sequencing

1. After harvesting the cells, RNA extraction with Trizol (Thermo Fisher) is performed according to the manufacturer's instructions.
2. The extracted RNA is reverse-transcribed using ProtoScript II reverse transcriptase (NEB) with a specific primer annealing to the downstream sequence of substrate reporter segments to produce single-stranded cDNA of the target fragment.
3. The reaction is performed in a volume of 20 μL including 1 μg of total RNA, 0.6 μL of 100 μM of reverse primer, ProtoScript II reverse transcriptase buffer, 1 μL of 10 mM dNTP, 1 μL of 0.1 M DTT, 8 U RNase inhibitor, and 0.2 μL of ProtoScript II reverse transcriptase (NEB) for 1 h at 42 °C.
4. Then the cDNA is amplified using Phusion High-Fidelity DNA Polymerase (NEB) for 30 cycles [98 °C × 2.5 min − (98 °C × 20 s, 71.7 °C × 20 s, 72 °C × 30 s) × 30 cycles −72 °C × 5 min] by using a forward primer that anneals to the junction region (JUNC, Fig. 2), where the AAV intron is spliced out. This ensures that only spliced RNA is amplified. The PCR product is then cleaned up using a spin column PCR cleanup kit (Thermo Fisher) to remove the free primers.
5. The PCR clean-up product is sent to Genewiz for Sanger Sequencing using the junction forward primer to precheck the editing activity.
6. At this stage, editing could be quantified from the Sanger trace using MultiEditR (Kluesner et al., 2021).

5. Circular RNAseq assay

5.1 Cell culture, transfection and RNA extraction

1. The detailed cell culture and maintenance protocol is described above.
2. In summary, HEK293T cells are cultured in DMEM medium containing 10% FBS, streptomycin (100 μg/mL), and penicillin (100 U/mL) and incubated at 37 °C, 5% CO_2.

3. The cell suspension solution is seeded one day before transfection at an approximate concentration of 250,000 cells/mL in 10 mL media into two 10 cm plates (CellVis).
4. The A1 only (control) and A1 + RBM46 editor vector (24 μg) are transfected in the HEK293T cells using Lipofectamine 3000 Transfection Reagent (Thermo Fisher) and incubated for 48 h, according to the manufacturer's recommended instructions.
5. After harvesting the cells, RNA extraction using RiboPure™ RNA Purification Kit is performed to enrich mRNA according to the manufacturer's instructions.

5.2 Library construction and sequencing

1. Fragment 1100 ng of enriched mRNA with the NEBNext RNase III RNA Fragmentation Module (E6146S) for 25 min at 37 °C.
2. Purify RNA fragments using an Oligo Clean & Concentrator kit (D4061) by Zymo Research according to the manufacturer's recommendations, except that the columns are washed twice instead of once to increase purity from organic contaminants.
3. Circularize the fragmented RNA using RNA ligase 1 in 20 μL reactions (NEB, M0204S) for 2 h at 25 °C.
4. Purify the circularized RNA is purified with the Oligo Clean & Concentrator kit (D4061) by Zymo Research.
5. To reverse-transcribe the circular RNA templates in a rolling-circle fashion, first incubate the RNA with random hexamer primers for 10 min at 25 °C to allow the primers to bind to the templates.
6. Then, shift the reaction to 42 °C and incubate for 20 min to allow for primer extension and cDNA synthesis.
7. Second strand synthesis and the remaining steps for library preparation are then performed with the NEBNext Ultra RNA Library Prep Kit for Illumina (E7530L) and the NEBNext Multiplex Oligos for Illumina (E7335S, E7500S) according to the manufacturer's protocols.
8. Briefly, cDNA templates are purified with the Oligo Clean & Concentrator kit (D4061) by Zymo Research and incubated with the second strand synthesis kit from NEB (E6111S).
9. Double-stranded DNA is then entered into the end-repair module of RNA Library Prep Kit for Illumina from NEB, and size selected for 500–700 bp inserts using AMPure XP beads.

10. These molecules are then amplified with Q5 PCR enzyme using 11 cycles of PCR, using a two-step protocol with 65 °C primer annealing and extension and 95 °C melting steps.
11. The PCR product is applied to NGS sequencing.
12. Sequencing data was converted to industry-standard Fastq files using BCL2FASTQv1.8.4.

5.3 Error identification

1. Tandem repeats are identified within each read (minimum repeat size: 30 nt, minimum identity between repeats: 90 %), and a consensus sequence of the repeat unit is built.
2. Next, the position corresponding to the 5′ end of the RNA template is identified by searching for the longest continuous mapping region.
3. The consensus sequence is then reorganized to start from the 5′ end of the original RNA fragment, mapped against the genome with Tophat (version 2.1.0 with bowtie 2.1.0) and all non-perfect hits go through a refining algorithm to search for the location of the 5′ end before being mapped again.
4. Finally, every mapped nucleotide is inspected and must pass a 4 steps check to be retained and considered a real editing event: (1) it must be part of at least 3 repeats generated from the original RNA template; (2) all repeats must make the same base call; (3) the sum of all qualities scores of this base must be > 100; (4) it must be > 2 nucleotides away from both ends of the consensus sequence.

5.4 Safe-sequencing system (SSS)

The Safe-Sequencing System (SSS) is an established strategy to address the high error rates (10^{-2} to 10^{-3}) of NGS, enabling the detection of rare mutations or RNA editing events (Kinde et al., 2011). By using SSS, the average error frequency per base pair can be reduced to approximately 10^{-4} to 10^{-5}, significantly enhancing the ability to identify low-frequency mutations. The basic experimental design and detailed analysis of SSS is described in the original paper (Eboreime et al., 2016). SSS is particularly effective for detecting rare mutations within specific DNA or RNA segments. As an example, we have employed it to successfully detect APOBEC-mediated RNA editing activities at various levels in the SARS-CoV-2 genome (Kim et al., 2022).

The key step of the SSS method happens during library preparation, when a library of $\sim 10^{12}$ different randomized nucleotide fragments called the Unique IDentifiers (UID) is added to the population of DNA fragments before the two initial PCR cycles. Each of these unique UIDs is ligated to a unique DNA fragment for sequencing, allowing it to be identified from the pool. After UID ligation, a short universal sequence required for Illumina sequencing is added. The process begins with two initial PCR cycles, followed by a subsequent round of PCR that specifically amplifies the target population using primers complementary to the Illumina adapter sequence, resulting in the formation of the SSS library (Fig. 3A). To achieve distinction between original target DNA or RNA molecules, each descendant of a particular original target molecule will bear the same UID sequence (Fig. 3B). If a minority of the final sequencing reads sharing the same UID display a mutation, it is likely attributable to errors from the library preparation or other errors associated with NGS (yellow or blue dots in Fig. 3B). Conversely, a high proportion of reads with identical mutations suggests that these mutations were present in the original genomic DNA or RNA molecule (red dots in Fig. 3B) (Kinde et al., 2011).

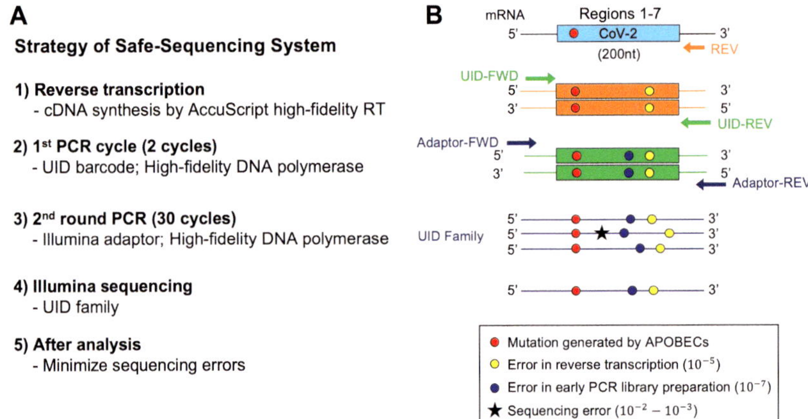

Fig. 3 Critical Steps of the Safe-Sequencing-System (SSS). (A) Schematic representation of the Safe-Sequencing-System (SSS) strategy, developed to minimize errors from reverse transcription, PCR amplification, and sequencing. After reverse transcription of SARS-CoV-2 RNA from cellular extracts, the cDNA is sequentially amplified to insert a unique identifier (UID) barcode in the initial 2 cycles and the Illumina adapter in the following 30 cycles. This method facilitates distinguishing genuine C-to-U editing events attributable to APOBECs from errors introduced during the protocol process. (B) Diagram illustrating the stepwise workflow of the Safe-Sequencing-System (SSS), with different types of errors marked by distinct symbols.

6. Materials and reagents

6.1 Cell culture reagents

- DMEM (Dulbecco's Modified Eagle's Medium) (Corning, 10-013-CV)
- Fetal Bovine Serum (Sigma-Aldrich, F2442)
- Penicillin-Streptomycin (10,000 U/mL) (Gibco, 15140122)
- Trypsin–EDTA (0.25) (Gibco, 25200072)
- DPBS (Gibco, 14190144)
- 10 cm plates (VWR, 10062-890)
- Lipofectamine 3000 Transfection Reagent (Thermo Fisher, L3000015)

6.2 Library preparation and sequencing reagents

- Trizol (Thermo Fisher, 15596026)
- Accuscript High-Fidelity Reverse Transcriptase (Agilent, 200820)
- All primers are ordered from IDT
- Phusion® High-Fidelity DNA Polymerase (NEB, M0530L)
- GeneJET PCR Purification Kit (ThermoFisher, K0702)
- full HiSeq Lane (PE150, 370 M paired reads, Novogene)

7. Methods

7.1 The cell-based RNA editing system

The cell-based RNA editing system is adapted from a previously published report (Wolfe et al., 2019), details of which are described in Chapter 1. In summary, reporter vectors containing DNA corresponding to the different RNA segments of SARS-CoV-2 (NC_045512.2) (Fig. 4A and B) and the APOBEC (A1 +A1CF, A3A or A3G) editor vectors (Fig. 4C) are constructed. A1 + A1CF is constructed as one open reading frame (ORF) with a self-cleaving peptide T2A inserted between A1 and A1CF (A1-T2A-A1CF), allowing the production of single A1 and A1CF proteins in a 1:1 ratio (Liu et al., 2017; Ryan et al., 1991). HEK293T cells are cultured in DMEM medium supplemented with 10% FBS, streptomycin (100 μg/mL), and penicillin (100 U/mL) and maintained at 37 °C, 5% CO_2. One day before transfection, the cells (250 μL) are seeded at an approximate concentration of 250,000 cells/mL into an 8-well glass chamber (CellVis). The cells are then transfected with a transfection mixture (25 μL) of an APOBEC editor vector (500 ng) and a SARS-CoV-2 reporter vector (50 ng) and 1.5 μL of X-tremeGENE 9 transfection reagent (Sigma) and then incubated for 48 h. After harvesting the cells, RNA extraction with

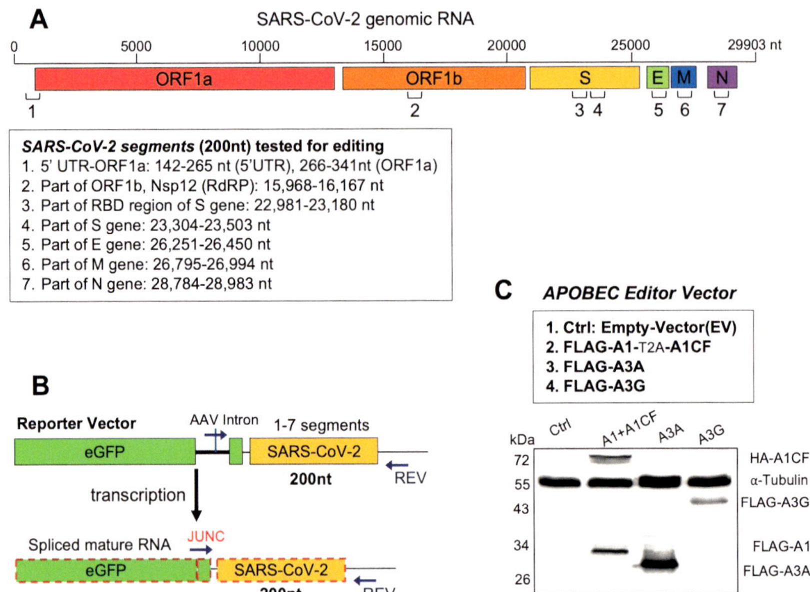

Fig. 4 APOBEC-mediated editing test of SARS-CoV-2 RNA. (A) Schematic diagram of the SARS-CoV-2 genomic RNA, highlighting the positions (boxed regions) of the seven RNA segments (1–7) selected for analyzing APOBEC-mediated RNA editing. (B) Reporter vectors (top) incorporate each of the seven viral RNA segments transcribed into RNA with an AAV intron inserted between the eGFP and the viral RNA sequence. Splicing of the AAV intron generates a mature mRNA containing a novel spliced junction sequence (JUNC), which distinguishes it from its corresponding coding DNA plasmid. This design allows selective amplification of either the mature mRNA or its coding DNA for sequencing the RNA transcript or its original DNA coding sequence. (C) Overview of three APOBEC editor constructs (top: A1–2A-A1CF, A3A, and A3G) and Western blot analysis confirming their expression in HEK293T cells (bottom). The A1–2A-A1CF construct is designed as a single open reading frame (ORF) with a self-cleaving T2A peptide inserted between A1 and A1CF, enabling the production of A1 and A1CF proteins in a 1:1 ratio.

Trizol (Thermo Fisher) and DNA extraction with QuickExtract (Epi-Centre) are performed, respectively, according to the manufacturer's recommended instructions.

7.2 Safe-sequencing system (SSS) sequencing library preparation

1. The extracted RNA is reverse transcribed with Accuscript High-Fidelity Reverse Transcriptase (Agilent) to produce single-stranded cDNA using a specific primer annealing to the downstream sequence of SARS-CoV-2 reporter segments.

2. The reaction is performed in a volume of 20 μL containing 1 μg of total RNA, 1 μL of 100 μM reverse primer, 1 × Accuscript buffer, 1 μL of 10 mM dNTP, 1 μL of 0.1 M DTT, 8 U RNase Inhibitor, and 1 μL of Accuscript High-Fidelity Reverse Transcriptase (Agilent) for 1 h at 42 °C.
3. The cDNA is then amplified for 2 cycles by adding a forward primer annealing to the junction region (JUNC, Fig. 4B), where the AAV intron is spliced out.
4. In this first 2-cycle PCR amplification, the forward and reverse primers are attached to barcodes consisting of 15 randomized nucleotides as the Unique Identifier (UID), plus four trinucleotides designating four different experimental conditions: TGA for A1 + A1CF; CAT for A3A; GTC for A3G; and ACG for Ctrl.
5. Phusion® High-Fidelity DNA Polymerase (NEB) is used for this PCR reaction: 98 °C × 5 min – (98 °C × 30 s, 71.4 °C × 30 s, 72 °C × 1 min) × 2 cycles − 72 °C × 5 min
6. The PCR product (330 bp) is then cleaned up using a spin column PCR cleanup kit (ThermoFisher) to remove the free first-round barcode primers.
7. The second-round PCR is performed for 30 cycles with Illumina flowcell adaptor primers using Phusion® High-Fidelity DNA Polymerase (NEB): 98 °C x 5 min − (98 °C × 30 s, 72 °C × 1 min) × 30 cycles − 72 °C x 5 min
8. All 28 (4 editors × 7 different SARS-CoV-2 substrates) of the different pooled PCR products (399 bp) are combined in equal amounts for the final libraries.
9. The final libraries are subjected to a full HiSeq Lane (PE150, 370 M paired reads, Novogene).
10. The primers for the sequencing library preparation are listed in Table 1.

7.3 Analysis of safe-sequencing-system (SSS)

1. To distinguish a true mutation/editing from random PCR and sequencing errors, the approach as reported in Ref (Kinde et al., 2011) is followed. The details of our implementation of the method are described in Ref (Eboreime et al., 2016).
2. Python scripts were written to analyze the sequencing data. We only considered those sequencing reads such that (1) at least 85% of the bases matched the reference sequence, and (2) the quality scores for all the UID bases were 30 or greater (probability of a sequencing error < 0.001).
3. We clustered reads with the same UID and barcode into UID families. We only considered those families with at least three reads with the same UID and barcode.

Table 1 The primers for the sequencing library preparation.

Primer	Purpose	Sequence (5′–3′)	Feature
First round Ctrl-Forward	Sequencing library	CGACGCTCTTCCGATCTNNNNNNNNNNNNNNNA CGCTGCTGCCCGACAACCACTAC	15 UID, trinucleotide ACG for Ctrl
First round Ctrl-Reverse	Sequencing library	TGAACCGCTCTTCCGATCTNNNNNNNNNNNNNN NACGACTAAAGGGAAGCGGCCTCATTA	15 UID, trinucleotide ACG for Ctrl
First round A1 +A1CF-Forward	Sequencing library	CGACGCTCTTCCGATCTNNNNNNNNNNNNNNNT GACTGCTGCCCGACAACCACTAC	15 UID, trinucleotide TGA for A1 +A1CF
First round A1 +A1CF-Reverse	Sequencing library	TGAACCGCTCTTCCGATCTNNNNNNNNNNNNNNN TGAACTAAAGGGAAGCGGCCTCATTA	15 UID, trinucleotide TGA for A1 +A1CF
First round A3A-Forward	Sequencing library	CGACGCTCTTCCGATCTNNNNNNNNNNNNNNNCA TCTGCTGCCCGACAACCACTAC	15 UID, trinucleotide CAT for A3A
First round A3A-Reverse	Sequencing library	TGAACCGCTCTTCCGATCTNNNNNNNNNNNNNN NCATACTAAAGGGAAGCGGCCTCATTA	15 UID, trinucleotide CAT for A3A
First round A3G-Forward	Sequencing library	CGACGCTCTTCCGATCTNNNNNNNNNNNNNNN GTCCTGCTGCCCGACAACCACTAC	15 UID, trinucleotide GTC for A3G
First round A3G-Reverse	Sequencing library	TGAACCGCTCTTCCGATCTNNNNNNNNNNNNNNN GTCACTAAAGGGAAGCGGCCTCATTA	15 UID, trinucleotide GTC for A3G
Second round universal-Forward	Sequencing library	AATGATACGGCGACCACCGAGATCTACACTCTTTC CCTACACGACGCTCTTCCGATC*T*	*=phosphorothioate bond, Illumina flowcell sequence in green
Second round universal-Reverse	Sequencing library	CAAGCAGAAGACGGCATACGAGATCGGTCTCG GCATTCCTGCTGAACCGCTCTTCCGA*T*	*=phosphorothioate bond, Illumina flowcell sequence in green

4. At each nucleotide site, the mutation frequency is calculated by dividing a numerator by a denominator.
5. The denominator is the number of UID families that, at this particular nucleotide site, have at least three reads with quality scores of at least 20 (probability of a sequencing error < 0.01; because of this quality restriction, the denominator may be different at different sites).
6. The numerator is the number of UID families that, at this particular site, (1) have at least three reads with quality scores of at least 20, and (2) 95% of these reads have the same base, which is different than the reference.
7. For instance, if 100 UID families are identified in step 5, and at a specific site, following the step 6 calculation, 85 of these families are found to have base T (which differs from the reference base C), the mutation frequency is calculated as $85/100 \times 100\% = 85\%$.
8. The probability that three out of three reads will all have the same sequencing error at a site is then 10^{-7} $(= (0.01^3)/(3^2))$.

8. Conclusions

Both the Circular RNAseq assay and the Safe-Sequencing System (SSS) are powerful tools for detecting low-frequency DNA mutations and RNA edits mediated by APOBECs with significantly reduced error rates. Moreover, these two methods can also be used to detect and quantify A-to-I editing by ADARs and editing activities by CEBs and ABEs. However, the choice between these assays depends on specific research conditions and purposes. For instance, the Circular RNAseq assay is better suited for whole transcriptome sequencing. This assay is ideal for screening and examining RNA editing activity across the entire transcriptome, particularly for identifying novel C-to-U mutations or edits mediated by APOBEC proteins. It enables high-throughput sequencing and provides comprehensive coverage of DNA/RNA editing activity while maintaining reduced error rates compared to conventional library preparation methods.

On the other hand, if the goal is to precisely quantify C-to-U editing activity by APOBEC proteins within a specific gene or transcript fragment that has already been identified, the Safe-Sequencing System (SSS) offers a superior level of accuracy. By minimizing technical errors to an average frequency of $\sim 10^{-4}$ to 10^{-5} per base pair, SSS provides highly reliable measurements of editing rates, making it the method of choice for targeted and quantitative studies.

Acknowledgements

This work was supported by NIH grant R01AI150524 to X.S.C.

References

Acevedo, A., & Andino, R. (2014). Library preparation for highly accurate population sequencing of RNA viruses. *Nature Protocols, 9*(7), 1760–1769. https://doi.org/10.1038/nprot.2014.118.

Eboreime, J., Choi, S. K., Yoon, S. R., Arnheim, N., & Calabrese, P. (2016). Estimating exceptionally rare germline and somatic mutation frequencies via next generation sequencing. *PLoS One, 11*(6), e0158340. https://doi.org/10.1371/journal.pone.0158340.

Eisen, T. J., Eichhorn, S. W., Subtelny, A. O., Lin, K. S., McGeary, S. E., Gupta, S., & Bartel, D. P. (2020). The dynamics of cytoplasmic mRNA metabolism. *Molecular Cell,* 77(4), 786–799.e710. https://doi.org/10.1016/j.molcel.2019.12.005.

Fritsch, C., Gout, J. P., & Vermulst, M. (2018). Genome-wide surveillance of transcription errors in eukaryotic organisms. *Journal of Visualized Experiments (139).* https://doi.org/10.3791/57731.

Gout, J. F., Li, W., Fritsch, C., Li, A., Haroon, S., Singh, L., ... Vermulst, M. (2017). The landscape of transcription errors in eukaryotic cells. *Science Advances, 3*(10), e1701484. https://doi.org/10.1126/sciadv.1701484.

Gout, J. F., Thomas, W. K., Smith, Z., Okamoto, K., & Lynch, M. (2013). Large-scale detection of in vivo transcription errors. *Proceedings of the National Academy of Sciences of the United States of America, 110*(46), 18584–18589. https://doi.org/10.1073/pnas.1309843110.

Kim, K., Calabrese, P., Wang, S., Qin, C., Rao, Y., Feng, P., & Chen, X. S. (2022). The roles of APOBEC-mediated RNA editing in SARS-CoV-2 mutations, replication and fitness. *Research Square.* https://doi.org/10.21203/rs.3.rs-1524060/v1.

Kinde, I., Wu, J., Papadopoulos, N., Kinzler, K. W., & Vogelstein, B. (2011). Detection and quantification of rare mutations with massively parallel sequencing. *Proceedings of the National Academy of Sciences of the United States of America, 108*(23), 9530–9535. https://doi.org/10.1073/pnas.1105422108.

Kluesner, M., Tasakis, R. N., Lerner, T., Arnold, A., Wüst, S., Binder, M., ... Pecori, R. (2021). MultiEditR: The first tool for detection and quantification of multiple RNA editing sites from Sanger sequencing demonstrates comparable fidelity to RNA-seq. *Molecular Therapy – Nucleic Acids.* https://doi.org/10.1016/j.omtn.2021.07.008.

Liu, Z., Chen, O., Wall, J., Zheng, M., Zhou, Y., Wang, L., ... Liu, J. (2017). Systematic comparison of 2A peptides for cloning multi-genes in a polycistronic vector. *Scientific Reports,* 7(1), 1–9.

Liu, Z., Yang, L., Deng, R., & Tian, J. (2017). An effective approach with feasible space decomposition to solve resource-constrained project scheduling problems. *Automation in Construction, 75,* 1–9.

Ryan, M. D., King, A. M., & Thomas, G. P. (1991). Cleavage of foot-and-mouth disease virus polyprotein is mediated by residues located within a 19 amino acid sequence. *Journal of General Virology, 72*(11), 2727–2732.

Stark, R., Grzelak, M., & Hadfield, J. (2019). RNA sequencing: The teenage years. *Nature Reviews. Genetics, 20*(11), 631–656. https://doi.org/10.1038/s41576-019-0150-2.

Wang, S., Kim, K., Gelvez, N., Chung, C., Gout, J. F., Fixman, B., ... Chen, X. S. (2023). Identification of RBM46 as a novel APOBEC1 cofactor for C-to-U RNA-editing activity. *Journal of Molecular Biology, 435*(24), https://doi.org/10.1016/j.jmb.2023.168333.

Wolfe, A. D., Arnold, D. B., & Chen, X. S. (2019). Comparison of RNA editing activity of APOBEC1-A1CF and APOBEC1-RBM47 complexes reconstituted in HEK293T cells. *Journal of Molecular Biology, 431*(7), 1506–1517. https://doi.org/10.1016/j.jmb.2019.02.025.

CHAPTER THREE

Purification of enzymatically active APOBEC proteins from an insect cell expression system

Linda Chelico[a,*] and Madison B. Adolph[b,*]
[a]Department of Biochemistry, Microbiology, and Immunology, University of Saskatchewan, Saskatoon, SK, Canada
[b]Department of Biochemistry and Molecular Biology, Saint Louis University School of Medicine, St. Louis, MO, United States
*Corresponding authors. e-mail address: linda.chelico@usask.ca; madison.adolph@health.slu.edu

Contents

Methods in Enzymology, Volume 713
ISSN 0076-6879, https://doi.org/10.1016/bs.mie.2024.11.035
Copyright © 2025 Elsevier Inc. All rights are reserved, including those for text and data mining, AI training, and similar technologies.

Abstract

The APOBEC cytidine/deoxycytidine deaminase family of enzymes has 11 members in humans. These enzymes carry out essential developmental, metabolic, and immunological functions through the deamination of cytosine to form uracil in RNA or single-stranded DNA. The known physiological functions relate to lipid absorption (APOBEC1), immunoglobulin gene diversification (AID), virus restriction (APOBEC3A-H, excluding E), and muscle differentiation (APOBEC2). The ability to characterize in vitro how APOBEC enzymes interact with and catalyze cytidine/deoxycytidine deamination of their substrate has provided key insights and understanding of their physiological functions. Having the most highly active and soluble enzyme to carry out in vitro experiments is essential. For APOBEC enzymes this requires purification from a mammalian or insect cell system. Since mammalian cell expression is lower than robustly engineered recombinant systems such as the *Spodoptera frugiperda 9* (*Sf9*) and baculovirus systems, we have developed recombinant baculovirus expression and purification methods for APOBEC enzymes from *Sf9* cells. The yield for all family members is suitable for biochemical assays, with some enzymes yielding milligram amounts (suitable for structural studies). Here we describe the expression and purification of APOBEC3A, APOBEC3B, APOBEC3C, APOBEC3F, APOBEC3G, APOBEC3H (Haplotypes II, V, VII), and APOBEC1 using existing molecular biology reagents. We also describe how to clone a novel gene into the system for expression and purification. Due to different expression levels and solubility, three purification methods are detailed that enable high, medium, and low expressing APOBECs to be purified.

1. Introduction

The apolipoprotein B mRNA editing complex (APOBEC) family of enzymes, named after the first member to be discovered, APOBEC1, are single stranded (ss) cytidine/deoxycytidine deaminases that modify cytidines in mRNA and deoxycytidines in single-stranded DNA (ssDNA) to form uracils (Adolph, Love, & Chelico, 2018; Feng, Baig, Love, & Chelico, 2014; Salter, Bennett, & Smith, 2016; Sharma et al., 2015; Sharma, Patnaik, Taggart, & Baysal, 2016). This 11-member family of enzymes has diverse roles on ssDNA substrates including affinity maturation of antibodies in B cells (AID, activation induced cytidine deaminase), suppression of retrotransposons (APOBEC1, AID, APOBEC3 subfamily A-H, excluding E), suppression of viral replication (APOBEC3 subfamily), and introduction of promutagenic uracil lesions in genomic DNA (Cheng et al., 2021; Conticello, 2008; Harris, 2015). In addition to deoxycytidine deamination, the APOBEC family also exhibits deamination-independent functions due to their ability to bind RNA and ssDNA with high affinity (Bishop, Holmes, & Malim, 2006; Iwatani, Takeuchi, Strebel, & Levin, 2006; Koito & Ikeda, 2013). APOBEC enzymes can have one or two zinc-dependent deaminase domains (ZDD) (Fig. 1). APOBEC enzymes deaminate cytidines/deoxycytidines to uracil through a zinc-dependent hydrolytic reaction coordinated within a conserved catalytic pocket using an activating water molecule (Marx, Galilee, & Alian, 2015). The deaminated cytidines introduced into mRNA by APOBEC1 typically occur at 5'AUC sites, cause a direct change to the mRNA coding sequence, and can change mRNA function (Greeve, Navaratnam, & Scott, 1991). There is much more diversity in the deoxycytidines that are deaminated in ssDNA, with motifs such as 5′AGC, 5′TTC, 5′ATC, 5′CTC, or 5′CCC being deaminated by specific AID or APOBEC3 (A3) enzymes. Such deamination can result in C-to-T transition mutations in the DNA if the uracil is used as a template during replication or mutation to any base that is dependent on the fate of DNA repair acting on the uracil (Adolph, Love, Feng, & Chelico, 2017; Buisson et al., 2019; Liddament, Brown, Schumacher, & Harris, 2004).

In vitro mechanistic and structural studies of A3 enzymes require highly pure and active proteins. Numerous groups have obtained purified recombinant A3s from a variety of hosts, including bacteria and eukaryotic organisms. Early studies of A3 enzymes utilized bacterial expression systems

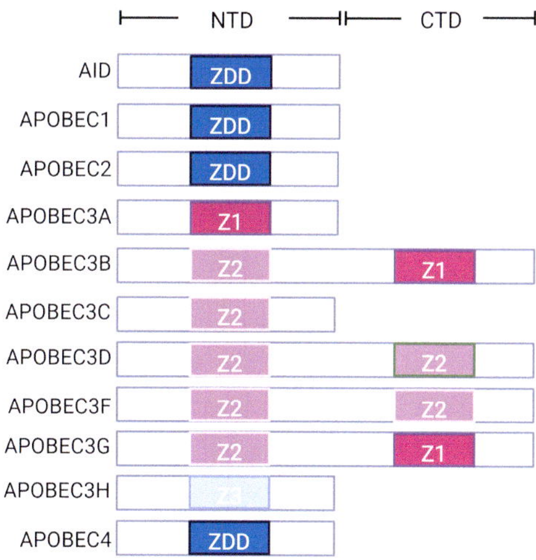

Fig. 1 Overview of APOBEC family protein domains. The eleven human APOBEC enzymes have either a single or double Zinc-dependent deaminase domain (ZDD). In the double domain enzymes, the N-terminal domain (NTD) mediates processivity and the C-terminal domain (CTD) catalyzes cytosine deaminations. For the APOBEC3 enzymes, the ZDD domains are organized into Z1, Z2, or Z3 types across the paralogs (LaRue et al., 2008). *Figure modified from Wong, Sami, & Chelico (2022).*

due to production of high protein yields at low cost (Beale et al., 2004; Harris et al., 2003). Production of A3 proteins from bacterial systems produced sufficient soluble protein for many initial structural studies (Bohn et al., 2013; Chen et al., 2008; Holden et al., 2008; Kitamura et al., 2012); however, it has been demonstrated that only a small percentage of the total protein folds correctly and bacterial systems are unable to produce protein that is reproducibly enzymatically active (Iwatani et al., 2006; Larijani et al., 2007; Zhang et al., 2003). Mammalian expression systems produce properly folded proteins with appropriate post-translation modifications that may be required for enzymatic activity (Borzooee & Larijani, 2019). However, scaling up cell numbers to produce a sufficient yield of protein is time consuming and cost prohibitive. Therefore, insect cell expression systems present an attractive alternative. Expression of recombinant proteins using *Spodoptera frugiperda 9* (*Sf9*) cells, a derivative of *Sf21*, is achieved through infection of the cells with recombinant baculovirus at a specific multiplicity of infection (MOI) to obtain high levels of expression of individual proteins or multi-protein complexes (Gradia et al., 2017; Li et al., 2023; Schneider & Seifert, 2010). An advantage of *Sf9* expression is that most post-translational modifications from mammalian cells are preserved. To date, numerous studies have used insect cell expression systems to purify high quality, active APOBEC proteins (Adolph, Ara, et al., 2017; Adolph, Love, et al., 2017; Ara, Love, & Chelico, 2014; Chelico, Pham, Calabrese, & Goodman, 2006; Chelico, Sacho, Erie, & Goodman, 2008; Chelico, Prochnow, Erie, Chen, & Goodman, 2010; Feng et al., 2015; Iwatani et al., 2006; Love, Xu, & Chelico, 2012; Sharma et al., 2015; Wong, Vizeacoumar, Vizeacoumar, & Chelico, 2021). Multiple *Sf9*/baculovirus systems can be purchased. This method describes the Bac-to-Bac Baculovirus Expression Vector Systems (BEVS) with specific modifications optimized for APOBEC expression.

When using Bac-to-Bac BEVS there is a significant amount of time dedicated to producing the recombinant baculovirus containing the gene of interest that will be overexpressed in the *Sf9* cells (Fig. 2A). Once the virus is made, it must be serially passaged to reach a high titer stock (Fig. 2B). This high titer stock can be stored at −80 °C (years) or at 4 °C (0.5–1 year) for use when recombinant protein production is needed. This process of preparing the virus stock takes 3–4 weeks. Infection of *Sf9* cells occurs over 48–72 h after which time cells are harvested and lysed to purify the recombinant protein (Fig. 2C). Initially, a baculovirus transfer vector was used to insert the gene of interest into a plasmid that would recombine with a linear baculovirus genome after transfection into *Sf9* cells. This inefficient recombination resulted in a

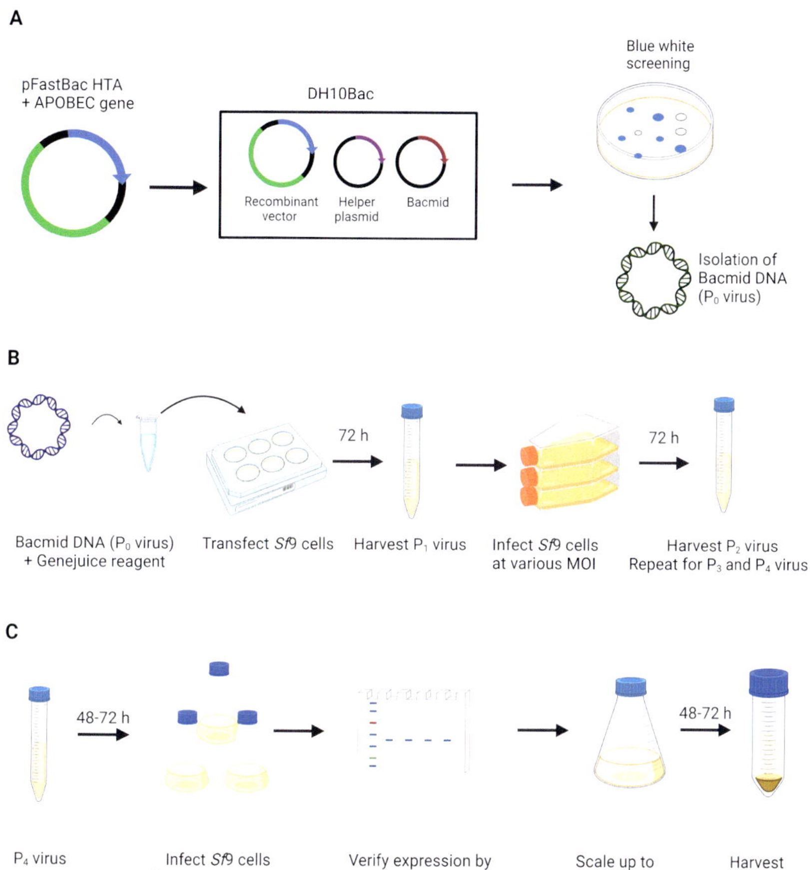

Fig. 2 Production and amplification of recombinant baculovirus to infect *Sf9* cells for protein expression. (A) Production of recombinant bacmid DNA. (B) Amplification of recombinant baculovirus. (C) Infection of *Sf9* cells to prepare protein pellets for APOBEC enzyme purification.

very low titer of recombinant virus. The next generation is the Bac-to-Bac system that uses a bacmid that recombines the baculovirus genome and the gene of interest in *Escherichia coli* (Fig. 2A). The bacmid DNA is then purified and transfected into *Sf9* cells to begin the baculovirus amplification process (Fig. 2B). The bacmid method is now used and recommended since the initial virus stock is much more concentrated which decreases the time required to amplify the virus stock to high titers. This method describes how we use the Bac-to-Bac system to produce high titer stocks of APOBEC-expressing baculovirus for infection of *Sf9* cells and subsequent protein purification.

A unique challenge of purifying A3 proteins is the variability in solubility, activity, and stability during purification and storage. APOBEC enzymes have charged surfaces and therefore can promote non-specific protein-protein and protein-RNA/DNA interactions. A3s strongly co-purify with RNA, and therefore elute as high molecular weight aggregates during purification. Treatment with RNase during purification reduces the aggregation of A3s, improves solubility and promotes a significant increase in enzymatic activity of the purified protein (Chelico et al., 2006; Iwatani et al., 2006). However, the RNase treatment does not completely remove the RNA and labs have used either high salt washes, urea treatment, or both during purification to remove residual RNA (Chelico et al., 2008; Iwatani et al., 2006; Salter, Polevoda, Bennett, & Smith, 2019). Both of these treatments cause partial denaturation of the protein and are only recommended for high expressing APOBECs. If these treatments are used for APOBECs that have low recombinant expression levels, then there may be no protein yield. We have adapted salt washes to each type of APOBEC expression, which is detailed in this chapter (Fig. 3). Similarly, addition of purification tags such as Glutathione S-transferase (GST) or Maltose Binding Protein (MBP) at the N- or C- terminus serves to facilitate purification, improve solubility, and stabilize protein folding (Chelico et al., 2006; Ito et al., 2018; Iwatani et al., 2006). The protein purification protocols described in this chapter use a GST tag (Fig. 4A). It is best practice to cleave the tag during purification to ensure there is no effect of the tag on the function of the protein. Nonetheless, depending on the protease cleavage site there still may be additional amino acids left on the protein, but this is less likely to affect protein function compared to a full protein tag (Fig. 4A). We have found that the GST tag does not affect specific activity of most A3s, except GST-A3B which is not active and GST-A3F that loses processivity (Adolph, Love, et al., 2017; Ara et al., 2014). Due to poor solubility of some APOBEC family enzymes such as AID, the GST tag is usually not cleaved (and in this case does not affect processivity or activity) (Bransteitter, Pham, Scharff, & Goodman, 2003). In these cases where the tag cannot be cleaved, a GST tag that is 20 kDa is preferable over the MBP that is 42 kDa. Since single catalytic domain APOBECs are approximately 20 kDa, in these instances, it is preferable not to have a tag that exceeds the size of the protein of interest, although it has worked for MBP-A3H (Ito et al., 2018). Nonetheless, using a GST tag in *Sf9* cells has the drawback that *Sf9* cells also produce GST, which binds to the Glutathione Sepharose affinity resin (Bichet, Mollat, Capdevila, & Sarubbi, 2000). When there is robust expression of APOBEC enzymes, purifying the

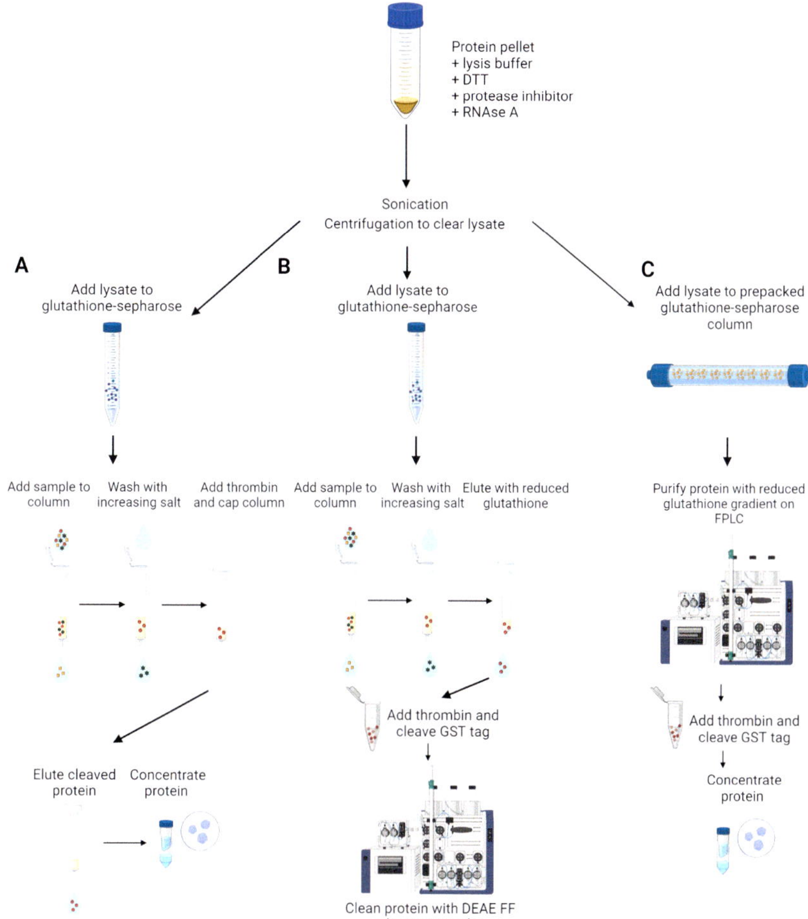

Fig. 3 Schematic of protein purification of APOBEC3 enzymes based on expression level. (A) Purification of APOBEC3 proteins using on column cleavage. (B) Purification of APOBEC3 using off column cleavage followed by Fast Protein Liquid Chromatography (FPLC) using a DEAE Fast Flow (FF) column. (C) Purification of APOBEC proteins using a prepacked GST column and FPLC.

APOBEC away from the GST tag and *Sf9* GST is not an issue, but for low expressing APOBECs this is a contaminant that purification steps are designed to remove. The issue is that when there is low GST-APOBEC expression the *Sf9* GST can be more abundant than the GST-APOBEC. However, this is certainly not the case for the majority of APOBEC purifications, but it is something to be aware of during the interpretation of protein bands on an SDS-PAGE gel post-purification.

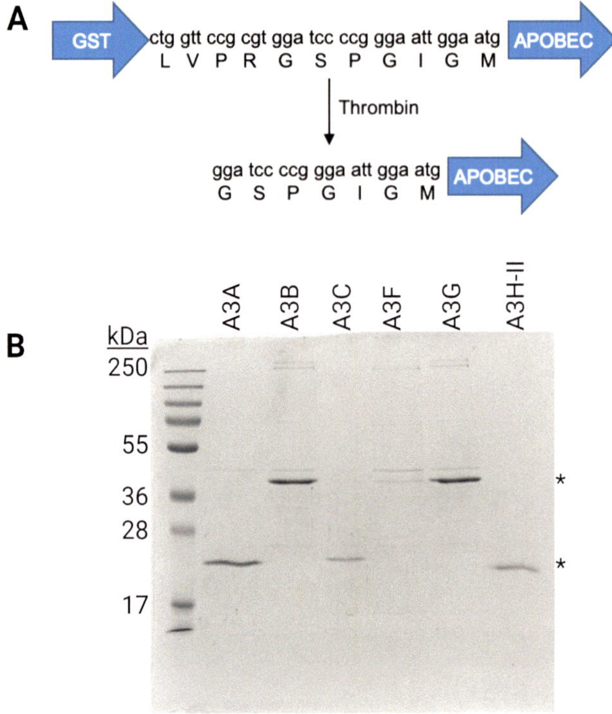

Fig. 4 Expected purity of APOBEC3 enzymes. (A) Schematic of thrombin cleavage site for GST-APOBEC proteins described in this chapter. Using the plasmids described in this chapter, the thrombin cleavage site will leave a six amino acid linker at the N-terminus of the protein. (B) GST-A3A, GST-A3B, GST-A3C, GST-A3F, GST-A3G and GST-A3H-II (A3H Haplotype II) expression plasmids were constructed by using the pAcG2T (BD Biosciences) or pFAST-bac1 (Invitrogen) vectors. Using the methods described in this chapter, recombinant baculovirus was used to infect *Sf*9 cells, with cells harvested 72 h post-infection. Cells were lysed, and the GST-tagged proteins were purified to obtain protein that was cleaved from the GST tag. The asterisks (*) denote the different A3 protein apparent molecular weights by SDS-PAGE and Coomassie staining. *Figure reprinted from (McDaniel et al., 2020).*

Additional considerations are related to the storage of purified APOBEC enzymes. We have found that some APOBEC enzymes require storage in liquid nitrogen to avoid loss of activity after 3–6 months when stored at −80 °C (e.g., A3B, A3C, A3H, APOBEC1). Other APOBEC enzymes are stable indefinitely at −80 °C (e.g., A3A, A3F, A3G). All the enzymes also have different abilities to resist freeze-thaw cycles and so post-purification, aliquots must be carefully prepared to avoid activity loss from multiple freeze-thaw cycles. A3 enzymes also vary widely in recombinant protein expression and stability during purification. While a minimum of 1 mg of A3A or A3G can be expected from 1 L of recombinant baculovirus infected *Sf9* cells (infected at 2×10^6 cells/mL), the yield can be at least 10-fold lower for A3B or A3C (Table 1). Therefore, there are multiple purification strategies based on these unique characteristics.

Here, we describe methods to produce APOBEC proteins from *Sf9* insect cells tagged with GST on the N-terminus. We present three purification schemes dependent on the expression level of the APOBEC protein. The GST tag is cleaved from the APOBEC protein with thrombin protease to yield highly active proteins that are > 95 % pure for robust expressing enzymes. A3 proteins purified from *Sf9* expression systems are suitable for all downstream biochemical assays to study substrate binding or catalytic mechanisms and structure.

2. Methods to prepare for protein purification

The general method is outlined in Figs. 2 and 3 and has the following steps: (1) production and amplification of recombinant baculovirus, (2) infection of the *Sf9* cells (3) harvesting *Sf9* cells and purifying the APOBEC enzyme. For information on how to handle and subculture *Sf9* cells refer to information available from Invitrogen. The *Sf9* cells are simple to maintain in culture and can go between adherent and non-adherent cell culture. The *Sf9* cells are grown at 27 °C with non-adherent cells agitated at 120 rpm. Best practices are to grow non-adherent cells in a range of 5×10^5–2×10^6 cells/mL. The allowable but undesired *Sf9* cell range is 4×10^5–3×10^6 cells/mL. *Sf9* cells should be maintained in culture for no more than 25–30 passages. Flask size is important for aeration. Do not exceed a 1:2 media to flask size ratio.

2.1 Production of recombinant baculovirus

2.1.1 Equipment

Shaker incubator (37 °C)
Stationary incubator (37 °C)
Heat block or thermomixer
Benchtop microcentrifuge
Nano spectrophotometer
Microcentrifuge tubes (1.5 mL)
Sterile round bottom polypropylene tubes (15 mL)
Sterile pipette tips

Table 1 Recombinant baculovirus infection conditions, expected yield and storage considerations.

APOBEC	Infection conditions	Expected yield	Storage and stability
APOBEC3A	• Infect with 10 mL of high titer virus for 72 h • Make 1 L at a time	1–2 mg/1 L	• Active at −80 °C indefinitely • Avoid more than 2–3 freeze-thaw cycles or there will be complete activity loss
APOBEC3B	• Infect with 20 mL high titer virus for 72 h • Make 4 L at a time	0.1–0.5 mg/4 L	• Active at −80 °C for 6 months, recommend long term storage in liquid nitrogen • Avoid more than 5 freeze-thaw cycles
APOBEC3C	• Infect with 20 mL of high titer virus for 72 h • Make 2 L at a time	0.5–1 mg/2 L	• Active at −80 °C for 6 months, recommend long term storage in liquid nitrogen • Avoid more than 5 freeze-thaw cycles
APOBEC3F	• Infect with 10 mL of high titer virus for 72 h • Make 3–6 L at a time	0.05 mg/4 L	• Active at −80 °C indefinitely • Avoid more than 5 freeze-thaw cycles
APOBEC3G	• Infect with 1 mL of high titer virus for 72 h • Make 1 L at a time	1 mg/1 L	• Active at −80 °C indefinitely • Avoid more than 5 freeze-thaw cycles
APOBEC3H (Haplotypes II, V, VII)	• Infect with 1 mL of high titer virus for 72 h • Make 1–2 L at a time	0.5–1 mg/1 L	• Active at −80 °C for 6 months, recommend long term storage in liquid nitrogen • Avoid more than 5 freeze-thaw cycles
APOBEC1	• Infect with 1 mL of high titer virus for 72 h • Make 2 L at a time	0.05 mg/2 L	• Active at −80 °C for 6 months, recommend long term storage in liquid nitrogen • Avoid more than 2–3 freeze-thaw cycles

2.1.2 Reagents

pFAST-Bac-HTA plasmid (Invitrogen Catalog #10584027)
E.coli DH5α competent cells, prepared in house or purchased (ThermoFisher Catalog #18258012)
E. coli DH10B chemically competent cells prepared in house or purchased (ThermoFisher Catalog #12331013)
Ampicillin (ThermoFisher Catalog #BP1760-25)
Gentamycin (ThermoFisher Catalog #BP918-1)
Tetracycline HCl (BioShop Catalog #TET701.10)
X-gal (Invitrogen Catalog #15520034) or Bluo-Gal (Invitrogen Catalog #15519028)
IPTG (Invitrogen Catalog #15520034)
SOC media (Invitrogen Catalog #15544034) or prepared in house (see recipe in 2.1.3 Buffers)
M13 Forward primer (5′ GTAAAACGACGGCCAG 3′)
M13 Reverse Primer (5′ CAGGAAACAGCTATGAC 3′)
QIAprep Spin Miniprep and Midiprep Kit (Qiagen Catalog #27104 and 12243 or equivalent from other supplier)

2.1.3 Buffers

LB-Ampicillin plates (500 mL).
5 g NaCl
5 g Tryptone
2.5 g Yeast Extract
7.5 g Agar
100 μg/mL Ampicillin

LB-Bac plates (500 mL).
5 g NaCl
5 g Tryptone
2.5 g Yeast Extract
7.5 g Agar
50 μg/mL Kanamycin sulfate
7 μg/mL Gentamicin
10 μg/mL Tetracycline
100 μg/mL X-Gal or Bluo-Gal
40 μg/mL IPTG

SOC media.
2% w/v tryptone

0.5 % w/v yeast extract
10 mM NaCl
2.5 mM KCl
10 mM $MgCl_2$
10 mM $MgSO_4$
20 mM glucose

2.1.4 Procedure

1. Clone gene of interest (GST-APOBEC) into pFAST-Bac-HTA vector using a method of choice (see **Note 1**).
2. Transform product into DH5α cells and grow on LB-Ampicillin plate. Screen clones for correct insert by plasmid purification using a commercial miniprep kit, determine the concentration using a nano spectrophotometer, and confirm the correct sequence by Sanger sequencing.
3. After the correct sequence has been confirmed by sequencing, transform 100 ng of pFAST-Bac-HTA vector into DH10Bac cells.
   - **a.** Gently mix the cells and place the tube on ice.
   - **b.** Incubate on ice for 30 min.
   - **c.** Heat shock at 42 °C for 45 s. Transfer to ice and chill for 2 min.
   - **d.** Transfer transformed DH10Bac cells into a pre-chilled 15 mL round-bottom polypropylene tube.
   - **e.** Add 900 μL of SOC media and shake at 37 °C for at least 4 h at 225 rpm.
4. Plate 100 μL of cells into the LB-Bac plates. Do 10-fold and 100-fold serial dilutions in SOC media for optimal plating results.
5. Incubate at 37 °C for at least 48 h. Longer incubation times allow for white and blue colonies to be distinguished.
6. Pick white colonies and restreak on LB-Bac plates. True white colonies tend to be larger. Choose the largest, most isolated colonies and avoid picking colonies that appear gray and are dark in the center.
7. Incubate at 37 °C for 24 h to confirm that the re-streaked colony is truly white.
8. Inoculate a white colony into a 5 mL LB broth culture that contains kanamycin (50 μg/mL), gentamicin (7 μg/mL), tetracycline (10 μg/mL). Incubate overnight at 37 °C at 225 rpm.
9. Isolate recombinant bacmid DNA using a commercial midi prep kit. Bacmid DNA can be stored at 4 °C for up to 2 weeks or at −20 °C indefinitely. If frozen, it must be frozen in aliquots and not freeze-thawed.
10. Verify successful transposition and insertion of the GST-APOBEC gene by Sanger sequencing using M13 forward and M13 reverse

primers. Make a glycerol stock of the DH10Bac cells containing the sequence verified recombinant bacmid DNA for future use.

11. This bacmid DNA is P_0 virus for transfecting *Sf9* cells.

2.2 Amplification of baculovirus

2.2.1 Equipment

27 °C incubator (refrigeration capacity best to avoid temperature fluctuations)
27 °C shaking incubator (refrigeration capacity best to avoid temperature fluctuations)
Biosafety cabinet
Tissue Culture Microscope
Compound Microscope
Hemocytometer
Microcentrifuge tubes (1.5 mL)
Vortex
Benchtop refrigerated centrifuge
Nano spectrophotometer
Pipette controller and serological pipettes
6 well cell culture dish
T75 cm^2 flasks
15 mL conical tubes
50 mL conical tubes
Polycarbonate flasks 125 mL-2 L volume (Fisherbrand™ Shaker Flasks, plain bottom with vented caps, Catalog #PBV125, PBV250, PBV500, PBV500, PBV1000). See **Note 2**.
Plastic container with sterile distilled water and a microcentrifuge rack to be used as a support
Resealable sandwich bag

2.2.2 Reagents

Bacmid DNA
Sf9 cells (ThermoFisher Catalog #11496015)
SF900-II SFM media (ThermoFisher Catalog #10902096)
Insect GeneJuice® Transfection Reagent (Sigma Catalog #71259)

2.2.3 Procedure

All *Sf9* cell work is conducted in a biosafety cabinet. Common biosafety protocols for a BSL2 level laboratory must be followed, ensuring personal protective equipment is worn while performing the experiments.

Cell preparation for a 6-well plate:

1. Make 15 mL of a fresh dilution of *Sf9* cells from an exponentially growing non-adherent culture. Dilute the cells in prewarmed (27 °C) SF900-II SFM media to 0.5×10^6 cells/mL. Use a compound microscope and hemocytometer to count the cells.
2. One hour before transfection, add 2.0 mL (1×10^6 cells) of the cell suspension to each well of a 6-well plate (or as needed) to allow adherence. Check the cell adherence using a tissue culture microscope.

Transfection:

1. Dilute 1 μg bacmid DNA in 100 μL SF900-II SFM media in a sterile 1.5 mL tube. Also, dilute 10 μL of Insect GeneJuice Transfection Reagent in 100 μL SF900-II SFM media in a separate sterile 1.5 mL tube. The Insect GeneJuice Transfection Reagent is stored at 4 °C and should be mixed by inversion before use. See **Note 3**.
2. Slowly add the diluted DNA *dropwise* to the diluted Insect GeneJuice Transfection Reagent. Mix immediately by gentle vortexing to avoid precipitation. Use vortex setting at half strength for ~1 s.
3. Incubate the Insect GeneJuice/DNA mixture at room temperature for 15 min.
4. After the 15 min incubation, add 800 μL SF900-II SFM media to the Insect GeneJuice/DNA transfection mixture.
5. Aspirate the SF900-II SFM media from the cells and add the transfection mixture *dropwise* to the cells. Adding the mixture dropwise ensures the cells are not disrupted by forceful pipetting.
6. Add the cover to the plate and carefully transfer it to a plate support, e.g., microcentrifuge rack, within a flat-bottomed covered storage container containing sterile distilled water (to avoid evaporation). Ensure the plate support is raised above the water level. Incubate the cells for 4 h (or up to 24 h) at 27 °C.
7. Remove the transfection mixture and replace it with 2 mL SF900-II SFM media. Place the plate in a partially open resealable sandwich bag (to avoid evaporation). Incubate the plate at 27 °C.
8. Harvest the supernatant 72 h later by aspiration into a sterile 15 mL conical tube.

Harvest P_1 recombinant baculovirus:

1. Collect the medium containing virus from each well (~2 mL) and transfer to a sterile 15 mL conical tube.

2. Centrifuge the tube at 500 × *g* for 5 min to remove cells and large debris.
3. Transfer the clarified supernatant to a fresh 15 mL tube. This is the P_1 virus stock. Store an aliquot at −80 °C (two 500 μL aliquots) for when virus reamplification is needed. This prevents going back to the bacmid transfection step. Until recombinant protein expression is confirmed, also store the remaining P1 virus at 4 °C, protected from light.
4. (Optional) Check virus titer using the Baculovirus qPCR method detailed in Hitchman et al. (Hitchman, Siaterli, Nixon, & King, 2007). See **Note 4**.

Amplify P_1 virus to make high titer stock:

1. Plate 8.4 × 10^6 *Sf9* cells in a T75 cm^2 flask using 10 mL SF900-II SFM media. Use a compound microscope and hemocytometer to count the cells. Do this in triplicate. After cells have adhered (around 15 min, check with a tissue culture microscope), replace media with 15 mL fresh SF900-II SFM media.
2. Add to the flask the flowing and incubate for 2-3 days at 27 °C.
   a. 100 μL of P_1.
   b. 500 μL of P_1.
   c. Nothing (cells only).
3. After 2-3 days, visually inspect the cells for cytopathic effects of infected cells compared to cells only. Common cytopathic effects are cell rounding (cells look more plump and will not line up with adjacent cells) and cell detachment (floating, rounded cells). Choose the lowest amount of virus (100 μL or 500 μL) where cells have cytopathic effects. If no cytopathic effects are observed, choose the 500 μL condition. For the chosen condition, transfer the virus-containing medium to a 15 mL tube and centrifuge at 500 × g for 5 min. Transfer the clarified supernatant to a fresh 15 mL tube. This is the P_2 virus stock. Store an aliquot at −80 °C (multiple 500 μL aliquots) for when reamplification of the virus is needed. This prevents going back to the bacmid transfection step. Until recombinant protein expression is confirmed, also store the remaining P_2 virus at 4 °C, protected from light.
4. Repeat the above steps to generate P_3 virus stock.
5. The P_3 virus stock is then amplified in suspension culture to make a high volume P_4 stock for infecting larger volumes of *Sf9* cells. In suspension, prepare in a 125 mL polycarbonate flask a 50 mL cell suspension at 2 × 10^6 cells/mL in SF900-II SFM media. Use a compound microscope

and hemocytometer to count the cells. Add 100 μL of P_3 (if from a T75 flask infected with 500 μL) or 50 μL of P_3 (if from a T75 flask infected with 100 μL).

6. Incubate the P_4 flask with shaking for 48 h.
7. After 48 h, transfer cells to a sterile 50 mL conical tube and centrifuge the tube at 500 x g for 5 min to remove cells and large debris.
8. Pour the supernatant (this is the high titer P_4 virus stock) into a sterile 50 mL conical tube. The high titer stock is not frozen at −80 °C, but kept at 4 °C, protected from light and used for up to 1 year (at which time viral titers usually drop). This virus stock is used for expression testing and infecting for protein production.
9. (Optional) Check virus titer using the Baculovirus qPCR method detailed in Hitchman et al. (Hitchman et al., 2007). See **Note 4**.

2.3 Infection of the *Sf9* cells for expression of APOBEC proteins

The infection conditions established for each APOBEC are detailed in Table 1. However, when starting to use this protocol, it is best to conduct an expression test to determine what passage of virus and conditions produce the highest protein yield, as there may be lab specific differences.

2.3.1 Equipment

27 °C shaking incubator (refrigeration capacity best to avoid temperature fluctuations)
Biosafety cabinet
Compound Microscope
Hemocytometer
Benchtop refrigerated centrifuge
High speed centrifuge with high volume capacity rotor and bottles
50 mL conical tubes
Pipette controller and serological pipettes
Polycarbonate flasks 125 mL-2 L volume (Fisherbrand™ Shaker Flasks, plain bottom with vented caps, Catalog #PBV125, PBV250, PBV500, PBV500, PBV1000). See **Note 2**.
Vertical gel electrophoresis system

2.3.2 Reagents

Sf9 cells (ThermoFisher Catalog #11496015)
SF900-II SFM media (ThermoFisher Catalog #10902096)

Glutathione-sepharose 4B resin (Cytvia Catalog #17075605)
RNase A 100 mg/mL (Qiagen Catalog #19101)
cOmplete Mini, EDTA-free Protease Inhibitor Cocktail (Sigma Catalog #11836170001)
1X PBS, made from 10X PBS (see recipe in 2.3.2 Buffers)
2X Laemmli Buffer (BioRad Catalog #1610737)
SDS-PAGE gels (made in house or purchased)
SDS-PAGE Coomassie stain solution (see recipe in 2.3.3 Buffers)
SDS-PAGE de-stain solution (see recipe in 2.3.3 Buffers)

2.3.3 Buffers

Lysis Buffer.
20 mM HEPES pH 7.5
150 mM NaCl
1 % (w/v) Triton-X 100
10 mM NaF
10 mM $NaH_2PO_4{\cdot}H_2O$ (MW=137.99)
10 mM $Na_4P_2O_7.10H_2O$ (MW=446.06)
1 mM EDTA
100 μM $ZnCl_2$ (see **Note 5**)
10 % (v/v) glycerol
10 mM DTT (added immediately before use)
EDTA-free Protease Inhibitor Cocktail (added immediately before use)
RNase A (added immediately before use, amount described in procedure)

1 % TritonX-100 + 500 mM NaCl.
1 % TritonX-100
500 mL NaCl
Bring up to final volume with 1X PBS

10X PBS.
27 mM KCl
1.37 M NaCl
100 mM $Na_2HPO_4.7H_20$
18 mM KH_2PO_4
pH to 7.5

SDS-PAGE Coomassie stain solution.
0.1 % Coomassie R250 (BioShop Catalog #CB0037)
10 % Acetic acid, glacial
50 % Methanol

SDS-PAGE de-stain solution.
10 % Acetic Acid, glacial
40 % Methanol

2.3.4 Expression test (optional)

1. Prepare four 125 mL flasks with 15 mL of *Sf9* cells at 2×10^6 cells/mL in SF900-II SFM media. Use a compound microscope and hemocytometer to count the cells.
2. Add recombinant P_4 baculovirus to cells with the following volumes: 25 μL (equivalent to 1 mL/L), 125 μL (equivalent to 5 mL/L), 250 μL (equivalent to 10 mL/L), and 500 μL (equivalent to 20 mL/L).
3. After 48 and 72 h, remove a 10 mL sample and place it into a sterile 15 mL conical tube. Centrifuge the sample for 15 min at 100 × g. Pour off the supernatant and freeze the pellet at −20 °C.
4. After all samples are collected, defrost pellets for 15 min on ice. While pellets are defrosting, prepare lysis buffer. Add fresh DTT to 10 mM, a protease inhibitor tablet (or portion of a tablet according to manufacturer instructions), and 30 μL of 100 mg/mL RNase A to the volume of lysis buffer (1 mL per pellet).
5. Resuspend each pellet in 1 mL lysis buffer with additives. Rock suspension at 4 °C for 30 min. Then, clarify lysate by centrifuging for 30 min at 12,000 × g at 4 °C.
6. During the incubation and centrifugation time, wash 50 μL (per pellet) of Glutathione-Sepharose 4B resin twice with 300 μL 1X PBS in a 1.5 mL tube. Resin can settle by gravity or be spun for 1 min at 400 × g.
7. Apply 1 mL of the clarified lysate to the washed resin. Incubate clarified lysate with resin for 2 h on a rocker in a cold room.
8. Collect resin by centrifuging for 1 min at 800 × g.
9. Wash resin with 500 μL 1 % TritonX-100 + 500 mM NaCl by resuspending resin by inversion and then by centrifuging for 1 min at 800 × g.
10. Resuspend resin in 20 μL 2X Laemmli Buffer and heat tubes for 2 min at 95 °C.
11. Resolve 20 μL of the sample by SDS-PAGE in 1X Laemmli buffer and visualize proteins by Coomassie staining.

2.3.5 Expression of APOBECs in Sf9 cells

1. Express the GST-tagged APOBEC protein in *Sf9* cells in SF900-II SFM media by adding the appropriate amount of P_4 recombinant baculovirus to the *Sf9* cells. Typically, 1 L of cells or more are infected at 2×10^6 cells/

mL and are harvested 72 h post infection. Use a compound microscope and hemocytometer to count the cells. Typical number of liters, infection times and recombinant baculovirus volumes for all APOBECs tested is in Table 1.

2. At the time of harvest, remove cells from incubator and pour into centrifuge bottles (250 mL or 500 mL preferred).
3. Centrifuge the cells for 15 min at 100 × g.
4. Carefully pour off the supernatant. The *Sf9* cells do not form a stable pellet. Place bottles on ice until all cell pellets are ready for next step.
5. Resuspend the first cell pellet in 30 mL of ice cold 1X PBS. And use this PBS/cells suspension to collect all the cells. Pour into a 50 mL conical tube.
6. Centrifuge the cells for 15 min at 100 × g.
7. Carefully pour off the supernatant. The *Sf9* cells do not form a stable pellet. Proceed to protein purification or store the pellet at −80 °C.

3. Purification of APOBEC3 proteins using on column cleavage (high expression: A3A, A3G, A3H)

GST-A3 proteins with high expression and high solubility are purified using on column cleavage directly from the Glutathione-Sepharose 4B resin (Fig. 3A). This allows for GST free A3 protein with > 95 % purity to be obtained by thrombin protease cleavage from the tag while the GST remains bound to the resin. This purification method eliminates the need for downstream Fast Protein Liquid Chromatography (FPLC) purification.

3.1 Equipment

Branson Sonifier 450 Sonicator
High speed refrigerated centrifuge
High speed centrifuge tubes (50 mL)
Benchtop refrigerated centrifuge
Pipette controller and serological pipettes
50 mL conical tubes
Orbital rotary shaker
Econo-Pac Chromatography Columns (Bio-Rad Catalog #7321010)
Spectrophotometer
Vertical gel electrophoresis system

3.2 Reagents

Glutathione-Sepharose 4B resin (Cytvia Catalog #17075605) (See **Note 6**)
Bradford Reagent (Bio-Rad Catalog #5000205)
Amicon Ultra Centrifugal Filters, 10 kDa (Sigma Catalog #UFC901008)
RNase A 100 mg/mL (Qiagen Catalog #19101)
cOmplete Mini, EDTA-free Protease Inhibitor Cocktail (Sigma Catalog #11836170001)
Thrombin protease (ThermoFisher Catalog #T6884)
Parafilm
6X Laemmli Buffer (ThermoFisher Catalog #J61337-AD) or made in house (see recipe in 3.3 Buffers)
SDS-PAGE gels (made in house or purchased)
SDS-PAGE Coomassie stain solution (see recipe in 3.3 Buffers)
SDS-PAGE de-stain solution (see recipe in 3.3 Buffers)

3.3 Buffers

Lysis Buffer.
20 mM HEPES pH 7.5
150 mM NaCl
1 % (w/v) Triton-X 100
10 mM NaF
10 mM $NaH_2PO_4{\cdot}H_2O$ (MW=137.99)
10 mM $Na_4P_2O_7.10H_2O$ (MW=446.06)
1 mM EDTA
100 μM $ZnCl_2$ (See **Note 5**)
10 % (v/v) glycerol
10 mM DTT (added immediately before use)
EDTA-free Protease Inhibitor Cocktail (added immediately before use)
RNase A (added immediately before use, amount describe in procedure)

Digestion buffer.
50 mM HEPES, pH 7.4
250 mM NaCl
10 % glycerol
1 mM DTT (added immediately before use)

10X PBS.
27 mM KCl
1.37 M NaCl
100 mM $Na_2HPO_4.7H_20$

18 mM KH_2PO_4
pH to 7.5

1 % TritonX-100 + 500 mM NaCl.
1 % TritonX-100
500 mL NaCl
Bring up to final volume with 1X PBS

1 M NaCl solution.
1 M NaCl
Bring up to final volume with 1X PBS

500 mM NaCl solution.
500 mM NaCl
Bring up to final volume with 1X PBS

250 mM NaCl solution.
250 mM NaCl
Bring up to final volume with 1X PBS

6X Laemmli Buffer.
375 mM Tris-HCl pH 6.8
6 % SDS
4.8 % Glycerol
9 % 2-Mercaptoethanol
0.03 % Bromophenol blue

SDS-PAGE Coomassie stain solution.
0.1 % Coomassie R250 (BioShop Catalog #CB0037)
10 % Acetic acid, glacial
50 % Methanol

SDS-PAGE de-stain solution.
10 % Acetic Acid, glacial
40 % Methanol

3.4 Procedure

1. Partially defrost a 1 L pellet (or more depending on final use) of *Sf9* cells on ice for 30–45 min (if stored at −80 °C).
2. During this time prepare lysis buffer. For each 1 L pellet use 20 mL lysis buffer. Add fresh DTT to 10 mM, a protease inhibitor tablet, and 30 μL of 100 mg/mL RNase A to the lysis buffer.

3. Add the lysis buffer to the partially defrosted pellet. Rock at 4 °C for 30–45 min.
4. During the incubation/lysis time, use a 20 mL serological pipette to break up remaining frozen cells and ensure cells are thoroughly suspended in the lysis buffer.
5. Sonicate lysate for 1.5 min on 50 % output and Setting 5 using Branson Sonifier 450 sonicator. Settings for alternative instruments must be determined empirically.
6. Transfer to a 50 mL high speed centrifuge tube and clear lysate at 27,000 × g for 1 h at 4 °C in a high speed centrifuge.
7. During the centrifugation step, prepare Glutathione-Sepharose 4B resin in a 50 mL conical tube. Wash 1 mL of the 75 % Glutathione-Sepharose 4B resin twice with 25 mL of 1X PBS. Spin for 1 min at 500 × g between each wash in a benchtop refrigerated centrifuge. Typically, 1 mL of packed resin binds 8 mg of GST, so this recipe allows the purification of 6 mg of protein.
8. When the lysate centrifugation step is complete, transfer the supernatant to the 50 mL conical tube with the washed Glutathione-Sepharose 4B resin. Add another 30 μL of RNase A to the cleared lysate with resin. Incubate cleared lysate with resin overnight on a rocker in a cold room in the 50 mL conical tube. *All downstream steps from this point are performed at 4 °C unless indicated otherwise.*
9. The next morning, spin lysate and resin mixture for 15 min at 500 × g in a refrigerated benchtop centrifuge (See **Note 7**). Discard supernatant and add 10 mL of 1 % TritonX-100 + 500 mM NaCl and pour resin into 20 mL disposable Bio-Rad econo column and allow liquid to drain by gravity flow. Once all liquid has passed through, close the drain with an end cap. Be mindful to not let the resin dry out during subsequent steps.
10. Wash once with 10 mL 1 % TritonX-100 + 500 mM NaCl by adding liquid to column with end cap. Gently swirl column to mix resin and then remove end cap of column and allow liquid to drain by gravity flow.
11. Wash once with 10 mL 1 M NaCl solution by adding liquid to column with end cap. Gently swirl column to mix resin and then remove end cap of column and allow liquid to drain by gravity flow.
12. Wash once with 10 mL 500 mM NaCl solution by adding liquid to column with end cap. Gently swirl column to mix resin and then remove end cap of column and allow liquid to drain by gravity flow.
13. Wash once with 10 mL 250 mM NaCl solution by adding liquid to column with end cap. Gently swirl column to mix resin and then remove end cap of column and allow liquid to drain by gravity flow.

14. Wash once with 10 mL digestion buffer by adding liquid to column with end cap. Gently swirl column to mix resin and then remove end cap of column and allow liquid to drain by gravity flow.
15. Replace end cap and add 2 column volumes (CV) digestion buffer + 50 μL Thrombin (stock at 1 U/μL). Place top cap on column. Cover both caps with parafilm to ensure a stable closure. Incubate with gentle shaking (i.e., column on side with slight invert) on a rotary shaker at *room temperature* for 12–16 h.
16. Working at 4 °C, elute digestion buffer that will contain cleaved APOBEC protein. This is the first elution (E1). Add successive 1 CV or 0.5 CV volumes of digestion buffer to an end capped column, mix and allow liquid to drain to elute protein (E2, E3, E4). Repeat elutions until a Bradford reagent reaction indicates no more protein is coming off; usually, 3 or 4 elutions are sufficient. Keep elutions on ice. Perform Bradford reagent reaction according to manufacturer's instructions.
17. (Optional) Pool elutions with the highest amounts of protein (>0.5 mg/mL) and concentrate using an Amicon centrifugal filter with a 10 kDa MW cutoff. Spin concentrator at 3000 × g (half recommended maximum) at 4 °C and stop every 5-10 min to mix the protein solution by pipetting or gentle inversion to prevent it from attaching to walls of concentrator. A Bradford reagent reaction can be done periodically to determine if the appropriate concentration has been reached. The Amicon centrifugal filter should be kept at 4 °C during this process. Determine protein concentration of pooled elutions by Bradford reagent reaction according to manufacturer's instructions.
18. Resolve protein ssamples o by SDS-PAGE in 1X Laemmli buffer with Coomassie staining to confirm yield and determine purity. Freeze elutions/aliquots of elutions at −80 °C or liquid nitrogen. Typical purity is shown in Fig. 4B. Follow appropriate storage of APOBEC proteins detailed in Table 1.

4. Purification of APOBEC3 using on column cleavage followed by DEAE chromatography (medium expression: A3B and A3C)

On column cleavage of A3 proteins is only efficient when there are high yields of protein. For A3 enzymes with medium expression, proteins are eluted from Glutathione-Sepharose 4B resin with reduced glutathione and the GST tag is cleaved in solution by thrombin protease (Fig. 3B). The native A3 protein can then be purified away from GST and thrombin by Diethylaminoethyl (DEAE) Sepharose chromatography to yield > 95 % pure A3 protein.

4.1 Equipment

Branson Sonifier 450 sonicator
High speed refrigerated centrifuge
High speed centrifuge tubes (50 mL)
Benchtop refrigerated centrifuge
50 mL conical tubes
Pipette controller and serological pipettes
Orbital Rotary shaker
Econo-Pac Chromatography Columns (Bio-Rad Catalog #7321010)
Bio-Rad DuoFlow FPLC
DynaLoop 25 (Bio-Rad Catalog #7500451)
Vertical gel electrophoresis system
Chemidoc-MP imaging system (Bio-Rad)

4.2 Reagents

Glutathione-Sepharose 4B resin (Cytvia Catalog #17075605) (See **Note 6**)
Bradford Reagent (Bio-Rad Catalog #5000205)
RNase A 100 mg/mL (Qiagen Catalog #19101)
cOmplete Mini, EDTA-free Protease Inhibitor Cocktail (Sigma Catalog #11836170001)
Reduced Glutathione (ThermoFisher Catalog #BP2521)
Thrombin protease (ThermoFisher Catalog #T6884)
HiTrap DEAE Sepharose FF (Cytvia Catalog #17505501) (see **Note 8**)
Bovine Serum Albumin (Sigma)
6X Laemmli Buffer (ThermoFisher Catalog #J61337-AD) or made in house (see recipe in 4.3 Buffers)
SDS-PAGE gels (made in house or purchased)
SDS-PAGE Coomassie stain solution (see recipe in 4.3 Buffers)
SDS-PAGE de-stain solution (see recipe in 4.3 Buffers)

4.3 Buffers

Lysis Buffer.
20 mM HEPES pH 7.5
150 mM NaCl
1 % (w/v) TritonX 100
10 mM NaF

10 mM $NaH_2PO_4 \cdot H_2O$ (MW=137.99)
10 mM $Na_4P_2O_7.10H_2O$ (MW=446.06)
1 mM EDTA
100 μM $ZnCl_2$ (See **Note 5**)
10 % glycerol
10 mM DTT (added immediately before use)
EDTA-free Protease Inhibitor Cocktail (added immediately before use)
RNase A (added immediately before use, amount describe in procedure)

10X PBS.
27 mM KCl
1.37 M NaCl
100 mM $Na_2HPO_4.7H_20$
18 mM KH_2PO_4
pH to 7.5

1 % TritonX-100 + 500 mM NaCl.
1 % TritonX-100
500 mL NaCl
Bring up to final volume with 1X PBS

0.5 % TritonX-100 + 250 mM NaCl.
0.5 % TritonX-100
250 mM NaCl
Bring up to final volume with 1X PBS

1 M NaCl solution.
1 M NaCl
Bring up to final volume with 1X PBS

500 mM NaCl solution.
500 mM NaCl
Bring up to final volume with 1X PBS

250 mM NaCl solution.
250 mM NaCl
Bring up to final volume with 1X PBS

Elution buffer.
100 mM Tris-HCl pH 8.8
250 mM NaCl
10 % glycerol
50 mM reduced glutathione (added immediately before use)

DEAE dilution buffer A.
10 % glycerol

DEAE dilution buffer B.
50 mM Tris HCl pH 8.0 at room temperature
10 % glycerol

DEAE Buffer A.
50 mM Tris HCl pH 8.0 at room temperature
50 mM NaCl
10 % glycerol
1 mM DTT (added immediately before use)

DEAE Buffer B.
50 mM Tris-HCl pH 8.0 at room temperature
1 M NaCl
10 % glycerol
1 mM DTT (added immediately before use)

6X Laemmli Buffer.
375 mM Tris-HCl pH 6.8
6 % SDS
4.8 % Glycerol
9 % 2-Mercaptoethanol
0.03 % Bromophenol blue

SDS-PAGE Coomassie stain solution.
0.1 % Coomassie R250 (BioShop Catalog #CB0037)
10 % Acetic acid, glacial
50 % Methanol

SDS-PAGE de-stain solution.
10 % Acetic Acid, glacial
40 % Methanol

4.4 Procedure

1. Partially defrost two to four 1 L pellets of A3B or A3C (use more or less depending on end use) of *Sf9* cells on ice for 30–45 min (if stored at −80 °C).
2. During this time prepare lysis buffer. For each 1 L pellet use 20 mL lysis buffer. Add fresh DTT to 10 mM, a protease inhibitor tablet, and 30 μL of 100 mg/mL RNase A to the lysis buffer.

3. Add the lysis buffer to the partially defrosted pellet. Rock at 4 °C for 30–45 min.

   During the incubation/lysis time, use a 20 mL serological pipette to break up remaining frozen cells and ensure cells are thoroughly suspended in the lysis buffer.
4. Sonicate lysate for 1 min on 50 % output and Setting 5 using Branson Sonifier 450 sonicator. Settings for alternative instruments must be determined empirically.
5. Transfer lysates to 50 mL high speed centrifuge tubes and clear lysates at 27,000 × g for 1 h at 4 °C in a high speed centrifuge.
6. During the centrifugation step, prepare Glutathione-Sepharose 4B resin in a 50 mL conical tube. Wash 1 mL of the 75 % Glutathione-Sepharose 4B resin twice with 25 mL of 1X PBS. Spin for 1 min at 500 × g between each wash in a benchtop refrigerated centrifuge. Typically, 1 mL of packed resin binds 8 mg of GST; therefore, this recipe allows the purification of 6 mg of protein. Use 1 mL of the 75 % Glutathione-Sepharose 4B resin per A3, not per pellet. After washing, aliquot the resin into two to four 50 mL conical tubes so that each 50 mL lysate can be separately incubated with resin.
7. When the lysate centrifugation step is complete, transfer the supernatants to a 50 mL conical tube with the washed Glutathione-Sepharose 4B resin. Add another 30 μL of RNase A to the cleared lysate with resin. Incubate the cleared lysate with resin overnight on a rocker in a cold room in the 50 mL conical tube. *All downstream steps from this point are performed at 4 °C unless indicated otherwise.*
8. Spin down resin at 500 × g in a refrigerated benchtop centrifuge and pour off lysate (See **Note 7**). Add 10 mL 1.0 % TritonX-100 + 500 mM NaCl to one of the 50 mL conical tubes, swirl to mix and use this volume to collect the other resins from the other 50 mL conical tubes. When all the resin is in one 50 mL conical tube, swirl to mix into a homogenous slurry. Then, pour the resin solution into a 20 mL Bio-Rad chromatography econo column and allow the liquid to drain by gravity flow.
9. Wash once with 10 mL 1 % TritonX-100 + 500 mM NaCl by adding liquid to column with end cap. Gently swirl column to mix resin and then remove end cap of column and allow liquid to drain by gravity flow.

10. Wash once with 10 mL 1 M NaCl solution by adding liquid to column with end cap. Gently swirl column to mix resin and then remove end cap of column and allow liquid to drain by gravity flow.
11. Wash once with 10 mL 500 mM NaCl solution by adding liquid to column with end cap. Gently swirl column to mix resin and then remove end cap of column and allow liquid to drain by gravity flow.
12. Wash once with 10 mL 250 mM NaCl solution by adding liquid to column with end cap. Gently swirl column to mix resin and then remove end cap of column and allow liquid to drain by gravity flow.
13. Place end cap on the column and add 500 μL of elution buffer. Ensure that the stock of reduced glutathione is prepared fresh on same day of purification. Gently swirl column to mix resin and allow elution buffer to incubate with resin for 30–45 min. Collect the elution by gravity flow into a 1.5 mL tube. Repeat 2–4 more times until Bradford reaction indicates protein is no longer eluting. Typically, only elutions 1 and 2 have high yields of protein.
14. Digest each elution with thrombin protease. Add 30 μL thrombin per 500 μL of elution and digest for 6 h at room temperature with gentle rocking.
15. To purify the A3 away from free GST and thrombin, A3 proteins are applied to a DEAE FF column connected to an FPLC with an automated fraction collector. The FPLC should be in a cold room. Equilibrate column into DEAE equilibration buffer with 15 mL Buffer A (flow rate 0.5 mL/min).
16. A3 proteins must be diluted to the same buffer that the DEAE column is equilibrated with. First dilute A3 elution 1:1 with DEAE dilution buffer A, and then dilute further 1:1 with DEAE dilution buffer B. This equalizes the conductivity between the A3 sample and the column and ensures proper binding to the column. Load sample onto column using a DynaLoop 25. From the load to the linear gradient, 1 mL fractions are collected.
17. The program is then set to wash the bound proteins with 15 mL (0.5 mL/min) of Buffer A, then a linear gradient is started that progresses from 100 % to 0 % Buffer A (0–100 % Buffer B) over 15 mL (0.5 mL/min). A3 fractions typically elute at 300 mM NaCl. Fractions of 500 μL are collected during the linear gradient. A final 5 mL (0.5 mL/min) wash with 100 % Buffer A removes any remaining proteins from the column. Columns should be washed and stored according to manufacturer instructions.

18. Determine protein concentration by resolving samples from peak fractions by SDS-PAGE in 1X Laemmli buffer with known standards (BSA at 250, 500, and 1000 ng/μL) with subsequent Coomassie staining. It is advisable to obtain a digital image of the gel and analyze the protein concentration based on a standard curve from the BSA standards. The SDS-PAGE will also determine purity. Freeze aliquots of elutions at −80 °C or liquid nitrogen. Typical purity is shown in Fig. 4B. Follow the appropriate storage of APOBEC proteins detailed in Table 1.

5. Purification of proteins using prepacked GST column (low expression: A3F and APOBEC1)

For APOBEC proteins with low expression, lysates can be loaded directly onto a prepacked 1 mL GST column through a dynamic loop and eluted in reduced glutathione (Fig. 3C). Eluted proteins are then cleaved by thrombin protease in solution to obtain protein.

5.1 Equipment

Branson Sonifier 450 sonicator
High speed refrigerated centrifuge
High speed centrifuge tubes (50 mL)
Benchtop refrigerated centrifuge
50 mL conical tubes
Pipette controller and serological pipettes
Orbital Rotary shaker
Bio-Rad DuoFlow FPLC
Vacuum Flask
Glass Vacuum Filter Holder Kit, 47 mm (Sigma Catalog #XX1014720)
DynaLoop 90 (Bio-Rad Catalog #7500452) or DynaLoop 25 (Bio-Rad Catalog #7500451)
Vertical gel electrophoresis system
Chemidoc-MP imaging system (Bio-Rad)

5.2 Reagents

GST 1 mL prepacked column (Bio-Rad Catalog #12009295)
Bradford Reagent (Bio-Rad Catalog #5000205)
RNase A 100 mg/mL (Qiagen Catalog #19101)

cOmplete Mini, EDTA-free Protease Inhibitor Cocktail (Sigma Catalog #11836170001)
Thrombin protease (ThermoFisher Catalog # T6884)
PVDF Membrane Filter, 0.45 μm Pore Size (Sigma Catalog #HVLP04700)
Slide-a-lyzer mini dialysis cups, 10 kDa MW cutoff (ThermoFisher Catalog #88401)
6X Laemmli Buffer (ThermoFisher Catalog #J61337-AD) or made in house (see recipe in 5.3 Buffers)
SDS-PAGE gels (made in house or purchased)
SDS-PAGE Coomassie stain solution (see recipe in 5.3 Buffers)
SDS-PAGE de-stain solution (see recipe in 5.3 Buffers)
Bovine Serum Albumin (ThermoFisher Catalog #23210)

5.3 Buffers

Lysis Buffer.
20 mM HEPES pH 7.5
150 mM NaCl
1 % (w/v) Triton-X 100
10 mM NaF
10 mM $NaH_2PO_4 \cdot H_2O$ (MW=137.99)
10 mM $Na_4P_2O_7.10H_2O$ (MW=446.06)
1 mM EDTA
100 μM $ZnCl_2$ (See **Note 5**)
10 % (v/v) glycerol
10 mM DTT (added immediately before use)
EDTA-free Protease Inhibitor Cocktail (added immediately before use)
RNase A (added immediately before use, amount described in procedure)

10X PBS.
27 mM KCl
1.37 M NaCl
100 mM $Na_2HPO_4.7H_20$
18 mM KH_2PO_4
pH to 7.5

Buffer A.
100 mM Tris-HCl pH 8.8
200 mM NaCl
10 % glycerol

Buffer B1 (100 % Glutathione Wash).
100 mM Tris-HCl pH 8.8
200 mM NaCl
10 % glycerol
50 mM reduced glutathione (Add fresh on day of purification)

Buffer B2.
0.5 % v/v Triton-X 100
250 mM NaCl
Bring up to final volume with 1X PBS

Buffer B3.
250 mM NaCl
Bring up to final volume with 1X PBS

Dialysis Buffer.
100 mM Tris-HCl pH 8.8
200 mM NaCl
10 % glycerol

6X Laemmli Buffer.
375 mM Tris-HCl pH 6.8
6 % SDS
4.8 % Glycerol
9 % 2-Mercaptoethanol
0.03 % Bromophenol blue

SDS-PAGE Coomassie stain solution.
0.1 % Coomassie R250 (BioShop Catalog #CB0037)
10 % Acetic acid, glacial
50 % Methanol

SDS-PAGE de-stain solution.
10 % Acetic Acid, glacial
40 % Methanol

5.4 Procedure

1. Partially defrost two 1 L *Sf9* cell pellets of APOBEC1 or three 1 L *Sf9* cell pellets of A3F to obtain a suitable yield. If stored at −80 °C, defrost the *Sf9* cell pellets on ice for 30–45 min.
2. During this time prepare lysis buffer. For each 1 L pellet use 20 mL lysis buffer. Add fresh DTT to 10 mM, a protease inhibitor tablet, and 30 μL of 100 mg/mL RNase A to the lysis buffer.

3. Add the lysis buffer to the partially defrosted pellet. Rock at 4 °C for 30–45 min.

   During the incubation/lysis time, use a 20 mL serological pipette to break up remaining frozen cells and ensure cells are thoroughly suspended in the lysis buffer.
4. Sonicate lysate for 1 min on 50 % output and Setting 5 using Branson Sonifier 450 sonicator. Settings for alternative instruments must be determined empirically.
5. Transfer to a 50 mL high speed centrifuge tube and clear lysate at 27,000 × g for 1 h at 4 °C in a high speed centrifuge. For A3F, proceed to step 8.
6. Recombinant APOBEC1 expression produces lysates that require additional clarification to avoid clogging the column. Transfer lysate to a new high speed centrifuge tube and clear lysate a second time centrifuging at 27,000 × g for 30 min at 4 °C.
7. Due to the thick lysates produced after recombinant APOBEC1 expression, lysates are additionally cleared by filtering through a 0.45 μm filter disc under vacuum with vacuum flask and filter clamp apparatus.
8. During step 3–4 or before starting the FPLC purification portion, attach a Bio-Rad prepacked 1 mL GST column and wash with 5 mL water and 5 mL Buffer B3 at a flow rate of 0.5 mL/min. After this step, prepare the FPLC for the purification run by priming the lines with appropriate buffers. An FPLC in a cold room that is able to switch between four buffers is needed. When attaching each buffer and priming lines do it in the following order: Buffer A, Buffer B1, Buffer B2, Buffer B3. Run the Buffer B3 on purge for 2 min. This ensures that the pump head is full of the buffer than is set to go on the column first.
9. Load the clarified lysate onto the column using a DynaLoop 90.
   a. For APOBEC1, the ~60 mL of clarified lysate from the two 1 L cell pellets must be diluted with lysis buffer up to 90 mL due to the thickness of the lysate. If this is not done, the column will go over pressure and the run will stop. The load is done in intervals of 5 mL from the DynaLoop (0.2 mL/min) followed by 3 mL Buffer B3 (0.3 mL/min) and continued until the entire lysate is loaded. The volume loaded from the DynaLoop can be adjusted based on the actual volume of clarified lysate.

b. For APOBEC3F the ~90 mL of clarified lysate from the three 1 L cell pellets can be directly loaded onto the column without dilution. The load is done in intervals of 25 mL from the DynaLoop (0.2 mL/min) and 10 mL Buffer B3 (0.3 mL/min) and continued until the entire lysate is loaded. The volume loaded from the DynaLoop can be adjusted based on the actual volume of clarified lysate.

10. Wash bound proteins with 20 mL of Buffer B2 (0.3 mL/min).
11. Wash bound proteins with 20 mL of Buffer B3 (0.3 mL/min).
12. Elution steps. Starting at this step, collect 500 μL fractions.
    a. For APOBEC1, run 10 mL (0.3 mL/min) of 2-20 % Buffer B1 to remove free GST and *Sf9* contaminant proteins. The free GST results from *Sf9* GST that binds to the resin but has a different binding affinity than the recombinant GST protein (Bichet et al., 2000). Depending on the age and source of reduced glutathione, we have found that 2–20 % Buffer B1/98–80 % Buffer A is needed. The best amount of Buffer B1 to remove free GST must be determined empirically using a linear gradient and then a step gradient can be designed for best results. Then, run 1 mL of 100 % Buffer B1 (0.2 mL/min) to saturate column with elution buffer and pause program for 60 min. Repeat this 100 % Buffer B1/pause cycle two more times to elute the protein.
    b. For APOBEC3F, run a 10 mL (0.3 mL/min) linear gradient of 0 % to 100 % Buffer B1 (100 % to 0 % Buffer A).
13. Finish the run with 20 mL (0.3 mL/min) of Buffer B1 to elute any protein remaining bound to the column. Column should be washed and stored according to manufacturer instructions.
14. After the run, resolve samples of peak fractions by SDS-PAGE in 1X Laemmli Buffer and visualize bands by Coomassie staining to confirm where the APOBEC1 or A3F were eluted. APOBEC1 will be in the 100 % Buffer B1 peak fractions and A3F will elute between 65 % and 100 % Buffer B1. For peak fractions, add 30 μL thrombin per 500 μL fraction. Digest for 6 h at room temperature with gentle rocking.
15. After thrombin digestion, dialyze peak fractions overnight in a cold room in 2 L cold dialysis buffer, changing buffer once after 2 h. Use of Slide-a-lyzer mini dialysis cups is recommended due to the small volume of peak fractions (two to four 500 μL fractions). The free thrombin and GST are left in solution with the protein since there is usually insufficient protein to proceed through an additional purification step.

16. Determine protein concentration by resolving a protein sample by SDS-PAGE in 1X Laemmli buffer with known standards (BSA at 250, 500, and 1000 ng/μL) with subsequent Coomassie staining. It is advisable to obtain a digital image of the gel and analyze the protein concentration based on a standard curve from the BSA standards. The SDS-PAGE will also determine purity. Freeze aliquots of elutions at −80 °C or liquid nitrogen. Typical purity is shown in Fig. 4B. Follow the appropriate storage of APOBEC proteins detailed in Table 1.

6. Summary and conclusions

These methods detail the purification of GST tagged APOBEC proteins from *Sf9* insect cells. Each APOBEC protein has unique expression, solubility, stability, and enzymatic activity profiles and, therefore, must be purified and stored by different methods. This chapter highlights three separate purification schematics optimized based on the expression profile of the APOBEC proteins. After cleavage from the GST tag by thrombin protease, these methods result in enzymatically active APOBEC enzymes that are suitable for use in all biochemical assays. For high and medium expression level APOBECs the method results in > 95 % pure untagged protein. Catalytic activity can be determined using methods detailed in Chelico and Feng, Chapter 7.

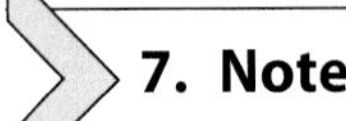

7. Notes

Note 1: APOBECs cloned into the pFAST-Bac-HTA vectors are available from the Chelico lab for A3B, A3C (human S188I, non-human primate), A3H (Haplotype II, V, VII), and APOBEC1 (Adolph, Ara, et al., 2017; Adolph, Love, et al., 2017; Feng et al., 2015; Wong et al., 2021). These vectors contain a custom GST and thrombin cut site preceding the APOBEC gene that is derived from pAcG2T (BD Biosciences) (Fig. 4A). This is due to the discontinuation of the BD Biosciences insect cell reagents and to maintain the same tag and linker region, a portion of the pAcG2T was inserted into pFAST-Bac-HTA. A3A, A3C (human common), A3F, and A3G are in pAcG2T (Adolph, Ara, et al., 2017; Ara et al., 2014; Chelico et al., 2006; Love et al., 2012).

Note 2: Glass flasks can be used during normal *Sf9* cell growth. However, we advise against using glass flasks during protein expression. We have found that glass flasks cause greater shear force on the cells and the resulting stress on the *Sf9* cells decreases protein yield.

Note 3: While any insect cell transfection reagent can be used, we have found that many of the reagents stop working within a short time frame. In contrast, the Insect GeneJuice Transfection Reagent has a long shelf-life.

Note 4: Checking the baculovirus titer when harvesting the P_1 ensures that the bacmid produced viable baculovirus. Knowing the exact titer against a known concentration is not required. A Ct value of 22 to 25 of the P_1 virus represents a productive infection and we have calculated that to be 2×10^4–2×10^5 infectious units/mL using a standard curve. Checking the virus titer after the P_4 will give an exact titer only if a control sample is used to make a standard curve. However, the Bac-to-Bac Baculovirus Expression System estimates a log fold increase at each amplification step, thus resulting in a final P_4 titer of 2×10^8–2×10^9 infectious units/mL. Conducting an expression test (Section 2.3.4) precludes the need to know the exact titer since the optimal milliliters of the virus for infection will be determined empirically.

Note 5: $ZnCl_2$ added to the lysis buffer results in higher yields of A3 proteins as the SF900-II SFM media often does not contain sufficient Zn^{2+} to be fully retained in the A3 active site.

Note 6: This specific Glutathione-Sepharose 4B resin is resistant to 1 M salt washes. If conducting a purification with 1 M salt washes, this exact reagent is recommended. If purchasing another reagent, carefully inspect the column stability in the presence of different reagents, because not all others function in the presence of high salt.

Note 7: It is advised to take samples of the lysate before addition of the resin, of the flow through during washing, and of the resin after elution to examine the efficiency of your purification. These samples can be examined by SDS-PAGE and Coomassie staining.

Note 8: This exact column is recommended. Use of other DEAE columns may not produce the same results.

Acknowledgments

We thank Yuqing Feng for thoughtful comments on the manuscript. Figures presented in this chapter were created with BioRender.com with permissions. This work was supported by CIHR (L.C.) PJT-162407 and PJT-159560.

References

Adolph, M. B., Ara, A., Feng, Y., Wittkopp, C. J., Emerman, M., Fraser, J. S., & Chelico, L. (2017). Cytidine deaminase efficiency of the lentiviral viral restriction factor APOBEC3C correlates with dimerization. *Nucleic Acids Research, 45*(6), 3378–3394. https://doi.org/10.1093/nar/gkx066.

Adolph, M. B., Love, R. P., & Chelico, L. (2018). Biochemical basis of APOBEC3 deoxycytidine deaminase activity on diverse DNA substrates. *ACS Infectious Diseases, 4*(3), 224–238. https://doi.org/10.1021/acsinfecdis.7b00221.

Adolph, M. B., Love, R. P., Feng, Y., & Chelico, L. (2017). Enzyme cycling contributes to efficient induction of genome mutagenesis by the cytidine deaminase APOBEC3B. *Nucleic Acids Research, 45*(20), 11925–11940. https://doi.org/10.1093/nar/gkx832.

Ara, A., Love, R. P., & Chelico, L. (2014). Different mutagenic potential of HIV-1 restriction factors APOBEC3G and APOBEC3F is determined by distinct single-stranded DNA scanning mechanisms. *PLoS Pathogens, 10*(3), e1004024. https://doi.org/10.1371/journal.ppat.1004024.

Beale, R. C., Petersen-Mahrt, S. K., Watt, I. N., Harris, R. S., Rada, C., & Neuberger, M. S. (2004). Comparison of the differential context-dependence of DNA deamination by APOBEC enzymes: Correlation with mutation spectra in vivo. *Journal of Molecular Biology, 337*(3), 585–596. https://doi.org/10.1016/j.jmb.2004.01.046.

Bichet, P., Mollat, P., Capdevila, C., & Sarubbi, E. (2000). Endogenous glutathione-binding proteins of insect cell lines: Characterization and removal from glutathione S-transferase (GST) fusion proteins. *Protein Expression and Purification, 19*(1), 197–201. https://doi.org/10.1006/prep.2000.1239.

Bishop, K. N., Holmes, R. K., & Malim, M. H. (2006). Antiviral potency of APOBEC proteins does not correlate with cytidine deamination. *Journal of Virology, 80*(17), 8450–8458. https://doi.org/10.1128/JVI.00839-06.

Bohn, M. F., Shandilya, S. M., Albin, J. S., Kouno, T., Anderson, B. D., McDougle, R. M., ... Schiffer, C. A. (2013). Crystal structure of the DNA cytosine deaminase APOBEC3F: The catalytically active and HIV-1 Vif-binding domain. *Structure (London, England: 1993), 21*(6), 1042–1050. https://doi.org/10.1016/j.str.2013.04.010.

Borzooee, F., & Larijani, M. (2019). Pichia pastoris as a host for production and isolation of mutagenic AID/APOBEC enzymes involved in cancer and immunity. *New Biotechnology, 51*, 67–79. https://doi.org/10.1016/j.nbt.2019.02.006.

Bransteitter, R., Pham, P., Scharff, M. D., & Goodman, M. F. (2003). Activation-induced cytidine deaminase deaminates deoxycytidine on single-stranded DNA but requires the action of RNase. *Proceedings of the National Academy of Sciences of the United States of America, 100*(7), 4102–4107. https://doi.org/10.1073/pnas.0730835100.

Buisson, R., Langenbucher, A., Bowen, D., Kwan, E. E., Benes, C. H., Zou, L., & Lawrence, M. S. (2019). Passenger hotspot mutations in cancer driven by APOBEC3A and mesoscale genomic features. *Science (New York, N. Y.), 364*(6447), https://doi.org/10.1126/science.aaw2872.

Chelico, L., Pham, P., Calabrese, P., & Goodman, M. F. (2006). APOBEC3G DNA deaminase acts processively 3′ −> 5′ on single-stranded DNA. *Nature Structural & Molecular Biology, 13*(5), 392–399. https://doi.org/10.1038/nsmb1086.

Chelico, L., Prochnow, C., Erie, D. A., Chen, X. S., & Goodman, M. F. (2010). Structural model for deoxycytidine deamination mechanisms of the HIV-1 inactivation enzyme APOBEC3G. *The Journal of Biological Chemistry, 285*(21), 16195–16205. https://doi.org/10.1074/jbc.M110.107987.

Chelico, L., Sacho, E. J., Erie, D. A., & Goodman, M. F. (2008). A model for oligomeric regulation of APOBEC3G cytosine deaminase-dependent restriction of HIV. *The Journal of Biological Chemistry, 283*(20), 13780–13791. https://doi.org/10.1074/jbc.M801004200.

Chen, K. M., Harjes, E., Gross, P. J., Fahmy, A., Lu, Y., Shindo, K., ... Matsuo, H. (2008). Structure of the DNA deaminase domain of the HIV-1 restriction factor APOBEC3G. *Nature, 452*(7183), 116–119. https://doi.org/10.1038/nature06638.

Cheng, A. Z., Moraes, S. N., Shaban, N. M., Fanunza, E., Bierle, C. J., Southern, P. J., ... Harris, R. S. (2021). APOBECs and herpesviruses. *Viruses, 13*(3), https://doi.org/10.3390/v13030390.

Conticello, S. G. (2008). The AID/APOBEC family of nucleic acid mutators. *Genome Biology, 9*(6), 229. https://doi.org/10.1186/gb-2008-9-6-229.

Feng, Y., Baig, T. T., Love, R. P., & Chelico, L. (2014). Suppression of APOBEC3-mediated restriction of HIV-1 by Vif. *Frontiers in Microbiology, 5*, 450. https://doi.org/10.3389/fmicb.2014.00450.

Feng, Y., Love, R. P., Ara, A., Baig, T. T., Adolph, M. B., & Chelico, L. (2015). Natural polymorphisms and oligomerization of human APOBEC3H contribute to single-stranded DNA scanning ability. *The Journal of Biological Chemistry, 290*(45), 27188–27203. https://doi.org/10.1074/jbc.M115.666065.

Gradia, S. D., Ishida, J. P., Tsai, M. S., Jeans, C., Tainer, J. A., & Fuss, J. O. (2017). MacroBac: New technologies for robust and efficient large-scale production of recombinant multiprotein complexes. *Methods in Enzymology, 592*, 1–26. https://doi.org/10.1016/bs.mie.2017.03.008.

Greeve, J., Navaratnam, N., & Scott, J. (1991). Characterization of the apolipoprotein B mRNA editing enzyme: No similarity to the proposed mechanism of RNA editing in kinetoplastid protozoa. *Nucleic Acids Research, 19*(13), 3569–3576. https://doi.org/10.1093/nar/19.13.3569.

Harris, R. S. (2015). Molecular mechanism and clinical impact of APOBEC3B-catalyzed mutagenesis in breast cancer. *Breast Cancer Research: BCR, 17*(1), 8. https://doi.org/10.1186/s13058-014-0498-3.

Harris, R. S., Bishop, K. N., Sheehy, A. M., Craig, H. M., Petersen-Mahrt, S. K., Watt, I. N., ... Malim, M. H. (2003). DNA deamination mediates innate immunity to retroviral infection. *Cell, 113*(6), 803–809. https://doi.org/10.1016/s0092-8674(03)00423-9.

Hitchman, R. B., Siaterli, E. A., Nixon, C. P., & King, L. A. (2007). Quantitative real-time PCR for rapid and accurate titration of recombinant baculovirus particles. *Biotechnology and Bioengineering, 96*(4), 810–814. https://doi.org/10.1002/bit.21177.

Holden, L. G., Prochnow, C., Chang, Y. P., Bransteitter, R., Chelico, L., Sen, U., ... Chen, X. S. (2008). Crystal structure of the anti-viral APOBEC3G catalytic domain and functional implications. *Nature, 456*(7218), 121–124. https://doi.org/10.1038/nature07357.

Ito, F., Yang, H., Xiao, X., Li, S. X., Wolfe, A., Zirkle, B., ... Chen, X. S. (2018). Understanding the structure, multimerization, subcellular localization and mC selectivity of a genomic mutator and anti-HIV factor APOBEC3H. *Scientific Reports, 8*(1), 3763. https://doi.org/10.1038/s41598-018-21955-0.

Iwatani, Y., Takeuchi, H., Strebel, K., & Levin, J. G. (2006). Biochemical activities of highly purified, catalytically active human APOBEC3G: Correlation with antiviral effect. *Journal of Virology, 80*(12), 5992–6002. https://doi.org/10.1128/JVI.02680-05.

Kitamura, S., Ode, H., Nakashima, M., Imahashi, M., Naganawa, Y., Kurosawa, T., ... Iwatani, Y. (2012). The APOBEC3C crystal structure and the interface for HIV-1 Vif binding. *Nature Structural & Molecular Biology, 19*(10), 1005–1010. https://doi.org/10.1038/nsmb.2378.

Koito, A., & Ikeda, T. (2013). Intrinsic immunity against retrotransposons by APOBEC cytidine deaminases. *Frontiers in Microbiology, 4*, 28. https://doi.org/10.3389/fmicb.2013.00028.

Larijani, M., Petrov, A. P., Kolenchenko, O., Berru, M., Krylov, S. N., & Martin, A. (2007). AID associates with single-stranded DNA with high affinity and a long complex half-life in a sequence-independent manner. *Molecular and Cellular Biology, 27*(1), 20–30. https://doi.org/10.1128/MCB.00824-06.

LaRue, R. S., Jonsson, S. R., Silverstein, K. A., Lajoie, M., Bertrand, D., El-Mabrouk, N., ... Harris, R. S. (2008). The artiodactyl APOBEC3 innate immune repertoire shows evidence for a multi-functional domain organization that existed in the ancestor of placental mammals. *BMC Molecular Biology, 9*, 104. https://doi.org/10.1186/1471-2199-9-104.

Li, Y. L., Langley, C. A., Azumaya, C. M., Echeverria, I., Chesarino, N. M., Emerman, M., ... Gross, J. D. (2023). The structural basis for HIV-1 Vif antagonism of human APOBEC3G. *Nature, 615*(7953), 728–733. https://doi.org/10.1038/s41586-023-05779-1.

Liddament, M. T., Brown, W. L., Schumacher, A. J., & Harris, R. S. (2004). APOBEC3F properties and hypermutation preferences indicate activity against HIV-1 in vivo. *Current Biology: CB, 14*(15), 1385–1391. https://doi.org/10.1016/j.cub.2004.06.050.

Love, R. P., Xu, H., & Chelico, L. (2012). Biochemical analysis of hypermutation by the deoxycytidine deaminase APOBEC3A. *The Journal of Biological Chemistry, 287*(36), 30812–30822. https://doi.org/10.1074/jbc.M112.393181.

Marx, A., Galilee, M., & Alian, A. (2015). Zinc enhancement of cytidine deaminase activity highlights a potential allosteric role of loop-3 in regulating APOBEC3 enzymes. *Scientific Reports, 5*, 18191. https://doi.org/10.1038/srep18191.

McDaniel, Y. Z., Wang, D., Love, R. P., Adolph, M. B., Mohammadzadeh, N., Chelico, L., & Mansky, L. M. (2020). Deamination hotspots among APOBEC3 family members are defined by both target site sequence context and ssDNA secondary structure. *Nucleic Acids Research, 48*(3), 1353–1371. https://doi.org/10.1093/nar/gkz1164.

Salter, J. D., Bennett, R. P., & Smith, H. C. (2016). The APOBEC protein family: United by structure, divergent in function. *Trends in Biochemical Sciences, 41*(7), 578–594. https://doi.org/10.1016/j.tibs.2016.05.001.

Salter, J. D., Polevoda, B., Bennett, R. P., & Smith, H. C. (2019). Regulation of antiviral innate immunity through APOBEC ribonucleoprotein complexes. *Sub-Cellular Biochemistry, 93*, 193–219. https://doi.org/10.1007/978-3-030-28151-9_6.

Schneider, E. H., & Seifert, R. (2010). Sf9 cells: A versatile model system to investigate the pharmacological properties of G protein-coupled receptors. *Pharmacology & Therapeutics, 128*(3), 387–418. https://doi.org/10.1016/j.pharmthera.2010.07.005.

Sharma, S., Patnaik, S. K., Taggart, R. T., & Baysal, B. E. (2016). The double-domain cytidine deaminase APOBEC3G is a cellular site-specific RNA editing enzyme. *Scientific Reports, 6*, 39100. https://doi.org/10.1038/srep39100.

Sharma, S., Patnaik, S. K., Taggart, R. T., Kannisto, E. D., Enriquez, S. M., Gollnick, P., & Baysal, B. E. (2015). APOBEC3A cytidine deaminase induces RNA editing in monocytes and macrophages. *Nature Communications, 6*, 6881. https://doi.org/10.1038/ncomms7881.

Wong, L., Sami, A., & Chelico, L. (2022). Competition for DNA binding between the genome protector replication protein A and the genome modifying APOBEC3 single-stranded DNA deaminases. *Nucleic Acids Research, 50*(21), 12039–12057. https://doi.org/10.1093/nar/gkac1121.

Wong, L., Vizeacoumar, F. S., Vizeacoumar, F. J., & Chelico, L. (2021). APOBEC1 cytosine deaminase activity on single-stranded DNA is suppressed by replication protein A. *Nucleic Acids Research, 49*(1), 322–339. https://doi.org/10.1093/nar/gkaa1201.

Zhang, H., Yang, B., Pomerantz, R. J., Zhang, C., Arunachalam, S. C., & Gao, L. (2003). The cytidine deaminase CEM15 induces hypermutation in newly synthesized HIV-1 DNA. *Nature, 424*(6944), 94–98. https://doi.org/10.1038/nature01707.

CHAPTER FOUR

In vitro deamination assay to measure the activity and processivity of AID/APOBEC enzymes

Linda Chelico[a,*] and Yuqing Feng[b,*]
[a]Department of Biochemistry, Microbiology, and Immunology, University of Saskatchewan, Saskatoon, SK, Canada
[b]Department of Biology, York University, Toronto, ON, Canada
*Corresponding authors. e-mail address: linda.chelico@usask.ca; yqfeng@yorku.ca

Contents

Abstract

The AID/APOBEC family of enzymes are cytidine/deoxycytidine deaminases that primarily catalyze the deamination of deoxycytidines (dCs) into deoxyuridines (dUs) on single-stranded DNA (ssDNA). In humans, there are 11 members within the family. AID and APOBEC3 (A3) enzymes have been extensively characterized for their ability to introduce promutagenic dUs during antibody gene diversification and intrinsic immune defenses against viruses and retrotransposons, respectively. In order to search for a local dC deamination target to effectively catalyze the deamination reaction, AID/APOBEC enzymes adopt facilitated diffusion as a mechanism to search for the target deamination sites on ssDNA substrates, which includes one-dimensional (1D) movements termed sliding, and three-dimensional (3D) movements termed

Methods in Enzymology, Volume 713
ISSN 0076-6879, https://doi.org/10.1016/bs.mie.2024.11.038
Copyright © 2025 Elsevier Inc. All rights are reserved, including those for text and data mining, AI training, and similar technologies.

jumping and intersegment transfer. This type of diffusional mechanism enables AID/APOBEC enzymes to processively scan ssDNA substrates and serves as a key determinant to the mutagenic potential of AID/APOBEC enzymes *in vivo*. The catalysis and processive ssDNA scanning behaviors of AID/APOBEC enzymes can be assessed using purified proteins and synthetic ssDNA through an *in vitro* deamination assay. In this Chapter, we describe how to perform deamination assays where DNA scanning mechanisms and processivity can be measured under single-hit conditions using a fluorescently labeled ssDNA substrate. The *in vitro* deamination assay can also be applied to determine AID/APOBEC activity in cell lysates or in kinetic reactions to determine the specific activity of purified enzymes.

1. Introduction

Locating a specific DNA sequence among a myriad of non-target sequences can be a formidable challenge for DNA binding proteins (Chelico, Pham, & Goodman, 2009; Wong, Vizeacoumar, Vizeacoumar, & Chelico, 2021). The initial finding in the 1970s that the *Escherichia coli* Lac repressor is able to find the target operator site on DNA at a rate that is 1000-fold faster than the upper limit of normal diffusion spurred interest to understand the mechanisms that enable an accelerated rate of target sequence search by DNA binding proteins (Riggs, Bourgeois, & Cohn, 1970). As this field of research developed, the movement of proteins such as the Lac repressor on DNA was defined as facilitated diffusion.

Facilitated diffusion has been identified as a Brownian motion-driven diffusional mechanism that allows DNA binding proteins to rapidly search for their target sequence on a DNA substrate in the absence of an energy source such as ATP (Berg, Winter, & von Hippel, 1981; Halford & Marko, 2004; von Hippel & Berg, 1989). Proteins that translocate themselves through facilitated diffusion are greatly influenced by the electrostatic interaction between the negatively charged DNA backbone and the protein itself, which usually has a positively charged patch that interacts with the DNA (Feng, Baig, Love, & Chelico, 2014). Such interaction allows the DNA binding protein to translocate on its substrate at a rate that far exceeds the free diffusion rate of a protein molecule and has been characterized for double-stranded (ds) and single-stranded (ss) DNA binding proteins (Chelico, Pham, Calabrese, & Goodman, 2006; Halford & Marko, 2004). Facilitated diffusion can occur through one-dimensional (1D) sliding and three-dimensional (3D) jumping or intersegment transfer (Berg et al., 1981; Halford & Marko, 2004; Stanford, Szczelkun, Marko, & Halford, 2000;

von Hippel & Berg, 1989). The 1D sliding motion allows for an in-depth local search for target sites that are closely spaced (less than 20 nt), whereas 3D jumping or intersegment transfer movement mediates longer-range enzyme translocation on the DNA substrate. The jumping motion is defined by microscopic dissociation and re-association with the same DNA molecule without diffusing into the bulk solution, whereas intersegmental transfer requires a protein with two DNA binding sites to reach a doubly bound intermediate state before transferring to one of the bound segments without completely dissociating from the DNA. The most efficient search on DNA involves a combination of sliding (1D) and jumping or inter-segment (3D) movements.

Facilitated diffusion enables a processive mode of DNA scanning (Adolph, Love, & Chelico, 2018; Feng et al., 2014). The term "processivity" is defined as the ability of an enzyme to catalyze multiple reactions on a single enzyme-substrate encounter without diffusing into the bulk solution (Berg et al., 1981; Chelico et al., 2009; Halford & Marko, 2004; von Hippel & Berg, 1989). Processive DNA scanning has been adopted by multiple DNA binding proteins, such as transcription factors, restriction endonucleases, DNA repair proteins such as DNA glycosylases and apurinic/apyrimidinic (AP) endonuclease, and AID/APOBEC cytidine/deoxycytidine deaminases (Bonnet et al., 2008; Brown et al., 2016; Carey & Strauss, 1999; Chelico et al., 2009; Dowd & Lloyd, 1990; Esadze & Stivers, 2018; Tafvizi, Huang, Fersht, Mirny, & van Oijen, 2011).

The AID/APOBEC family of cytidine/deoxycytidine deaminases has 11 family members in humans, including activation-induced cytidine deaminase (AID), the apolipoprotein B mRNA-editing enzyme, catalytic polypeptide-like (APOBEC) members APOBEC1 (A1), APOBEC2 (A2), seven APOBEC3s (referred to as A3A, A3B, A3C, A3D, A3F, A3G, A3H), and APOBEC4 (A4) (Adolph et al., 2018; Feng et al., 2014; Feng, Seija, Di Noia, & Martin, 2021). AID is required to initiate antibody affinity maturation, class switch recombination and gene conversion in B cells (Di Noia & Neuberger, 2007; Feng et al., 2021; Methot & Di Noia, 2017). A3 enzymes are innate immune defense proteins against several retroviruses, DNA viruses, and retrotransposons, and the protective roles of A3 enzymes against the human immunodeficiency virus type 1 (HIV-1) have been extensively characterized (Adolph et al., 2018; Cheng et al., 2021; Feng et al., 2014; Gaba, Flath, & Chelico, 2021). AID and A3 enzymes have also been characterized to have "off-target" activity in the human genome that has been associated with cancer (Leeman-Neill, Bhagat, & Basu, 2024; Swanton, McGranahan,

Starrett, & Harris, 2015; Wang, Schmitt, Guo, Santiago, & Stephens, 2016). A2 and A4 have no characterized deamination activity to date (Lorenzo et al., 2024; Marino et al., 2016). A1 has a physiological function that became the namesake for the enzyme family and involves an essential mRNA editing event at cytidine 6666 of the apolipoprotein B mRNA to create a shorter protein product that is essential for the function of the small intestine (Greeve, Navaratnam, & Scott, 1991). However, A1 can still deaminate ssDNA (Harris, Petersen-Mahrt, & Neuberger, 2002). Conversely, A3A and A3G have been reported to deaminate mRNA (Jalili et al., 2020; Sharma et al., 2015; Sharma, Patnaik, Taggart, & Baysal, 2016). Overall, the main physiological function of AID/APOBEC enzymes, excluding A1, is on ssDNA which is the focus of this Chapter.

While functioning in different physiological contexts, the biological functions of AID and A3 enzymes are both initiated by the deamination of deoxycytidine (dC) to deoxyuridine (dU) on ssDNA (Fig. 1). After antigen encounter, AID expression is induced in B cells and it introduces dU lesions during transcription events at the *immunoglobulin* (*Ig*) locus. Error-prone repair by the cellular base excision and mismatch repair machinery can lead to mutagenic outcomes such as mutations and DNA strand breaks, allowing high-affinity class-switched antibody variants to be selected during antibody affinity maturation (Di Noia & Neuberger, 2007; Feng & Martin, 2022; Feng et al., 2021; Methot & Di Noia, 2017). Certain A3 enzymes including A3D, A3F, A3G, and A3H can induce

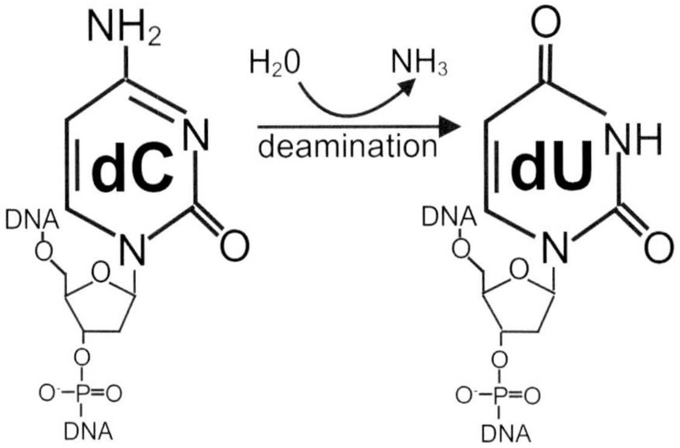

Fig. 1 Reaction scheme of deoxycytidine deamination catalyzed by AID/APOBEC family of enzymes. AID/APOBEC-induced mutagenesis on single-stranded DNA (ssDNA) is initiated by the deamination of deoxycytidine (dC) which forms deoxyuridine (dU). The cytosine deamination occurs through a zinc-mediated hydrolytic reaction.

mutations on HIV-1 proviral (–) cDNA through dC deamination if they escape proteasomal degradation mediated by the HIV-1 protein, Viral infectivity factor (Vif), allowing the inhibition of viral replication (Desimmie et al., 2014).

Processivity enables AID/APOBEC enzymes to introduce both an increased number and diverse types of mutations through dC deamination, therefore increasing the mutagenic capacity of these enzymes. A processive mode of ssDNA scanning mechanism has been identified to be essential for efficient deamination of HIV-1 proviral (–) cDNA by A3 enzymes (Ara, Love, & Chelico, 2014; Feng & Chelico, 2011; Feng, Love, & Chelico, 2013; Feng et al., 2014; Feng et al., 2015; Love, Xu, & Chelico, 2012). For AID, the processive behavior ensures that AID-induced dU lesions can persist into mutations even in the presence of highly efficient DNA repair (Chelico et al., 2009; Pham et al., 2019; Shen, Tanaka, Bozek, Nicolae, & Storb, 2006; Shen et al., 2009; Storb, Shen, & Nicolae, 2009; Xue, Rada, & Neuberger, 2006). For dsDNA viruses and for A3s that have been found to deaminate the human genome, the deamination occurs primarily during DNA replication or DNA repair where the DNA is transiently single-stranded (Cheng et al., 2019; Seplyarskiy et al., 2016; Wong, Sami, & Chelico, 2022). In contrast to the efficient processive scanning required for A3 enzymes to access HIV-1 ssDNA during reverse transcription or for AID to access *Ig* gene ssDNA during transcription, for genomic ssDNA undergoing replication or repair, the ability to undergo 3D rapid on/off cycling on ssDNA, either by a non-processive cycling mode (A3A) or intersegmental transfer (A1, A3B, A3H), is required to introduce the most dC deaminations and enables these A3 enzymes to compete with replication protein A (RPA) for available ssDNA (Adolph et al., 2018; Adolph, Love, Feng, & Chelico, 2017; Wong et al., 2022).

The biochemical activity and ssDNA scanning behaviors of AID/APOBEC can be quantified through an *in vitro* deamination assay using purified proteins or mammalian cell lysates and chemically synthesized and fluorescently labeled ssDNA oligonucleotides (Fig. 2A). The *in vitro* kinetic deamination assay is based on incubating the enzyme and substrate for a specific amount of time. Afterwards, the AID/APOBEC enzymes are inactivated and the substrate is incubated with *E. coli* Uracil-DNA Glycosylase (UDG), which removes the dU base creating an AP site (Fig. 2A). The AP site is sensitive to breakage when heated under alkaline conditions. The DNA fragments are then resolved on a urea polyacrylamide gel for visualization (Fig. 2A).

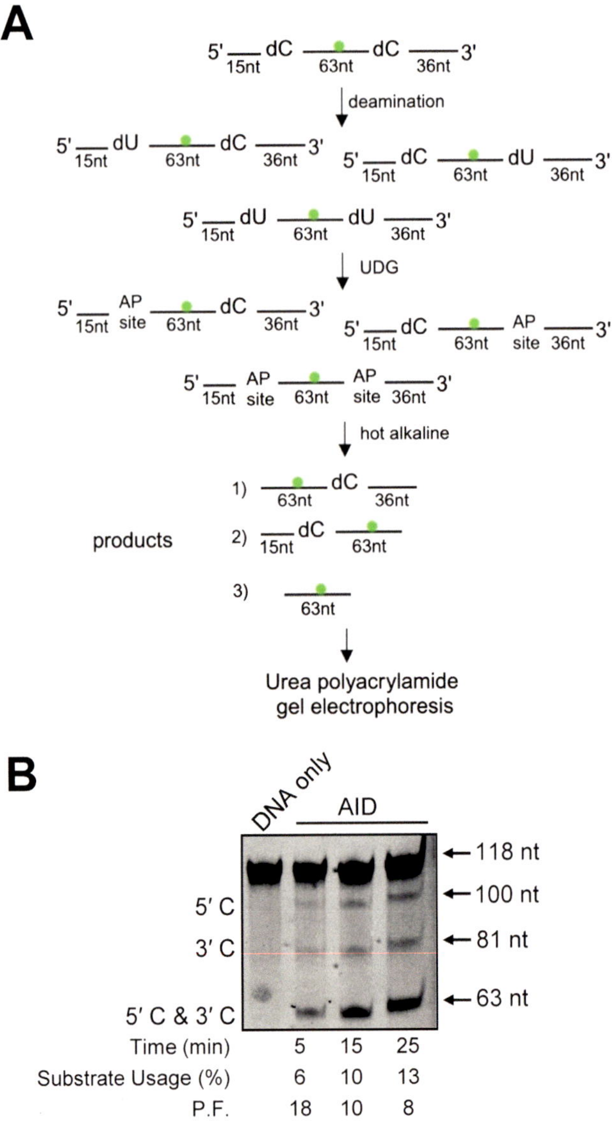

Fig. 2 Overview of processive deamination assay of AID/APOBEC enzymes. (A) A 118 nt ssDNA substrate containing two identical dC deamination targets embedded within the AID/APOBEC deamination hot-spot motifs (*e.g.*, 5′AGC̲ for AID or 5′CCC̲ for A3G) is used. The ssDNA substrate is fluorescently labeled between the two deamination motifs with a Fluorescein dT amidite (Fam-dT) to allow the direct and simultaneous measurements of double or single deamination on an individual ssDNA substrate. Incubation of the substrate with APOBECs generates 3 types of products, a single C-to-U deamination in the target dC motif located nearer the 5′ end (left scenario);

Commonly used substrates for deamination assays range from ~30 to 120 nucleotides (nt) and contain either one or two dC-containing motifs. Outside of the deamination motif, the DNA is composed of a random sequence of dA, dT, or dG (no other dC outside of the motif(s)). The processivity of AID/APOBEC enzymes is determined by placing two deamination motifs (Fig. 2A), such as 5′AGC for AID, 5′CCC for A3G, or 5′A/T/CTC for A1 and A3s other than A3G within the same ssDNA sequence (Adolph, Ara, et al., 2017; Ara et al., 2014; Bransteitter, Pham, Calabrese, & Goodman, 2004; Chelico et al., 2006; Feng & Chelico, 2011; Feng et al., 2013; Feng et al., 2015; Love et al., 2012; Pham, Bransteitter, Petruska, & Goodman, 2003). While a substrate with a single deamination motif directly measures the enzymatic activity, this measurement is accurate only on short substrates, *e.g.*, ~30–40 nt. This is because as the substrate length increases, so does the amount of nonspecific DNA that the enzymes must search through to find their deamination motif. This introduces multiple variables into the final calculation of "activity" which includes the actual catalytic efficiency when the dC is engaged in the active site and the time needed to search for that dC. Thus, when defining the substrate to use, it should be considered if the *in vitro* deamination reaction can provide data applicable to the biological system being studied, *e.g.*, if it is important to characterize enzyme processivity or determining only activity would be sufficient. Or, perhaps only a comparison of activity levels between A3 and

nearer the 3′ end (right scenario), or double deamination events occurring in both 5′ and 3′ deamination motifs (bottom scenario). Following the treatment of the ssDNA with UDG and hot alkaline, these three types of deamination products can be resolved through urea polyacrylamide gel electrophoresis. To determine processivity or specific activity, experiments must be carried out under single-hit condition to ensure that almost all deaminations, either double or single, take place during a single enzyme-ssDNA encounter. AP site, apurinic/apyrimidinic site; green circle, Fam-dT. This figure was created using BioRender with publication permission (https://biorender.com/). (B). Example of AID/APOBEC enzyme processivity determination using the deamination substrate depicted in panel (A) containing two 5′AGC deamination motifs. AID was incubated with the ssDNA substrate for 5, 15, or 25 min. Resolution of the bands on the urea polyacrylamide gel and subsequent analysis demonstrated that optimal substrate usage for calculation of a processivity factor (P.F.) occurred after 15 min (10% substrate usage) and gave a processivity factor of 10. Calculation of processivity when only 6% substrate was used gave a higher processivity factor since the calculation of the number of independent events that is used in the processivity calculation changes with substrate usage. For this reason, processivity calculations are recommended only for 10 ± 2% substrate usage where there is less variability in the processivity factor.

mutants is required. In the former, longer substrates and processivity determination are recommended since all natural substrates involve stretches of ssDNA with multiple deamination motifs. In the latter, a more controlled system may be used initially. However, if differences in activity are noticed after making mutations outside of the active site, then likely processivity has been altered and activity levels and processivity should be tested on longer substrates (Feng & Chelico, 2011).

The ssDNA substrate can be labeled multiple ways to enable detection. Originally the deamination assay used a radiolabeled DNA, but our current protocol uses a fluorescein label (Fam) either at the 5′-end for single deamination motif containing substrates or a Fam-dT modification (*i.e.*, fluorescein attached to position 5 of the thymine ring by a 6-carbon spacer arm) between the two cytosines for substrates with two deamination motifs (Fig. 2A). We and others have designed a variety of substrates that are recommended for use to enable comparison of activities between labs (Table 2). However, if designing substrates, certain basic principles should be followed. First, the substrate should be confirmed to not form stable secondary structures in the ionic conditions of the reaction and at 37 °C with a program such as DNA fold (Zuker, 2003). In addition, since the assay relies on a DNA break being introduced at a deamination site for substrates with two deamination motifs, both dC containing sites need to be off-set from the center to produce bands of different sizes (Fig. 2A). To measure enzyme processivity, a high DNA substrate:enzyme molar ratio is usually used in conjunction with a time course to ensure single-hit conditions for highly active enzymes (Creighton & Goodman, 1995). For less active enzymes, the DNA substrate:enzyme molar ratio can be equal to one, as long as the substrate usage remains within single-hit conditions. Single-hit conditions occur when less than 10 % of the substrate is used and means that each enzyme interacted with at most one DNA (Creighton & Goodman, 1995). As a result, a large majority of the ssDNA substrate remains unmodified (Fig. 2B, substrate usage 6–13 %). A processivity factor can then be calculated under these experimental conditions. The processivity factor is the ratio of double deaminations that resulted from one enzyme to the predicted double deamination events that would occur from two enzymes separately binding the DNA. Thus a processivity factor of ≤ 1 means that the enzyme is non-processive. Under single-hit conditions, non-processive enzymes such as A3A or A3C will not produce a double deamination band precluding analysis or the double deamination band will be faint and result in a processivity factor at or near 1 (Fig. 3A, 63 nt band

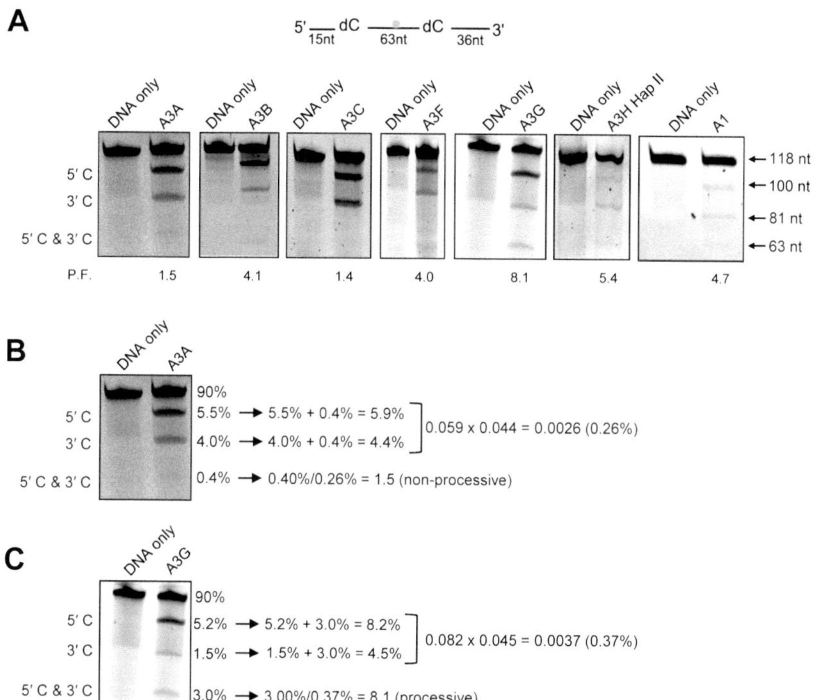

Fig. 3 Processive deamination analysis of APOBEC1 and APOBEC3 enzymes. (A) All A3 enzymes except A3D have been purified *in vitro* and have been characterized for processivity. Examples of the expected processivity factor (P.F.) for A3 enzymes and A1 on the L63 – 118 nt deamination substrate (see sketch) is shown. The ones used in this figure were TTC L63 – 118 nt (A3A, A3C, A3F, A1), ATC L63 – 118 nt (A3B), CCC L63 – 118 nt (A3G), CTC L63 – 118 nt (A3H). The DNA substrate sequences are shown in Table 2. The DNA substrate to enzyme ratios recommended for each enzyme are A3A, A3B, A3H – 1:0.5; A3C – 1:3.5; A3F – 1:2; A3G – 1: 0.2–0.5. (B) Example processivity calculation for A3A. (C) Example processivity calculation for A3G. Figures are reprinted with permission from (Adolph, Ara, et al., 2017; Adolph, Love, et al., 2017; Ara et al., 2014; Ara et al., 2017; Feng et al., 2015; Love et al., 2012; Wong et al., 2021).

and processivity factor for A3A and A3C) (Adolph, Ara, et al., 2017; Love et al., 2012). In contrast, highly processive APOBECs such as A3G will show a distinct double deamination band that can be easily visualized (Fig. 3A, 63 nt band and processivity factor for A3G) (Ara et al., 2014). Extending the reaction time enables non-processive enzymes to catalyze double deamination events, but these reactions are derived from multiple, separate DNA-enzyme encounters and do not satisfy the single-hit conditions required for calculating processivity. AID, A1, and A3 enzymes such A3B, A3F, A3G, and A3H are highly processive enzymes with a processivity factor of 4 and above (Feng et al., 2013; Pham et al., 2008) (Figs. 2B and 3A).

Different modes of ssDNA scanning can be dissected by altering the spacing between two deamination motifs or changing the assay conditions to favor one type of enzyme movement over the other. For example, a substrate with two dCs that are spaced at 5 to 15 nt apart can be used to measure 1D short-distance sliding because jumping motion only becomes effective when two deamination targets are spaced at least 20 nt apart (Ara et al., 2014; Halford & Marko, 2004). Substrates with deamination motifs 30 to 100 nt apart have been used to characterize jumping (Ara et al., 2014). A ssDNA substrate with a complementary DNA or RNA oligonucleotide annealed between the two deamination motifs can be used to confirm the jumping motion of the enzyme because AID/APOBEC enzymes do not bind or bind very weakly to dsDNA or RNA/DNA hybrids (Chelico et al., 2006; Chelico, Sacho, Erie, & Goodman, 2008; Feng et al., 2013). Attempts of an AID/APOBEC enzyme to slide over a duplex region induce dissociation of the enzyme from the substrate. As a result, when using a substrate with a duplex region between the motifs, only jumping motions enable the deamination of both deamination motifs in a single enzyme-substrate encounter (Ara et al., 2014; Ara et al., 2017). Intersegments transfer has been examined by increasing the concentration of the A3 and ssDNA substrate while keeping the ratio of the two components the same, thus allowing the reaction environment to become crowded and allowing the enzyme to become more likely to translocate to a different ssDNA than to translocate within the same ssDNA substrate (Adolph, Love, et al., 2017; Feng et al., 2015; Lieberman & Nordeen, 1997). The specific activity can be calculated in addition to processivity on any of these substrates by determining the picomoles of substrate deaminated per microgram of enzyme per minute.

The ionic strength of the buffer solution plays an essential role in properly characterizing AID/APOBEC enzyme activity and processivity on ssDNA (Chelico et al., 2006; Morse et al., 2017; Senavirathne et al., 2012; von Hippel & Berg, 1989; Wong et al., 2021). This is because electrostatic interaction between the positively charged AID/APOBEC enzymes and the negatively charged ssDNA backbone is a major determinant in driving the reaction rate above the diffusion rate during processive enzyme movement on ssDNA (Berg et al., 1981; Halford & Marko, 2004). As ions can shield the negative charges present on DNA and reduce the time that an enzyme remains in the domain of a DNA molecule, AID/APOBEC processivity on ssDNA can only be accurately observed at high salt conditions (such as at 5–10 mM $MgCl_2$) (Chelico et al., 2008; Feng & Chelico, 2011). Salt has also been shown to influence the polarity, oligomeric state, and specific activity of A3 enzymes on ssDNA (Chelico et al., 2008; Wong et al., 2021). Although AID/APOBEC enzymes can deaminate ssDNA in buffers with no salt, this would not give representative measurements since, within cells, there is a competitive electrostatic environment. The differences in buffer composition between labs has resulted the inability to compare specific activity and processivity results (Brown et al., 2021; Wong et al., 2021; Wong et al., 2022).

In this Chapter, we describe how to perform AID/APOBEC deamination assays. The methods detail how to measure AID/APOBEC catalysis and processive ssDNA scanning properties using purified enzymes and how to measure the deamination activity of AID/APOBEC enzymes using cell lysates. Note that while cell lysates are not recommended for detailed kinetic analyses, they omit the procedure of expressing and purifying recombination proteins and allow for the characterization of enzymatic behavior in a more physiologically relevant cellular context.

2. Kinetic *in vitro* deamination assay using purified enzymes

2.1 Equipment

1. Whatman #1 qualitative 9 cm circle filter (Sigma, Catalog #1001090).
2. Buchner funnel (10 cm diameter) fitted into a filter flask (used with Whatman #1 qualitative 9 cm circle filter).
3. Vacuum pump.
4. Heat block (37 °C and 85 °C).
5. Microcentrifuge.
6. Vertical gel electrophoresis system (Bio-rad, Mini-protean Tetra system) with glass plates having 0.75 mm spacers and a 10 well comb.
7. 5 mL syringe fitted with an 18-gauge needle.
8. Gel loading tips.
9. Chemidoc-MP imaging system (Bio-Rad).

2.2 Reagents

1. Urea.
2. 10x TBE (Tris-Borate, EDTA) buffer: for 1 L, add 108 g of Tris base, 56 g of Boric acid, 40 mL of 0.5 M EDTA, pH 8.0. Autoclave and store at room temperature.

3. Acrylamide: Bis-Acrylamide 19:1 (40 % Solution) (ThermoFisher, Catalog #BP1406-1). Store at 4 °C. The polyacrylamide is used to make different types of gels detailed in Table 1.
4. Tetra-methylethylenediamine (TEMED) (Sigma, Catalog #T9281).
5. 5 % (w/v) ammonium persulfate (APS) (Sigma, Catalog #A3678). Make fresh every 6 months and store at 4 °C in a container light protected or freeze into small aliquots and store at −20 °C for up to one year.
6. Phenol:chloroform:isoamyl alcohol 25:24:1 (Sigma, Catalog #P3803). Store at 4 °C.
7. Chloroform (Sigma, Catalog #NC2290834). Store at room temperature.
8. Uracil DNA glycosylase (UDG) and 10x UDG buffer (NEB, Catalog #M0280S).
9. 5 M NaOH.
10. MaXtract High Density tubes (Qiagen, Catalog #129046).
11. 10x deamination buffer.
    a. For activity or processivity assays (500 mM Tris-Cl, pH 7.5; 400 mM KCl; 100 mM $MgCl_2$; 10 mM DTT). Store aliquots at −20 °C. Avoid multiple freeze-thaw cycles since fresh DTT is important for enzyme activity.

Table 1 Components required to make 1 L of urea-acrylamide solution of various concentrations.

Polyacrylamide concentration	10 %	16 %	20 %
Acrylamide: Bis-Acrylamide 19:1 (40 % Solution)	250 mL	400 mL	500 mL
Urea	480 g	480 g	480 g
10x TBE buffer	100 mL	100 mL	100 mL
Deionized/distilled water (ddH_2O)	To 1 L (approximately 170 mL)	To 1 L (approximately 20 mL)	none required

Total volume can be adjusted to make more or less urea-polyacrylamide solution as necessary. Place solution in a beaker inside a basin filled with warm water. Place the basin and beaker on a magnetic stirrer. Stir beaker contents by magnetic stirring to ensure urea is completely dissolved. Filter through a Whatman #1 filter paper by vacuum before use. Store at room temperature for up to 6 months (**see Note 1**).

b. This solution is for activity assays only (500 mM Tris-Cl, pH 7.5; 500 mM NaCl; 10 mM DTT). Store aliquots at −20 °C. Avoid multiple freeze-thaw cycles, as fresh DTT is important for enzyme activity.

12. 1x Protein Dilution Buffer: 50 mM Tris, pH 7.5; 150 mM NaCl; 1 mM DTT; 10 % Glycerol.
13. Formamide-EDTA solution (clear): for 10 mL, mix 9.5 mL of 95 % formamide (ThermoFisher, Catalog #17899); 0.4 mL of 0.5 M EDTA; pH 8.0; and 0.1 mL of ddH_2O. Store at 4 °C until ready to use.
14. Formamide-EDTA solution (with dyes): mix 1 mL of clear formamide-EDTA solution with 0.1 mL of dye containing 1 % (w/v) bromophenol blue and 1 % (w/v) xylene cyanol. Store at 4 °C until ready to use.
15. Purified AID/APOBEC: purification is detailed in Chelico and Adolph, Chapter 6.
16. If purified AID/APOBEC was not treated with RNase during purification, then DNase-free RNase (Sigma, Catalog #11119915001) will need to be added to the reaction.
17. DNA substrate: DNA substrates up to 60 nt can be ordered from IDT Technologies. Longer substrates can be ordered from specialty companies such as Tri-Link Biotechnologies. If over 30 nt, the DNA must be HPLC or gel purified before use in the deamination assays. Previously characterized DNA substrates, their applications, and recommended urea polyacrylamide concentration for gel resolution are in Table 2.

2.3 Procedure

1. Thaw purified AID/APOBEC proteins on ice from the −80 °C freezer or liquid nitrogen (see **Note 2**). Dilute the AID/APOBEC proteins to an appropriate concentration using 1x Protein Dilution Buffer.
2. Thaw a ssDNA substrate sequence appropriate to your work at room temperature (see **Note 3**). After thawing, vortex briefly and place on ice.
3. Thaw 10x deamination buffer at room temperature. After thawing, vortex and place on ice.
4. Set up the reaction. Table 3 provides the recipe for one reaction, and the reaction can be scaled up accordingly, depending on the number of required time points for kinetic study. Although the sample volume taken in the end of the deamination assay is 10 μL (see step 6), the reaction is prepared to account for loss of sample volume during heating. Best practice is to prepare 5–10 μL extra than what is needed.

Table 2 DNA substrates.

DNA name	DNA sequence	Description	Urea polyacrylamide for band resolution	References
CCC L63 – 118 nt	5′ GAA TAT ATG TTG AGA CCC AAA GTA ATG AGA GAT TGA (Fam-dT) TAG ATG AGT GTA ATG TGA TAT ATG TGT ATG AAA GAT ATA AGA CCC AAA GAG TAA AGT TGT TAA TGT GTG TAG ATA TGT TAA	Two CCC motifs spaced 63 nt apart in a 118 nt substrate (deamination bands are 100 nt (5′C), 81 nt (3′C), and 63 nt (5′C&3′C)). Use for processivity (sliding and jumping), specific activity. Optimal substrate for A3G.	10 %	(Feng & Chelico, 2011)
AGC L63 – 118 nt	5′ GAA TAT ATG GTG AGA AGC GTA GTA ATG AGA GAT TGA TTA GAT TAG (Fam-dT) TTA ATG GGG TAT AGG GGT ATG AAA GGT ATA AGA AGC AGA GAG GAA AGG TGT TAA TTT GTG TAG ATA GGT TAA	Two AGC motifs spaced 63 nt apart in a 118 nt substrate (deamination bands are 100 nt (5′C), 81 nt (3′C), and 63 nt (5′C&3′C)). Use for processivity (sliding and jumping), specific activity, binding. Optimal substrate for AID.	10 %	Unpublished
ATC L63 – 118 nt	5′ GAA TAT ATG AGT TGA ATC AAA GTA ATG AGA GAG AAT (Fam-dT) TAG ATG AGT GTA ATG TGA TAT ATG TGT ATG AAA GAT ATA AGA ATC AAA GAG TAA AGT TGT TAA TGT GTG TAG ATA TGT TAA	Two ATC motifs spaced 63 nt apart in a 118 nt substrate (deamination bands are 100 nt (5′C), 81 nt (3′C), and 63 nt (5′C&3′C)). Use for processivity (sliding and jumping), specific activity, binding. Optimal substrate for A3B.	10 %	(Adolph, Love, et al., 2017)

CTC L63 – 118 nt	5′ GAA TAT AGT TTT TAG CTC AAA GTA AGT GAA GAT AAT (Fam-dT)TAG AGA GTT GTA ATG TGA TAT ATG TGT ATG AAA GAT ATA AGA CTC AAA GTG AAA AGT TGT TAA TGT GTG TAG ATA TGT TAA	Two CTC motifs spaced 63 nt apart in a 118 nt substrate (deamination bands are 100 nt (5′C), 81 nt (3′C), and 63 nt (5′C&3′C)). Use for processivity (sliding and jumping), specific activity, binding. Optimal substrate for A3H.	10 %	(Feng et al., 2015)
TTC L63 – 118 nt	5′ GAA TAT AGT TTT TAG TTC AAA GTA AGT GAA GAT AAT (Fam-dT) TAG AGA GTT GTA ATG TGA TAT ATG TGT ATG AAA GAT ATA AGA TTC AAA GAG TAA AGT TGT TAA TGT GTG TAG ATA TGT TAA	Two TTC motifs spaced 63 nt apart in a 118 nt substrate (deamination bands are 100 nt (5′C), 81 nt (3′C), and 63 nt (5′C&3′C)). Use for processivity (sliding and jumping), specific activity, binding. Optimal substrate for A3A, A3C, A3F, A1.	10 %	(Ara et al., 2014)
Complementary DNA to L63-118 nt	5′ CTT TCA TAC ACA TAT ATC AC	Can be heat annealed to any L63 – 118 nt to confirm jumping. Anneals between the two deamination motifs. Can be made as RNA.	10 % (for use with L63 – 118 nt)	(Feng & Chelico, 2011)

(continued)

Table 2 DNA substrates. (*cont'd*)

DNA name	DNA sequence	Description	Urea polyacrylamide for band resolution	References
CCC L30-85 nt	5′ AAA GAG AAA GTG ATA CCC AAA GAG TAA AGT (Fam-dT) AGA TAG AGA GTG ATA CCC AAA GAG TAA AGT TAG TAA GAT GTG TAA GTA TGT TAA	Two CCC motifs spaced 30 nt apart in an 85 nt substrate (deamination bands are 67 nt (5′C), 48 nt (3′C), and 30 nt (5′C&3′C)). Use for processivity (sliding and jumping), specific activity, binding. Optimal substrate for A3G.	16 %	(Feng & Chelico, 2011)
ATC L30-85 nt	5′ AAA GAG AAA GAG TAA ATC AAA GAG TAA AGT (Fam-dT) AAG TAG AGA GAT TAT ATC AAA GAG TAA AGT TAG TAA GAT GTG TAA GTA TGT TAA	Two ATC motifs spaced 30 nt apart in an 85 nt substrate (deamination bands are 67 nt (5′C), 48 nt (3′C), and 30 nt (5′C&3′C)). Use for processivity (sliding and jumping), specific activity, binding. Optimal substrate for A3B.	16 %	Unpublished
CTC L30-85 nt	5′ AAA GTG AAA GTG ATA CTC AAA TTT AAA AGT (Fam-dT) AGA TAG AAG GTG ATA CTC AAA TAT GAA AGT TAG TAA GAT GTG TAA GTA TGT TAA	Two CTC motifs spaced 30 nt apart in an 85 nt substrate (deamination bands are 67 nt (5′C), 48 nt (3′C), and 30 nt (5′C&3′C)). Use for processivity (sliding and jumping), specific activity, binding. Optimal substrate for A3H.	16 %	Unpublished

TTC L30-85 nt	5′ AAA GAG AAA GTG ATA TTC AAA GAG TAA AGT (Fam-dT) AGA TAG AGA GTG ATA TTC AAA GAG TAA AGT TAG TAA GAT GTG TAA GTA TGT TAA	Two TTC motifs spaced 30 nt apart in an 85 nt substrate (deamination bands are 67 nt (5′C), 48 nt (3′C), and 30 nt (5′C&3′C)). Use for processivity (sliding and jumping), specific activity, binding. Optimal substrate for A3A, A3C, A3F, A1.	16 %	(Ara et al., 2014)
CCC L14-69 nt	5′ AAA GAG AAA GTG AGA CCC AAA GAA (Fam-dT) GA AGA CCC AAA TGT TAG AAT TGT TAA TGT GTG TGA TGA TGT TGA	Two CCC motifs spaced 14 nt apart in a 69 nt substrate (deamination bands are 51 nt (5′C), 32 nt (3′C), and 14 nt (5′C&3′C)). Use for processivity (sliding), specific activity, binding. Optimal substrate for A3G.	16 %	(Ara et al., 2014)
ATC L14-69 nt	5′ AAA GAG TTA GGG TGA ATC AAA ATT (Fam-dT)AA AGA ATC AAA TGT TAG ATA TGT TAA TGT GTG TGA TGA TGT TGA	Two ATC motifs spaced 14 nt apart in a 69 nt substrate (deamination bands are 51 nt (5′C), 32 nt (3′C), and 14 nt (5′C&3′C)). Use for processivity (sliding), specific activity, binding. Optimal substrate for A3B.	16 %	(Adolph, Love, et al., 2017)

(continued)

Table 2 DNA substrates. *(cont'd)*

DNA name	DNA sequence	Description	Urea polyacrylamide for band resolution	References
TTC L14-69 nt	5′ AAA GAG TTA GTG AGA TTC AAA AT T (Fam-dT)AG AGA TTC AAA TGT TAG ATATGT TAA TGT GTG TGA TGA TGT TGA	Two TTC motifs spaced 14 nt apart in a 69 nt substrate (deamination bands are 51 nt (5′C), 32 nt (3′C), and 14 nt (5′C&3′C)). Use for processivity (sliding), specific activity, binding. Optimal substrate for A3A, A3C, A3F, A1.	16 %	(Ara et al., 2014)
CCC L5-60 nt	5′ AAA GAG AAA GTG ATA CCC A(Fam-dT)A CCC ATA GAG TAA AGT TAG TAA GAT GTG TAA GTA TGT TAA	Two CCC motifs spaced 5 nt apart in a 60 nt substrate (deamination bands are 42 nt (5′C), 23 nt (3′C), and 5 nt (5′C&3′C)). Use for processivity (sliding), specific activity, binding. Optimal substrate for A3G.	20 %	(Adolph, Love, et al., 2017)
ATC L5-60 nt	5′ AAA GAG AAA AGT ATA ATC A(Fam-dT)A ATC ATA GAG TAA AGT TAG TAA GAT GTG TAA GTA TGT TAA	Two ATC motifs spaced 5 nt apart in a 60 nt substrate (deamination bands are 42 nt (5′C), 23 nt (3′C), and 5 nt (5′C&3′C)). Use for processivity (sliding), specific activity, binding. Optimal substrate for A3B.	20 %	(Adolph, Love, et al., 2017)

TTC L5-60 nt	5′ AAA GAG AAA GTG ATA TTC A(Fam-dT)A TTC ATA GAG TAA AGT TAG TAA GAT GTG TAA GTA TGT TAA	Two TTC motifs spaced 5 nt apart in a 60 nt substrate (deamination bands are 42 nt (5′C), 23 nt (3′C), and 5 nt (5′C&3′C)). Use for processivity (sliding), specific activity, binding. Optimal substrate for A3A, A3C, A3F, A1.	20 %	(Ara et al., 2014)
CCC−43 nt	5′-Fam-ATT ATT ATT ACC CAA ATG GAT TTA TTT ATT TAT TTA TTT ATT T	One CCC motif in a substrate 43 nt in length. Produces a deamination band of 12 nt. Use for specific activity and binding. Optimal substrate for A3G.	16 %	Unpublished
TTC−43 nt	5′-Fam-ATT ATT ATT ATT CGA ATG GAT TTA TTT ATT TAT TTA TTT ATT T	One TTC motif in a substrate 43 nt in length. Produces a deamination band of 12 nt. Use for specific activity and binding. Optimal substrate for A3A, A3C, A3F and can also be used for A3B, A3H.	16 %	(Akre et al., 2016)

Fam-dT denotes Fluorescein amidite. Any DNA can be made into a UDG control by converting the 3′C in the deamination motif(s) to dU. Examples of how processivity substrates resolve on gels is shown in Fig. 3A.

Here we show an example with 5 μL of extra volume for a total reaction volume of 15 μL. Importantly, always include a no enzyme control in which the enzyme volume is replaced by ddH_2O. Initially, it is important to ensure that the amount of UDG used removes all possible dU by using a substrate synthesized with dU instead of dC. The dU oligonucleotide reaction can also be used as a size marker for the deamination product.

Add all the components, except for ssDNA, in a 0.5 mL microcentrifuge tube on ice and pipette mix. If the volume exceeds 30 μL, the reaction should be prewarmed at 37 °C for 30 s before moving to step 3 (see **Note 4**).

5. Start the reaction by adding ssDNA substrate, immediately pipette mixing and then incubating the reaction at 37 °C for 30 s to 30 min. Low percentage of substrate usage (8 %–12 %) is crucial for measuring processivity as it allows the reaction to be carried out under single-hit conditions (see **Note 5**). Conducting a time course enables determination of optimal substrate usage. Alternatively, the time course can be used to compare activity levels. Some A3s with lower activity may need different time courses and/or addition of enzyme at the upper range (Table 3).
6. Terminate the reaction by transferring 10 μL of reaction to a pre-spun MaXtract tube that contains 10 μL ddH_2O and 30 μL phenol:-chloroform:isoamyl alcohol. Vortex thoroughly for 20 s (see **Note 6**).
7. Spin down the MaXtract tube for 5 min at 12,000 x g using a microcentrifuge to separate the aqueous phase from the organic phase.
8. Remove any residual phenol by adding 30 μL of chloroform to samples in the same MaXtract tube. Vortex briefly, and spin the reaction for 5 min at 12,000 x g.

Table 3 Reaction set-up for a kinetic *in vitro* deamination assay.

Reagents	Stock concentration	Volume	Final concentration
Deamination buffer	10x	1.5 μL	1x
Enzyme	1 μM	x μL	20–350 nM
ssDNA substrate	1 μM	1.5 μL	100 nM
ddH_2O		To 15 μL total volume	

9. Set up the UDG reaction.
   9.1 While the MaXtract tubes are spinning, prepare reagents for the UDG assay by mixing 2.5 μL 10x UDG buffer and 2 μL of UDG enzyme in a clean 0.5 mL microcentrifuge tube.
   9.2 Carefully remove 20 μL solution containing protein-free ssDNA substrate from the aqueous phase of the MaXtract tube without disturbing the gel bed, and add to the UDG reaction. Mix the reaction by pipetting up and down a few times.
   9.3 Incubate the reaction for 45 to 60 min at 37 °C.
   9.4 After the reaction, spin down tubes briefly to collect any condensation (see **Note 7**).
10. Add 1 μL of 5 M NaOH and flick to mix. Heat the reaction for 10 min at 85 °C to hydrolyze the alkaline-sensitive AP site(s).
11. After the reaction, immediately place the tubes on ice to cool down. Centrifuge tubes briefly to collect any condensation.
12. Add an equal volume (~25 μL) of formamide-EDTA (clear) loading buffer to the reaction and flick until thoroughly mixed (see **Note 8**).
13. Resolve deamination products by following the "Denaturing urea polyacrylamide electrophoresis" protocol (Section 4).

3. Non-kinetic *in vitro* deamination assay using cell lysates

3.1 Equipment

1. T75 cm^2 flasks with vented cap (ThermoFisher, Catalog # 130190).
2. Cell scraper (Corning, Catalog #CLS3010).
3. Rocker (in a cold room).
4. Whatman #1 qualitative 9 cm circle filter (Sigma, Catalog #1001090).
5. Buchner funnel (10 cm diameter) fitted into a filter flask (used with Whatman #1 qualitative 9 cm circle filter) with vacuum.
6. Heat block (37 °C and 85 °C).
7. Microcentrifuge.
8. Vertical gel electrophoresis system (Bio-rad, Mini-protean Tetra system) with glass plates having 0.75 mm spacers and a 10 well comb.
9. 5 mL syringe fitted with an 18-gauge needle.
10. Gel loading tips.
11. Chemidoc-MP imaging system (Bio-Rad).

3.2 Reagents

1. 293 T cells (ATCC, Catalog #CRL-3216) or other preferred cell line.
2. GeneJuice® Transfection Reagent (Sigma, Catalog #70967).
3. 1x PBS (Cytvia, Catalog #SH3025601).
4. DNase-free RNase (Sigma, Catalog #11119915001).
5. Cell lysis buffer: 50 mM Tris-Cl pH 7.4, 1 % Nonidet-P40 substitute, 0.1 % sodium deoxycholate, 10 % glycerol, 150 mM NaCl, with the following components added fresh before use 1 mM DTT, 20 μg/mL RNase A, and cOmplete™, EDTA-free Protease Inhibitor Cocktail (Sigma, Catalog #11873580001).
6. Urea (ThermoFisher, Catalog #U15-3).
7. 10x TBE (Tris-Borate, EDTA) buffer: for 1 L, add 108 g of Tris base, 56 g of Boric acid, 40 mL of 0.5 M EDTA, pH 8.0. Autoclave and store at room temperature.
8. Acrylamide: Bis-Acrylamide 19:1 (40 % Solution) (ThermoFisher, Catalog #BP1406-1). Store at 4 °C. The polyacrylamide is used to make different types of gels detailed in Table 1.
9. Tetra-methylethylenediamine (TEMED) (Sigma, Catalog #T9281).
10. 5 % (w/v) ammonium persulfate (APS) (Sigma, Catalog #A3678). Make fresh every 6 months and store at 4 °C in a container impervious to light or freeze into small aliquots and store at −20 °C for up to one year.
11. Uracil DNA glycosylase (UDG) and 10x UDG buffer (NEB, Catalog #M0280S).
12. 5 M NaOH.
13. Formamide-EDTA solution (clear): for 10 mL, mix 9.5 mL of 95 % formamide (19:1); 0.4 mL of 0.5 M EDTA, pH 8.0; and 0.1 mL of ddH_2O. Store at 4 °C until ready to use.
14. Formamide-EDTA solution (with dyes): mix 1 mL of clear formamide-EDTA solution with 0.1 mL of dye that contains 1 % (w/v) bromophenol blue and 1 % (w/v) xylene cyanol. Store at 4 °C until ready to use.
15. DNA substrate: DNA substrates up to 60 nt can be ordered from IDT Technologies. Longer substrates can be ordered from specialty companies such as Tri-Link Biotechnologies. If over 30 nt the DNA must be HPLC or gel purified before use in the deamination assays. Previously characterized DNA substrates, their applications, and recommended polyacrylamide concentration are in Table 2.

3.3 Procedure

1. Express the AID/APOBEC by stable transduction or transfection in 293 T cells or other cell line in a T75 flask. The method used for AID/APOBEC expression in the mammalian cell line will not affect the results. Importantly, always include a "no enzyme" control in which cells not expressing AID/APOBEC are processed according to this procedure. An example experimental set-up would be to plate 1.0×10^6 293 T cells per 75 cm^2 flask. The next day transfect 1 μg of plasmid DNA containing the AID/APOBEC cDNA or empty plasmid (no enzyme) with 3 μL GeneJuice transfection reagent according to manufacturer's protocol.
2. Forty-eight hours after the transfection, remove media from cells, rinse with 1x cold PBS and then add 1 mL of cell lysis buffer.
3. Collect cells with a cell scraper and transfer to a 1.5 mL tube. Rock the tube on a rocker for 30 min at 4 °C to lyse cells.
4. Centrifuge tube (12,000 x g for 30 min) to pellet the insoluble fraction. Remove supernatant and place tube on ice. Prepare aliquots and freeze unneeded portions of the sample at −80 °C.
5. Thaw a ssDNA substrate sequence at room temperature appropriate to your work (see **Note 3**). After DNA thaws, vortex briefly and place on ice. Initially, it is important to ensure that UDG is removing all possible dU by using a substrate synthesized with dU instead of dC. The dU oligonucleotide reaction can also be used as a size marker for the deamination product.
6. Set up the deamination reaction with 2 μL UDG, 0.25 μL DNase-free RNase A, 2 μL 10x UDG buffer, 16.5 μL cell lysate. Incubate at 37 °C for 30 min to 2 h (**see Note 9**).
7. Add 1 μL of 5 M NaOH and flick to mix. Heat the reaction for 10 min at 85 °C to hydrolyze the alkaline-sensitive AP site(s).
8. Add an equal volume (~21 μL) of formamide-EDTA loading buffer to the reaction and flick until thoroughly mixed (see **Note 8**).
9. Resolve deamination products by following the "Denaturing urea polyacrylamide electrophoresis" protocol (Section 4).

4. Denaturing urea polyacrylamide electrophoresis

1. Prepare urea polyacrylamide solution (Table 1). One mini gel requires 5 mL of urea polyacrylamide solution. Place the urea polyacrylamide

solution in a small beaker. Add 100 μL of 5 % (w/v) APS and swirl to mix. Then immediately add 50 μL of TEMED and swirl to mix. Quickly pour or pipette the urea polyacrylamide solution between the glass plates. Insert the comb and allow gel to polymerize for at least 30 min (see **Note 10**).

2. After polymerization is completed, carefully remove the comb and place the gel into the gel box with 1x TBE running buffer. Flush the wells with 1x TBE running buffer using a syringe fitted with a needle. This will remove residual urea or polyacrylamide from the well and minimize sample smearing on the gel.
3. Pre-run the urea polyacrylamide gel in 1x TBE buffer at 150 V for 15 min
4. Before loading, heat all samples for 2 min at 85 °C to ensure the ssDNA is linear (or denature any dsDNA regions) and immediately place on ice afterwards.
5. Flush the wells a second time with 1x TBE running buffer using a syringe fitted with a needle. Load 10–15 μL of sample into a well using gel loading tips. Reserve a lane to load formamide-EDTA loading buffer with dye so that DNA migration can be tracked during electrophoresis. Dye is not used with samples since the dye may interfere with visualization of the Fam-labeled DNA.
6. Run gel at 160 V to resolve all reaction products. During sample migration, periodically examine the migration of the marker dye. The gel running can usually be stopped when the lower dye front reaches the end of the gel. However, duration of electrophoresis should be empirically decided, depending on the percentage of the polyacrylamide and size of the resolved bands. Tables for dye resolution can be found in Sambrook *et al.* or on-line resources (Green & Sambrook, 2012).
7. After gel resolution is complete, the glass plates are removed from the gel box but are not dissembled until reaching the Chemidoc-MP imaging system fitted with a tray for UV imaging (or any imaging system capable of imaging fluorescein (ex 488 nm/em 520 nm)). At the imaging system, the glass plates are opened and ddH_2O is used to coat the gel and avoid drying. Some ddH_2O is also sprayed onto the UV imaging tray. This will enable the user to lift the gel from one side and place it onto the UV imaging tray. The water on the UV imaging tray enables the user to smooth out the gel (wear clean gloves when touching the gel). Then, the gel is scanned using the fluorescein settings on the Chemidoc-MP imaging system.

8. The gel image can be exported as a TIFF file and analyzed by programs that determine the integrated gel band intensity, such as ImageQuant software (Molecular Dynamics), TotalLab Quant (TotalLab), Image Lab software (Bio-Rad).

5. Analysis of processive deamination by AID/APOBEC enzyme

1. Quantify the band intensities with a program that determines the integrated gel band intensity such as ImageQuant software (Molecular Dynamics), TotalLab Quant (TotalLab), Image Lab software (Bio-Rad). Draw boxes around each of the four possible bands and analyze one lane at a time to obtain the percentage of each band. Ensure that for processivity, lanes used only have 8–12 % total deamination. Use the "no A3" lane to draw boxes for background subtraction at the same place in the gel as where the expected deamination bands resolved.
2. Calculate the processivity factor. Examples of calculations are shown in Fig. 3B (non-processive) and Fig. 3C (processive).
   - **2.1** Calculate the sum of all deaminations at the 5′ end by combining the integrated band intensity of single deaminations at the 5′C (Fig. 3B and C, 100-nt band) and the integrated band intensity of the double (5′C & 3′C) deamination band (Fig. 3B and C, 63-nt band). Non-processive enzymes will not produce a double deamination band or the band will be very faint (Fig. 3A, A3A, A3C).
   - **2.2** Calculate the sum of all deamination at the 3′ end by combining the integrated band intensity of single deaminations at the 3′C (Fig. 3B and C, 81-nt band) and the integrated band intensity of double (5′C & 3′C) deamination band (Fig. 3B and C, 63-nt band). Non-processive enzymes will not produce a double deamination band or the band will be very faint (Fig. 3A, A3A, A3C).
   - **2.3** Calculate the predicted fraction of independent double deaminations (if two separate enzymes catalyzed both deamination events). This is calculated by multiplying the values obtained from steps 2.1 and 2.2 and is based on the probability rule that the probability of two events being random (independent) is their product.

2.4 The processivity factor is calculated as the ratio of the experimentally quantified double (5′C & 3′C) deaminations (Fig. 3B and C, 63-nt band) to the predicted fraction of independent double deaminations (value calculated in step 2.3).

6. Determine specific activity of an AID/APOBEC enzyme

1. Quantify the band intensities with a program that determines the integrated gel band intensity such as ImageQuant software (Molecular Dynamics), TotalLab Quant (TotalLab), Image Lab software (Bio-Rad). For substrates with more than one dC motif, draw boxes around each of the four possible bands individually (substrate and three product bands) and analyze one lane at a time to obtain the total percentage of deamination (add the three product bands, 5′C, 3′C and double (5′C & 3′C) deamination band). For substrates with one dC motif draw individual boxes around the substrate and product bands to determine the percent deamination (product). Ensure that for specific activity, lanes used for analysis are within the steady rise of the kinetic curve and not after the curve asymptotes. Use the "no A3" lane to draw boxes for background subtraction at the same place in gel as where expected deamination bands resolved.
2. Determine the percent of deamination by adding up the bands for the 5′C, 3′C, and double deamination band (if using a processivity substrate) or simply the one product band.
3. Calculate specific activity as picomoles of substrate used per microgram of enzyme per minute. The reaction volume used to calculate the picomoles substrate and micrograms of enzyme is 10 μL (volume taken from original reaction tube). For example, for 100 nM DNA in a 10 μL reaction, there is 1 pmole of substrate.

7. Notes

Note 1: Acrylamide is a neurotoxin. Appropriate personal protective equipment must be used when preparing polyacrylamide solution. After 6 months the urea polyacrylamide solution is still useful for electrophoresis, however, the color of the urea polyacrylamide becomes increasingly purple under the conditions needed to visualize the Fam labeled ssDNA substrate resulting in an unwanted background.

Note 2: The purification of APOBEC enzymes can be found in Chelico and Adolph, Chapter 6. The methods here assume that RNase A was added during purification. AID/APOBEC enzymes bind RNA in cells and for all except A3A and A1 this inhibits their catalytic activity (Bransteitter, Pham, Scharff, & Goodman, 2003; Chelico et al., 2006; Cortez et al., 2019; Wong et al., 2021). If RNase A was not added during purification, then DNase-free RNase A must be added to the deamination reaction. The activity of APOBEC enzymes is sensitive to repeated freeze-thaw cycles. Multiple aliquots of small volumes (10 μL to 20 μL) of purified enzymes should be stored at −80 °C or in liquid nitrogen after purification. APOBEC enzymes can be frozen after dilution in 1x Protein Dilution Buffer.

Note 3: Fluorescently labeled ssDNA can be synthesized from Integrated DNA Technologies or Tri-Link Biotechnologies. Ordered oligonucleotides should be HPLC-grade or gel purified if over 30 nt long. The concentration of DNA can be determined spectrophotometrically. Repeated freeze-thaw should be avoided to minimize the degradation of fluorescein.

Note 4: The concentration of enzyme will vary depending on which A3 enzymes are being used. The DNA substrate to enzyme ratios recommended for each enzyme are A3A, A3B, A3H – 1:0.5; A3C – 1:3.5; A3F – 1:2; A3G – 1: 0.2–0.5.

Note 5: The incubation time and the DNA to enzyme ratio will vary depending on the enzymatic activity of a specific deaminase. Use a broad time course initially to find the optimal time. Then, reactions can use only 3 time points surrounding ~10% substrate usage for subsequent experiments, *e.g.*, Fig. 2B. Processivity comparisons are only accurate if all samples had same substrate usage within ± 2% (Fig. 2B).

Note 6: MaXtract tubes can be prepared before starting the reaction. Prior to use, spin down the gel inside the MaXtract tubes in a microcentrifuge for 30 s at 12,000 x g. Perform all procedures involving phenol or chloroform in a chemical fume hood.

Note 7: Rather than proceeding directly to the UDG assay, DNA extracted from the MaXtract tube can be temporarily stored at −20 °C for the short-term, *e.g.*, 1–2 days.

Note 8: Reactions can now be saved at −20 °C until ready to proceed to electrophoresis.

Note 9: The amount of lysate and incubation time used needs to be determined empirically and can vary depending on the method of AID/APOBEC enzyme overexpression.

Note 10: Choose the concentration of urea polyacrylamide solution depending on your separation requirements and DNA substrate (Tables 1 and 2).

Acknowledgments

We thank Madison Adolph for thoughtful comments on the manuscript. This work was supported by CIHR (L.C.) PJT-162407 and PJT-159560.

References

Adolph, M. B., Ara, A., Feng, Y., Wittkopp, C. J., Emerman, M., Fraser, J. S., & Chelico, L. (2017). Cytidine deaminase efficiency of the lentiviral viral restriction factor APOBEC3C correlates with dimerization. *Nucleic Acids Research, 45*(6), 3378–3394. https://doi.org/10.1093/nar/gkx066.

Adolph, M. B., Love, R. P., & Chelico, L. (2018). Biochemical basis of APOBEC3 deoxycytidine deaminase activity on diverse DNA substrates. *ACS Infectious Diseases, 4*(3), 224–238. https://doi.org/10.1021/acsinfecdis.7b00221.

Adolph, M. B., Love, R. P., Feng, Y., & Chelico, L. (2017). Enzyme cycling contributes to efficient induction of genome mutagenesis by the cytidine deaminase APOBEC3B. *Nucleic Acids Research, 45*(20), 11925–11940. https://doi.org/10.1093/nar/gkx832.

Akre, M. K., Starrett, G. J., Quist, J. S., Temiz, N. A., Carpenter, M. A., Tutt, A. N., ... Harris, R. S. (2016). Mutation processes in 293-based clones overexpressing the DNA cytosine deaminase APOBEC3B. *PLoS One, 11*(5), e0155391. https://doi.org/10.1371/journal.pone.0155391.

Ara, A., Love, R. P., & Chelico, L. (2014). Different mutagenic potential of HIV-1 restriction factors APOBEC3G and APOBEC3F is determined by distinct single-stranded DNA scanning mechanisms. *PLoS Pathogens, 10*(3), e1004024. https://doi.org/10.1371/journal.ppat.1004024.

Ara, A., Love, R. P., Follack, T. B., Ahmed, K. A., Adolph, M. B., & Chelico, L. (2017). Mechanism of enhanced HIV restriction by virion coencapsidated cytidine deaminases APOBEC3F and APOBEC3G. *Journal of Virology, 91*(3), https://doi.org/10.1128/JVI.02230-16.

Berg, O. G., Winter, R. B., & von Hippel, P. H. (1981). Diffusion-driven mechanisms of protein translocation on nucleic acids. 1. Models and theory. *Biochemistry, 20*(24), 6929–6948. https://doi.org/10.1021/bi00527a028.

Bonnet, I., Biebricher, A., Porte, P. L., Loverdo, C., Benichou, O., Voituriez, R., ... Desbiolles, P. (2008). Sliding and jumping of single *Eco*RV restriction enzymes on non-cognate DNA. *Nucleic Acids Research, 36*(12), 4118–4127. https://doi.org/10.1093/nar/gkn376.

Bransteitter, R., Pham, P., Calabrese, P., & Goodman, M. F. (2004). Biochemical analysis of hypermutational targeting by wild type and mutant activation-induced cytidine deaminase. *The Journal of Biological Chemistry, 279*(49), 51612–51621. https://doi.org/10.1074/jbc.M408135200.

Bransteitter, R., Pham, P., Scharff, M. D., & Goodman, M. F. (2003). Activation-induced cytidine deaminase deaminates deoxycytidine on single-stranded DNA but requires the action of RNase. *Proceedings of the National Academy of Sciences of the United States of America, 100*(7), 4102–4107. https://doi.org/10.1073/pnas.0730835100.

Brown, A. L., Collins, C. D., Thompson, S., Coxon, M., Mertz, T. M., & Roberts, S. A. (2021). Single-stranded DNA binding proteins influence APOBEC3A substrate preference. *Scientific Reports, 11*(1), 21008. https://doi.org/10.1038/s41598-021-00435-y.

Brown, M. W., Kim, Y., Williams, G. M., Huck, J. D., Surtees, J. A., & Finkelstein, I. J. (2016). Dynamic DNA binding licenses a repair factor to bypass roadblocks in search of DNA lesions. *Nature Communications, 7*, 10607. https://doi.org/10.1038/ncomms10607.

Carey, D. C., & Strauss, P. R. (1999). Human apurinic/apyrimidinic endonuclease is processive. *Biochemistry, 38*(50), 16553–16560. https://doi.org/10.1021/bi9907429.

Chelico, L., Pham, P., Calabrese, P., & Goodman, M. F. (2006). APOBEC3G DNA deaminase acts processively 3′ – > 5′ on single-stranded DNA. *Nature Structural & Molecular Biology, 13*(5), 392–399. https://doi.org/10.1038/nsmb1086.

Chelico, L., Pham, P., & Goodman, M. F. (2009). Stochastic properties of processive cytidine DNA deaminases AID and APOBEC3G. *Philosophical Transactions of the Royal Society of London. Series B, Biological Sciences, 364*(1517), 583–593. https://doi.org/10.1098/rstb.2008.0195.

Chelico, L., Sacho, E. J., Erie, D. A., & Goodman, M. F. (2008). A model for oligomeric regulation of APOBEC3G cytosine deaminase-dependent restriction of HIV. *The Journal of Biological Chemistry, 283*(20), 13780–13791. https://doi.org/10.1074/jbc.M801004200.

Cheng, A. Z., Moraes, S. N., Shaban, N. M., Fanunza, E., Bierle, C. J., Southern, P. J., ... Harris, R. S. (2021). APOBECs and Herpesviruses. *Viruses, 13*(3), https://doi.org/10.3390/v13030390.

Cheng, A. Z., Yockteng-Melgar, J., Jarvis, M. C., Malik-Soni, N., Borozan, I., Carpenter, M. A., ... Harris, R. S. (2019). Epstein-Barr virus BORF2 inhibits cellular APOBEC3B to preserve viral genome integrity. *Nature Microbiology, 4*(1), 78–88. https://doi.org/10.1038/s41564-018-0284-6.

Cortez, L. M., Brown, A. L., Dennis, M. A., Collins, C. D., Brown, A. J., Mitchell, D., ... Roberts, S. A. (2019). APOBEC3A is a prominent cytidine deaminase in breast cancer. *PLoS Genetics, 15*(12), e1008545. https://doi.org/10.1371/journal.pgen.1008545.

Creighton, S., & Goodman, M. F. (1995). Gel kinetic analysis of DNA polymerase fidelity in the presence of proofreading using bacteriophage T4 DNA polymerase. *The Journal of Biological Chemistry, 270*(9), 4759–4774. https://doi.org/10.1074/jbc.270.9.4759.

Desimmie, B. A., Delviks-Frankenberrry, K. A., Burdick, R. C., Qi, D., Izumi, T., & Pathak, V. K. (2014). Multiple APOBEC3 restriction factors for HIV-1 and one Vif to rule them all. *Journal of Molecular Biology, 426*(6), 1220–1245. https://doi.org/10.1016/j.jmb.2013.10.033.

Di Noia, J. M., & Neuberger, M. S. (2007). Molecular mechanisms of antibody somatic hypermutation. *Annual Review of Biochemistry, 76*, 1–22. https://doi.org/10.1146/annurev.biochem.76.061705.090740.

Dowd, D. R., & Lloyd, R. S. (1990). Biological significance of facilitated diffusion in protein-DNA interactions. Applications to T4 endonuclease V-initiated DNA repair. *The Journal of Biological Chemistry, 265*(6), 3424–3431. https://www.ncbi.nlm.nih.gov/pubmed/2406255.

Esadze, A., & Stivers, J. T. (2018). Facilitated diffusion mechanisms in DNA base excision repair and transcriptional activation. *Chemical Reviews, 118*(23), 11298–11323. https://doi.org/10.1021/acs.chemrev.8b00513.

Feng, Y., Baig, T. T., Love, R. P., & Chelico, L. (2014). Suppression of APOBEC3-mediated restriction of HIV-1 by Vif. *Frontiers in Microbiology, 5*, 450. https://doi.org/10.3389/fmicb.2014.00450.

Feng, Y., & Chelico, L. (2011). Intensity of deoxycytidine deamination of HIV-1 proviral DNA by the retroviral restriction factor APOBEC3G is mediated by the noncatalytic domain. *The Journal of Biological Chemistry, 286*(13), 11415–11426. https://doi.org/10.1074/jbc.M110.199604.

Feng, Y., Love, R. P., Ara, A., Baig, T. T., Adolph, M. B., & Chelico, L. (2015). Natural polymorphisms and oligomerization of human APOBEC3H contribute to single-stranded DNA scanning ability. *The Journal of Biological Chemistry, 290*(45), 27188–27203. https://doi.org/10.1074/jbc.M115.666065.

Feng, Y., Love, R. P., & Chelico, L. (2013). HIV-1 viral infectivity factor (Vif) alters processive single-stranded DNA scanning of the retroviral restriction factor APOBEC3G. *The Journal of Biological Chemistry, 288*(9), 6083–6094. https://doi.org/10.1074/jbc.M112.421875.

Feng, Y., & Martin, A. (2022). Mutagenic repair during antibody diversification: Emerging insights. *Trends in Immunology, 43*(8), 604–607. https://doi.org/10.1016/j.it.2022.05.004.

Feng, Y., Seija, N., Di Noia, J. M., & Martin, A. (2021). AID in antibody diversification: There and back again: (Trends in Immunology 41, 586-600; 2020). *Trends in Immunology, 42*(1), 89. https://doi.org/10.1016/j.it.2020.10.011.

Gaba, A., Flath, B., & Chelico, L. (2021). Examination of the APOBEC3 barrier to cross species transmission of primate lentiviruses. *Viruses, 13*(6), https://doi.org/10.3390/v13061084.

Green, M. R., & Sambrook, J. (2012). *Molecular cloning: A laboratory manual, VOl. 3*. Cold Spring Harbor Laboratory Press.

Greeve, J., Navaratnam, N., & Scott, J. (1991). Characterization of the apolipoprotein B mRNA editing enzyme: No similarity to the proposed mechanism of RNA editing in kinetoplastid protozoa. *Nucleic Acids Research, 19*(13), 3569–3576. https://doi.org/10.1093/nar/19.13.3569.

Halford, S. E., & Marko, J. F. (2004). How do site-specific DNA-binding proteins find their targets? *Nucleic Acids Research, 32*(10), 3040–3052. https://doi.org/10.1093/nar/gkh624.

Harris, R. S., Petersen-Mahrt, S. K., & Neuberger, M. S. (2002). RNA editing enzyme APOBEC1 and some of its homologs can act as DNA mutators. *Molecular Cell, 10*(5), 1247–1253. https://doi.org/10.1016/s1097-2765(02)00742-6.

Jalili, P., Bowen, D., Langenbucher, A., Park, S., Aguirre, K., Corcoran, R. B., ... Buisson, R. (2020). Quantification of ongoing APOBEC3A activity in tumor cells by monitoring RNA editing at hotspots. *Nature Communications, 11*(1), 2971. https://doi.org/10.1038/s41467-020-16802-8.

Leeman-Neill, R. J., Bhagat, G., & Basu, U. (2024). AID in non-Hodgkin B-cell lymphomas: The consequences of on- and off-target activity. *Advances in Immunology, 161*, 127–164. https://doi.org/10.1016/bs.ai.2024.03.005.

Lieberman, B. A., & Nordeen, S. K. (1997). DNA intersegment transfer, how steroid receptors search for a target site. *The Journal of Biological Chemistry, 272*(2), 1061–1068. https://doi.org/10.1074/jbc.272.2.1061.

Lorenzo, J. P., Molla, L., Amro, E. M., Ibarra, I. L., Ruf, S., Neber, C., ... Papavasiliou, F. N. (2024). APOBEC2 safeguards skeletal muscle cell fate through binding chromatin and regulating transcription of non-muscle genes during myoblast differentiation. *Proceedings of the National Academy of Sciences of the United States of America, 121*(17), e2312330121. https://doi.org/10.1073/pnas.2312330121.

Love, R. P., Xu, H., & Chelico, L. (2012). Biochemical analysis of hypermutation by the deoxycytidine deaminase APOBEC3A. *The Journal of Biological Chemistry, 287*(36), 30812–30822. https://doi.org/10.1074/jbc.M112.393181.

Marino, D., Perkovic, M., Hain, A., Jaguva Vasudevan, A. A., Hofmann, H., Hanschmann, K. M., ... Munk, C. (2016). APOBEC4 enhances the replication of HIV-1. *PLoS One, 11*(6), e0155422. https://doi.org/10.1371/journal.pone.0155422.

Methot, S. P., & Di Noia, J. M. (2017). Molecular mechanisms of somatic hypermutation and class switch recombination. *Advances in Immunology, 133*, 37–87. https://doi.org/10.1016/bs.ai.2016.11.002.

Morse, M., Huo, R., Feng, Y., Rouzina, I., Chelico, L., & Williams, M. C. (2017). Dimerization regulates both deaminase-dependent and deaminase-independent HIV-1 restriction by APOBEC3G. *Nature Communications, 8*(1), 597. https://doi.org/10.1038/s41467-017-00501-y.

Pham, P., Bransteitter, R., Petruska, J., & Goodman, M. F. (2003). Processive AID-catalysed cytosine deamination on single-stranded DNA simulates somatic hypermutation. *Nature, 424*(6944), 103–107. https://doi.org/10.1038/nature01760.

Pham, P., Malik, S., Mak, C., Calabrese, P. C., Roeder, R. G., & Goodman, M. F. (2019). AID-RNA polymerase II transcription-dependent deamination of IgV DNA. *Nucleic Acids Research, 47*(20), 10815–10829. https://doi.org/10.1093/nar/gkz821.

Pham, P., Smolka, M. B., Calabrese, P., Landolph, A., Zhang, K., Zhou, H., & Goodman, M. F. (2008). Impact of phosphorylation and phosphorylation-null mutants on the activity and deamination specificity of activation-induced cytidine deaminase. *The Journal of Biological Chemistry, 283*(25), 17428–17439. https://doi.org/10.1074/jbc.M802121200.

Riggs, A. D., Bourgeois, S., & Cohn, M. (1970). The lac repressor-operator interaction. 3. Kinetic studies. *Journal of Molecular Biology, 53*(3), 401–417. https://doi.org/10.1016/0022-2836(70)90074-4.

Senavirathne, G., Jaszczur, M., Auerbach, P. A., Upton, T. G., Chelico, L., Goodman, M. F., & Rueda, D. (2012). Single-stranded DNA scanning and deamination by APOBEC3G cytidine deaminase at single molecule resolution. *The Journal of Biological Chemistry, 287*(19), 15826–15835. https://doi.org/10.1074/jbc.M112.342790.

Seplyarskiy, V. B., Soldatov, R. A., Popadin, K. Y., Antonarakis, S. E., Bazykin, G. A., & Nikolaev, S. I. (2016). APOBEC-induced mutations in human cancers are strongly enriched on the lagging DNA strand during replication. *Genome Research, 26*(2), 174–182. https://doi.org/10.1101/gr.197046.115.

Sharma, S., Patnaik, S. K., Taggart, R. T., & Baysal, B. E. (2016). The double-domain cytidine deaminase APOBEC3G is a cellular site-specific RNA editing enzyme. *Scientific Reports, 6*, 39100. https://doi.org/10.1038/srep39100.

Sharma, S., Patnaik, S. K., Taggart, R. T., Kannisto, E. D., Enriquez, S. M., Gollnick, P., & Baysal, B. E. (2015). APOBEC3A cytidine deaminase induces RNA editing in monocytes and macrophages. *Nature Communications, 6*, 6881. https://doi.org/10.1038/ncomms7881.

Shen, H. M., Poirier, M. G., Allen, M. J., North, J., Lal, R., Widom, J., & Storb, U. (2009). The activation-induced cytidine deaminase (AID) efficiently targets DNA in nucleosomes but only during transcription. *The Journal of Experimental Medicine, 206*(5), 1057–1071. https://doi.org/10.1084/jem.20082678.

Shen, H. M., Tanaka, A., Bozek, G., Nicolae, D., & Storb, U. (2006). Somatic hypermutation and class switch recombination in Msh6(-/-)Ung(-/-) double-knockout mice. *Journal of Immunology, 177*(8), 5386–5392. https://doi.org/10.4049/jimmunol.177.8.5386.

Stanford, N. P., Szczelkun, M. D., Marko, J. F., & Halford, S. E. (2000). One- and three-dimensional pathways for proteins to reach specific DNA sites. *The EMBO Journal, 19*(23), 6546–6557. https://doi.org/10.1093/emboj/19.23.6546.

Storb, U., Shen, H. M., & Nicolae, D. (2009). Somatic hypermutation: Processivity of the cytosine deaminase AID and error-free repair of the resulting uracils. *Cell Cycle (Georgetown, Tex.), 8*(19), 3097–3101. https://doi.org/10.4161/cc.8.19.9658.

Swanton, C., McGranahan, N., Starrett, G. J., & Harris, R. S. (2015). APOBEC enzymes: Mutagenic fuel for cancer evolution and heterogeneity. *Cancer Discovery, 5*(7), 704–712. https://doi.org/10.1158/2159-8290.CD-15-0344.

Tafvizi, A., Huang, F., Fersht, A. R., Mirny, L. A., & van Oijen, A. M. (2011). A single-molecule characterization of p53 search on DNA. *Proceedings of the National Academy of Sciences of the United States of America, 108*(2), 563–568. https://doi.org/10.1073/pnas.1016020107.

von Hippel, P. H., & Berg, O. G. (1989). Facilitated target location in biological systems. *The Journal of Biological Chemistry, 264*(2), 675–678. https://www.ncbi.nlm.nih.gov/pubmed/2642903.

Wang, Y., Schmitt, K., Guo, K., Santiago, M. L., & Stephens, E. B. (2016). Role of the single deaminase domain APOBEC3A in virus restriction, retrotransposition, DNA damage and cancer. *The Journal of General Virology, 97*(1), 1–17. https://doi.org/10.1099/jgv.0.000320.

Wong, L., Sami, A., & Chelico, L. (2022). Competition for DNA binding between the genome protector replication protein A and the genome modifying APOBEC3 single-stranded DNA deaminases. *Nucleic Acids Research, 50*(21), 12039–12057. https://doi.org/10.1093/nar/gkac1121.

Wong, L., Vizeacoumar, F. S., Vizeacoumar, F. J., & Chelico, L. (2021). APOBEC1 cytosine deaminase activity on single-stranded DNA is suppressed by replication protein A. *Nucleic Acids Research, 49*(1), 322–339. https://doi.org/10.1093/nar/gkaa1201.

Xue, K., Rada, C., & Neuberger, M. S. (2006). The in vivo pattern of AID targeting to immunoglobulin switch regions deduced from mutation spectra in msh2-/- ung-/- mice. *The Journal of Experimental Medicine, 203*(9), 2085–2094. https://doi.org/10.1084/jem.20061067.

Zuker, M. (2003). Mfold web server for nucleic acid folding and hybridization prediction. *Nucleic Acids Research, 31*(13), 3406–3415. https://doi.org/10.1093/nar/gkg595.

CHAPTER FIVE

Defining the genome-wide mutagenic impact of APOBEC3 enzymes

Eszter Németh[a,1], Rachel A. DeWeerd[b,c,1], Abby M. Green[b,c,*], and Dávid Szüts[a,*]

[a]Institute of Molecular Life Sciences, HUN-REN Research Centre for Natural Sciences, Budapest, Hungary
[b]Department of Pediatrics, Washington University School of Medicine, St. Louis, MO, United States
[c]Center for Genome Integrity, Siteman Cancer Center, Washington University School of Medicine, St. Louis, MO, United States
*Corresponding authors. e-mail address: abby.green@wustl.edu; szuts.david@ttk.hu

Contents

Abstract

Somatic mutations drive cancer initiation and tumor evolution. Therefore, the etiology of mutagenesis in cancer is important to preventative and treatment strategies. Somatic mutagenesis in cancer is a multifactorial process and includes both endogenous and exogenous sources of mutations. One recently recognized source of mutagenesis in cancer is the innate immune APOBEC3 family of enzymes, which catalyze cytosine deamination to restrict viral infection but can aberrantly act on the

[1] These authors contributed equally.

Methods in Enzymology, Volume 713
ISSN 0076-6879, https://doi.org/10.1016/bs.mie.2024.12.003
Copyright © 2025 Elsevier Inc. All rights are reserved, including those for text and data mining, AI training, and similar technologies.

cellular genome, resulting in mutations. Single base substitution (SBS) signatures, or mutational patterns, identified in cancer genomes have demonstrated widespread mutagenesis caused by APOBEC3 enzymes throughout human tumors. To comprehensively define the consequences of APOBEC3 mutagenesis, we developed an experimental pipeline for prospective analysis of genome-wide mutations caused by APOBEC3 activity. This pipeline can be adapted to analyze additional sources of mutagenesis across a spectrum of cells.

1. Introduction

Understanding the origins of mutations in cancer can uncover mutagenic processes and reveal clinically actionable etiologies of mutagenesis. The most common type of mutations in cancer are single-base substitutions (Vogelstein et al., 2013). While somatic base substitutions are commonly the result of DNA replication errors, several other factors, such as exogenous carcinogens, endogenous cellular processes, and defective DNA repair pathways, contribute to the genetic landscape of cancers.

Experimental assessment of mutagenicity has historically relied on phenotypic changes arising from inactivation of reporter genes; typical examples include β-galactosidase (*LacZ*) in *E. coli* and hypoxanthine phosphoribosyltransferase (*HPRT*) in mammalian cells (Liber & Thilly, 1982; Malling, 2004; O'Neill, Brimer, Machanoff, Hirsch, & Hsie, 1977; White et al., 2019; Yamamoto & Fujiwara, 1990). While these reporter assays are helpful in determining mutagenic potential, they provide only low-resolution data on mutation frequency and context.

With the meteoric rise in next-generation sequencing (NGS) technology, whole genome sequencing (WGS) offers an alternative, high-resolution opportunity to interrogate mutagenesis. WGS analysis by nonnegative matrix factorization (NMF) led to the identification of mutational patterns, or signatures, in cancer genomes that point to mutagenic processes currently or historically active in cancer cells. NMF is a computational method used to reduce the dimensionality of data to isolate significant characteristics (Berry, Browne, Langville, Pauca, & Plemmons, 2007; Lee & Seung, 1999). Several groundbreaking papers employed NMF to extract base substitution patterns from whole-genome and exome sequences of human cancers, resulting in the detection of single-base substitution (SBS) signatures (Alexandrov et al., 2013; Nik-Zainal et al., 2012a; Nik-Zainal et al., 2012b). SBS signatures account for the base change and trinucleotide context of mutations, resulting in 96 possible mutation classifications. The Catalog of Somatic Mutations in

Cancer (COSMIC) database includes over 80 SBS signatures, as well as signatures of other mutation types (indels, copy number variations, and others) (Tate et al., 2019).

Mutation patterns attributed to the AID/APOBEC family of cytosine deaminases were identified as one of the first five SBS signatures extracted from breast cancers (Nik-Zainal et al., 2012b). Since then, the APOBEC signatures, known as SBS2 and SBS13, have been noted across more than 50 % of human cancer types (Alexandrov et al., 2020). AID/APOBEC enzymes generate base substitutions by catalyzing the deamination of cytosine to uracil in single-stranded DNA and RNA. AID/APOBEC enzymes play multiple roles in biology, including inducing somatic hypermutation in developing B cells (AID) (Dickerson, Market, Besmer, & Papavasiliou, 2003; Muramatsu et al., 1999), generating ApoB isoforms for lipid metabolism (APOBEC1) (Teng, Burant, & Davidson, 1993), and restricting virus infection (APOBEC3) (Stavrou & Ross, 2015). However, aberrant APOBEC activity can lead to mutations in the host genome. Two members of the APOBEC3 family, APOBEC3A and APOBEC3B, can localize to the nucleus and act on genomic DNA (Bogerd, Wiegand, Doehle, & Cullen, 2007; Burns et al., 2013; Landry, Narvaiza, Linfesty, & Weitzman, 2011). Interestingly, APOBEC3 expression does not often correlate with APOBEC signature mutations (Chan et al., 2015). Therefore, identifying the SBS signatures caused by APOBEC3 expression can point to tumors in which APOBEC3 enzymes have been active, and can elucidate the mutagenic consequences of APOBEC3 activity.

Genome-wide evaluation of mutation patterns enables the prospective assessment of APOBEC3 activity under variable conditions. For example, deletion of REV1, a translesion synthesis polymerase, in cancer cell lines was shown to decrease APOBEC3-induced mutational burden and mutation clusters (Petljak et al., 2022). In addition, mutational signature analysis can be used to retrospectively interrogate cancer genomes to understand factors associated with APOBEC3 activity. For example, dysfunction of the Structural Maintenance of Chromosomes 5/6 complex (SMC5/6) in human cancers leads to a lack of APOBEC3 SBS signatures, an observation that is consistent with synthetic lethality between these two conditions (Fingerman et al., 2024). Thus, mutational signature analysis can be used for both mechanistic and clinically relevant studies of APOBEC3 enzymes in cancer.

As somatic mutation patterns in cancer (and other diseases) continue to be elucidated, methods to experimentally interrogate mutagen impact on a genome-wide scale have become increasingly useful and widespread. WGS provides information on mutations arising in all contexts and allows downstream analyses about the influence of sequence context, DNA structure, transcription, chromatin state, and other cellular features. Mutational processes are, therefore, ideally studied by WGS of cultured cells. To investigate genome-wide mutation patterns, we and others have taken advantage of cell lines with low spontaneous mutation rates. The chicken DT40 lymphoblastoma cell line has a near-normal karyotype and low spontaneous mutation rate, making it a useful model system for DNA repair and mutagenesis research (Molnar et al., 2014; Szikriszt et al., 2016). However, mutagenesis can be studied in any cell line using WGS, with consideration for the genetic properties of each cell line.

Here, we describe in detail the workflow to prospectively determine the mutagenic effects of APOBEC3A in DT40 cells as in our recent publication (DeWeerd et al., 2022). We use a doxycycline-inducible APOBEC3A construct to control APOBEC3A expression for the duration of the experiment. While the methods and analysis described here are tailored to APOBEC mutagenesis, we emphasize that this workflow can be adapted to study a variety of endogenous or exogenous mutagens.

2. Materials

2.1 Lentivirus production

1. pSLIK-A3A-HA lentivector (DeWeerd et al., 2022).
2. 3rd generation packaging plasmids. We use pMDL (gag/pol), pRSV-Rev (rev), and VSVg (env) available at Addgene (plasmid #12251, 12253, 12259, respectively). Other plasmids encoding lentiviral packaging components can be used.
3. Lipofectamine 2000 (Invitrogen #11668–019).
4. Opti-MEM (Gibco #31985–070).
5. 293 T cells (American Type Culture Collection #CRL-2316).
6. DMEM media (Gibco #10566–016) supplemented with 10% fetal bovine serum (FBS) (Sigma Aldrich #F0926).
7. Sterile 0.45 μM cell filters (Millipore #SLHVR33RS).
8. Sterile syringes.

2.2 DT40 cell culture, transduction, selection

1. DT40 cells (*Gallus gallus,* American Type Culture Collection #CRL-2111)
2. RPMI media (Gibco #11875-085), including the following supplements:

a. Tetracycline-free fetal bovine serum, heat-inactivated (FBS). 7% for normal cell passaging, 17% for single-cell isolation. (Sigma Aldrich #F2442) See **Note 1**.
b. 3% chicken serum (Sigma-Aldrich #C5405)
c. 1% penicillin/streptomycin (Corning #30-002-Cl)
d. 50 μM beta-mercaptoethanol (MP Biomedicals # 190242)

3. G418 (Sigma-Aldrich #G8168)

2.3 Genomic DNA extraction, library preparation

1. PureLink Genomic DNA Mini Kit (Invitrogen #K182002)
2. DNA Prep Kit (Illumina)

2.4 NGS sequencing and data analysis

1. NovaSeq X Plus Series (Illumina)
2. Raw data processing: Trimmomatic, BWA, GATK
3. Mutation calling: Isomut analysis pipeline, samtools, R studio

3. Methods

3.1 Overview

Experimental modeling of mutagenesis can be accurately detected if assayed between two single-cell cloning events (Fig. 1). The genome of the ancestral cell can be determined using WGS on the population of cells derived from it. Any mutations that arise after the cloning step will be

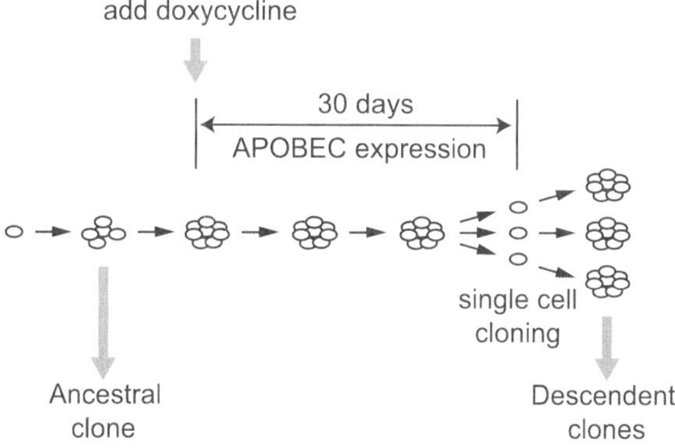

Fig. 1 Scheme of the long-term APOBEC expression experiment, to prepare ancestral and descendent samples for WGS.

subclonal and can be filtered out based on allele frequency. The genomes of descendent clones are determined similarly, and a comparison of the descendent and the matched ancestral clones reveals mutations that arise between the two cloning steps. Standard software packages such as the GATK Toolkit (McKenna et al., 2010) can be used for the primary processing of sequencing data, which involves quality control steps and alignment to the reference genome of a species. Pairwise comparisons with software tools (such as those used for finding somatic mutations in cancer samples with matched normal tissue from the same patient) can be used for mutation detection. However, the computational cost of analyzing multiple subclones simultaneously is high. Additionally, the noise of false-positive mutations may exceed the signal, which may be as low as a few mutations per 10^9 nucleotides. Also, even state-of-the-art mutation callers show varying accuracy (Xiao et al., 2021).

Mutational processes are generally cell-autonomous, regardless of the source of mutagenesis. Mutation detection can be improved by the simultaneous analysis of several isogenic genomes. In this setting, false positives tend to appear in the same position across several samples, for example, due to alignment uncertainties in repeat-containing genomic regions; these can be readily filtered out. In contrast, newly arising mutations are generally unique to individual samples and can be trusted. Cell line–based experiments can provide many isogenic samples minimizing the genomic differences between samples at baseline. We have, therefore, developed a dedicated mutation detection tool, IsoMut, specifically designed to efficiently detect clonal or near-clonal differences between isogenic samples (Pipek et al., 2017). IsoMut operates by simultaneously comparing alignment files from a substantial number of sequenced isogenic samples and using the read count numbers to calculate the probability at each genomic position that the sample most different from the reference (i.e., the signal) is identical to the sample that is the second most different from the reference (i.e., the highest noise). The probability scores are then used for mutation filtering by setting the filtering score such that the ancestral clone contains no or almost no unique mutations.

The method described here detects base substitutions and small insertions and deletions and it is therefore suitable for investigating the mutation patterns generated by APOBECs activity. Different software tools are required to optimally detect large-scale genomic changes (Li et al., 2020).

3.2 Lentiviral production and transduction

1. All steps should be performed in a dedicated tissue culture hood. Lentiviral production detailed below is adapted from standard assays (Benskey & Manfredsson, 2016).
2. Culture 293 T cells in DMEM with 10% FBS. The optimal seeding density for transfection is 60-70% confluency. Seed ~3 million 293 T cells in a 20 cm plate 24 h before transfection. Culture at 37 °C in a dedicated tissue culture incubator with 5% CO_2.
3. Mix 4:2:2:1 ratio of lentivector:pMDL:pVSVg:pREV. The total DNA quantity should be 45 μg. Add the DNA mix to 500 μL Opti-MEM in a microcentrifuge tube. In a separate microcentrifuge tube, add 40 μL of Lipofectamine to 500 μL of Opti-MEM and pipet to mix. Combine the contents of both tubes and incubate at room temperature for 20 min in a biosafety hood.
4. Add vector/Lipofectamine mix to the 20 cm plate of 293 T cells and culture for 48 h until confluent. Passage supernatant through a 0.45 μM filter using a sterile syringe, then add 1 mL of filtered supernatant to 100,000 DT40 cells in one well of a 12-well plate. A second well with an equivalent concentration of DT40s in RPMI media without viral supernatant serves as a control for antibiotic selection. Remaining filtered supernatant can be cryopreserved for future use.
5. Treat cells with 1 mg/mL G418 for 3-7 days until control cells are non-viable anymore.

3.3 Single-cell cloning and APOBEC3A induction

1. Isolate ancestral clones through single-cell sorting by flow cytometry. Sort single cells into a 96-well plate with RPMI media containing 17% tetracycline-free FBS, 3% chicken serum, and 1% penicillin/streptomycin, and culture at 37 °C under 5% CO_2. See **Note 2**.
2. Allow the clonal populations from single-cell sorting to grow in 96-well plates until clones are easily detected under a microscope (~1-2 weeks for DT40 cells). Move the clones by pipetting to progressively larger plates with more media to accommodate growth. See **Note 3**.
3. Once a satisfactory number of cells is obtained (>2 million), treat a sample of each clone with doxycycline to validate APOBEC3A expression and activity (see **Note 4**). A minimum of a single ancestral clone with confirmed APOBEC3A expression and activity after doxycycline should be selected for extended treatment.

4. Treat the ancestral clone with doxycycline every 3 days for a total of 30 days. See **Note 5**.
5. After 30 days, sort the pooled cells again to obtain descendent single-cell clones. See **Note 6**.
6. Culture and grow descendent clone populations as in Steps 3.3.2 and 3.3.3 until there are > 1 million cells in each well. See **Note 7**.

3.4 Genomic DNA extraction, library preparation, and sequencing

1. Extract genomic DNA (gDNA) using the PureLink Genomic DNA Mini Kit (Invitrogen) according to the manufacturer's instructions. We find 1-5 million cells to be sufficient for gDNA extraction.
2. One ancestral and at least three descendants from each group are necessary for downstream analysis. We recommend analyzing biological control groups in parallel to account for mutagenesis incurred during cell culture or those caused by doxycycline treatment. Examples include: sample without treatment, sample with no construct treated with doxycycline, sample with construct but no doxycycline treatment, and sample expressing a catalytically inactivated enzyme (i.e., A3A-C106S).
3. Quantitate gDNA by Nanodrop (Thermo Scientific) or equivalent before library preparation.
4. Perform library preparation to obtain ~350 bp fragments (Illumina DNA prep kit) according to manufacturer's instructions.
5. Sequence all libraries on Illumina NovaSeq X Plus Series (PE150) in 2×150 bp paired-end format to obtain ~30x coverage. In the case of DT40 cells, this corresponds to around 30 Gb of raw data per sample.

3.5 Raw sequence data processing

1. Prepare the reference genome file. The reference genome, such as Galgal4.73, should be supplemented with the lentiviral plasmid construct sequence.
2. Remove the adapter sequences using Trimmomatic (Bolger, Lohse, & Usadel, 2014).
3. Align raw sequencing reads to the reference genome using BWA-MEM (Li & Durbin, 2009).
4. Remove duplicated reads using Samblaster (Faust & Hall, 2014).
5. Perform indel realignment with the GATK indelrealigner tool (Gaujoux & Seoighe, 2010).

6. Check the alignment in a genome browser, such as IGV, including the plasmid which technically appears as an extra chromosome. To determine where the plasmid integrated into the genome, blast the soft clipped bases of the plasmid reads against the reference genome. Check the integration locus for possible problems, such as disruption of a transgene (see **Note 8**).
7. Perform general quality control on the aligned BAM file, such as assessment of mean coverage, GC content, and proportion of non-aligned reads.

3.6 Mutation calling

1. Use IsoMut (Pipek et al., 2017) to call mutations. 15−40 samples are required for one run, including ancestral and descendent clones. Use standard parameters for isomut: min_sample_freq = 0.21, min_other_ref_freq = 0.93, cov_limit = 5, base_quality_limit = 30, min_gap_dist_snv = 0, min_gap_dist_indel = 20. For reference genome sequence use the same as for raw read sequence alignment.
2. Process the raw list of possible mutations. IsoMut will assign a probability-based score for each mutation; this score will be used for filtering during postprocessing. Determine score thresholds for SNVs and insertions and deletions, separately. For SNVs, choose a value that allows a maximum of 5 false positive mutations in the starting clone (Fig. 2). Every mutation with a lower score than the threshold is to be disregarded.

3.7 Analysis of mutagenesis

1. Use R for all further analysis of the mutation data. These should include negative controls as defined in 3.4.2.
2. From the filtered SNV list from IsoMut, determine the 96-category triplet mutation spectrum for each sample (Fig. 3) using the same nomenclature and conventions as in the COSMIC database (Alexandrov et al., 2013; Alexandrov et al., 2020).
3. Perform nonnegative matrix factorization (NMF) using the R package nmf (Gaujoux & Seoighe, 2010) to determine *de novo* mutational signatures from the novel experimental data. Check the cosine similarity of the experimental spectra to the established mutation signatures (COSMIC) by deconstructing the experimental spectra to the NMF components.
4. Separately, deconstruct the experimental SBS spectrum to SBS signatures identified in cancer using MutationalPatterns (Manders et al., 2022) to get the ratio of signatures of interest (in this case, APOBEC signatures SBS2 and SBS13) and background mutagenesis.

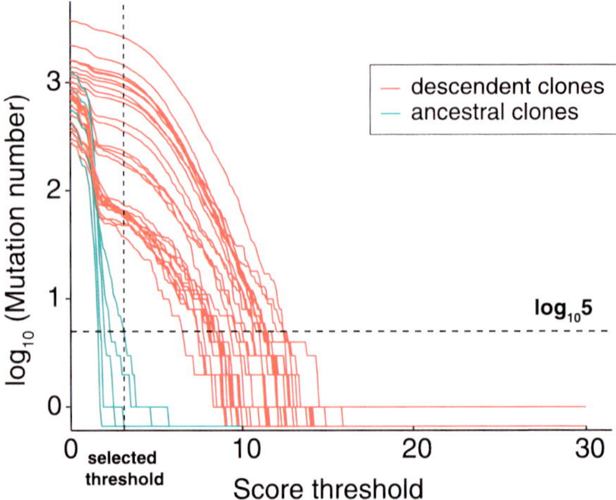

Fig. 2 Selection of SBS score threshold in the postprocessing of raw IsoMut output. A maximum of five false positives is allowed in any of the starting clones. Ideally, the threshold cuts the curves of descendent clones at a position with a nearly zero slope. Thus the final number of mutations is not sensitive to subtle changes in the score threshold.

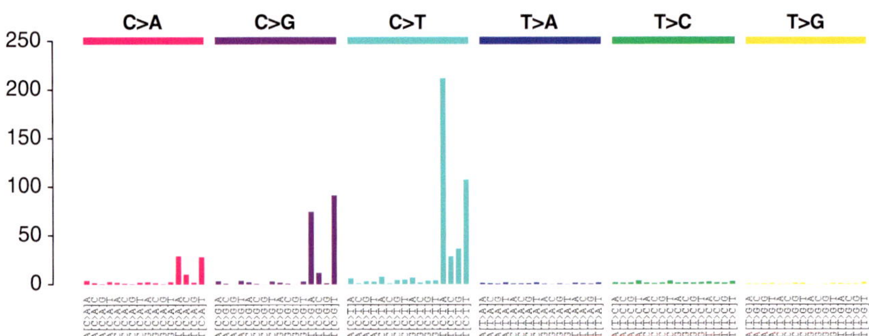

Fig. 3 Mutational spectrum from SBS mutations determined with IsoMut.

4. Notes

1. In **Step 2.2**, tetracycline-free fetal bovine serum is used to avoid tetracycline activation of the doxycycline-inducible APOBEC3A construct used in this protocol. Regular FBS can be used instead of tetracycline-free FBS with systems that do not require doxycycline induction.

2. In **Step 3.3.1**, the FBS content is 10% higher during single-cell sorting than in normal cell culture (17% compared to 7%). This supports single-cell growth post-sorting.
3. In **Step 3.3.2**, DT40 cells grow very rapidly (doubling time 10-12 h at 37 °C) and die when overgrown. Care should be taken during clonal growth to ensure healthy populations.
4. In **Step 3.3.3**, after single-cell sorting, characterization of ancestral clones by immunoblotting for APOBEC3A expression and deaminase activity (DeWeerd & Green, 2022) enables selection of clones with high or low APOBEC3A expression/activity as desired.
5. In **Step 3.3.4**, note that the timing of culture can be amended to study the acute (short-term) or chronic (long-term) effects of mutagenesis. All controls should follow the same sorting and culture timeline to account for mutations accrued during culture.
6. In **Step 3.3.5**, the descendent clones can again be tested for expression and activity (see Note 4). We find that chronic induction of a mutagen such as APOBEC3A for 30 days may lead to decreased expression or inactivation of the APOBEC3A transgene. Importantly, this does not affect the downstream computational analysis, as these experiments measure mutations caused by APOBEC3A at any time during the experiment. However, analysis may consequently lack complete information about the investigated mutagen. We recommend careful validation of inducible mutagens after treatment and propose adjustments in dosage and the total time of the experiment to avoid selection bias.
7. In **Step 3.3.6**, mutagenesis can also be assayed in cell lines without recourse to repeated cloning steps using duplex sequencing, which lowers the error rate below the expected mutation rates. Duplex sequencing of selected genomic regions is sufficient to determine mutation rates or spectra, but may not cover the effect of genomic attributes (Schmitt et al., 2012). Alternatively, the whole genome can be sequenced, which may become a widespread technology when costs decrease (Abascal et al., 2021).
8. In **Step 3.5.6**, we have observed disruption or deletion of integrated transgenes when they confer genotoxicity. If this occurs, the mutational data may be biased.

Funding

We thank the support from Department of Defense grant CA200867 (AMG) and the National Research, Development and Innovation Office of Hungary (grant PD134818 to EN, K142385 to DS).

References

Abascal, F., Harvey, L. M. R., Mitchell, E., Lawson, A. R. J., Lensing, S. V., Ellis, P., ... Wang, Y. (2021). Somatic mutation landscapes at single-molecule resolution. *Nature, 593*, 405–410.

Alexandrov, L. B., Kim, J., Haradhvala, N. J., Huang, M. N., Tian Ng, A. W., Wu, Y., ... Bergstrom, E. N. (2020). The repertoire of mutational signatures in human cancer. *Nature, 578*, 94–101.

Alexandrov, L. B., Nik-Zainal, S., Wedge, D. C., Aparicio, S. A., Behjati, S., Biankin, A. V., ... Borresen-Dale, A. L. (2013). Signatures of mutational processes in human cancer. *Nature, 500*, 415–421.

Benskey, M. J., & Manfredsson, F. P. (2016). Lentivirus production and purification. *Methods in Molecular Biology, 1382*, 107–114.

Berry, M. W., Browne, M., Langville, A. N., Pauca, V. P., & Plemmons, R. J. (2007). Algorithms and applications for approximate nonnegative matrix factorization. *Computational Statistics & Data Analysis, 52*, 155–173.

Bogerd, H. P., Wiegand, H. L., Doehle, B. P., & Cullen, B. R. (2007). The intrinsic antiretroviral factor APOBEC3B contains two enzymatically active cytidine deaminase domains. *Virology, 364*, 486–493.

Bolger, A. M., Lohse, M., & Usadel, B. (2014). Trimmomatic: A flexible trimmer for Illumina sequence data. *Bioinformatics (Oxford, England), 30*, 2114–2120.

Burns, M. B., Lackey, L., Carpenter, M. A., Rathore, A., Land, A. M., Leonard, B., ... Nikas, J. B. (2013). APOBEC3B is an enzymatic source of mutation in breast cancer. *Nature, 494*, 366–370.

Chan, K., Roberts, S. A., Klimczak, L. J., Sterling, J. F., Saini, N., Malc, E. P., ... Mieczkowski, P. A. (2015). An APOBEC3A hypermutation signature is distinguishable from the signature of background mutagenesis by APOBEC3B in human cancers. *Nature Genetics, 47*, 1067–1072.

DeWeerd, R. A., Nemeth, E., Poti, A., Petryk, N., Chen, C. L., Hyrien, O., ... Green, A. M. (2022). Prospectively defined patterns of APOBEC3A mutagenesis are prevalent in human cancers. *Cell Reports, 38*, 110555.

DeWeerd, R., & Green, A. M. (2022). Qualitative and quantitative analysis of DNA cytidine deaminase activity. *Methods in Molecular Biology, 2444*, 161–169.

Dickerson, S. K., Market, E., Besmer, E., & Papavasiliou, F. N. (2003). AID mediates hypermutation by deaminating single stranded DNA. *The Journal of Experimental Medicine, 197*, 1291–1296.

Faust, G. G., & Hall, I. M. (2014). SAMBLASTER: Fast duplicate marking and structural variant read extraction. *Bioinformatics (Oxford, England), 30*, 2503–2505.

Fingerman, D. F., O'Leary, D. R., Hansen, A. R., Tran, T., Harris, B. R., DeWeerd, R. A., ... Tennakoon, M. (2024). The SMC5/6 complex prevents genotoxicity upon APOBEC3A-mediated replication stress. *The EMBO Journal, 43*, 3240–3255.

Gaujoux, R., & Seoighe, C. (2010). A flexible R package for nonnegative matrix factorization. *BMC Bioinformatics, 11*, 367.

Landry, S., Narvaiza, I., Linfesty, D. C., & Weitzman, M. D. (2011). APOBEC3A can activate the DNA damage response and cause cell-cycle arrest. *EMBO Reports, 12*, 444–450.

Lee, D. D., & Seung, H. S. (1999). Learning the parts of objects by non-negative matrix factorization. *Nature, 401*, 788–791.

Li, H., & Durbin, R. (2009). Fast and accurate short read alignment with Burrows-Wheeler transform. *Bioinformatics (Oxford, England), 25*, 1754–1760.

Li, Y., Roberts, N. D., Wala, J. A., Shapira, O., Schumacher, S. E., Kumar, K., ... Haber, J. E. (2020). Patterns of somatic structural variation in human cancer genomes. *Nature, 578*, 112–121.

Liber, H. L., & Thilly, W. G. (1982). Mutation assay at the thymidine kinase locus in diploid human lymphoblasts. *Mutation Research, 94*, 467–485.

Malling, H. V. (2004). History of the science of mutagenesis from a personal perspective. *Environmental and Molecular Mutagenesis, 44*, 372–386.

Manders, F., Brandsma, A. M., de Kanter, J., Verheul, M., Oka, R., van Roosmalen, M. J., ... van Boxtel, R. (2022). MutationalPatterns: The one stop shop for the analysis of mutational processes. *BMC Genomics, 23*, 134.

McKenna, A., Hanna, M., Banks, E., Sivachenko, A., Cibulskis, K., Kernytsky, A., ... Daly, M. (2010). The Genome Analysis Toolkit: A MapReduce framework for analyzing next-generation DNA sequencing data. *Genome Research, 20*, 1297–1303.

Molnar, J., Poti, A., Pipek, O., Krzystanek, M., Kanu, N., Swanton, C., ... Szuts, D. (2014). The genome of the chicken DT40 bursal lymphoma cell line. *G3 (Bethesda), 4*, 2231–2240.

Muramatsu, M., Sankaranand, V. S., Anant, S., Sugai, M., Kinoshita, K., Davidson, N. O., & Honjo, T. (1999). Specific expression of activation-induced cytidine deaminase (AID), a novel member of the RNA-editing deaminase family in germinal center B cells. *The Journal of Biological Chemistry, 274*, 18470–18476.

Nik-Zainal, S., Alexandrov, L. B., Wedge, D. C., Van Loo, P., Greenman, C. D., Raine, K., ... Stebbings, L. A. (2012a). Mutational processes molding the genomes of 21 breast cancers. *Cell, 149*, 979–993.

Nik-Zainal, S., Van Loo, P., Wedge, D. C., Alexandrov, L. B., Greenman, C. D., Lau, K. W., ... Ramakrishna, M. (2012b). The life history of 21 breast cancers. *Cell, 149*, 994–1007.

O'Neill, J. P., Brimer, P. A., Machanoff, R., Hirsch, G. P., & Hsie, A. W. (1977). A quantitative assay of mutation induction at the hypoxanthine-guanine phosphoribosyl transferase locus in Chinese hamster ovary cells (CHO/HGPRT system): Development and definition of the system. *Mutation Research, 45*, 91–101.

Petljak, M., Dananberg, A., Chu, K., Bergstrom, E. N., Striepen, J., von Morgen, P., ... Alexandrov, L. B. (2022). Mechanisms of APOBEC3 mutagenesis in human cancer cells. *Nature, 607*, 799–807.

Pipek, O., Ribli, D., Molnar, J., Poti, A., Krzystanek, M., Bodor, A., ... Szuts, D. (2017). Fast and accurate mutation detection in whole genome sequences of multiple isogenic samples with IsoMut. *BMC Bioinformatics, 18*, 73.

Schmitt, M. W., Kennedy, S. R., Salk, J. J., Fox, E. J., Hiatt, J. B., & Loeb, L. A. (2012). Detection of ultra-rare mutations by next-generation sequencing. *Proceedings of the National Academy of Sciences of the United States of America, 109*, 14508–14513.

Stavrou, S., & Ross, S. R. (2015). APOBEC3 proteins in viral immunity. *Journal of Immunology, 195*, 4565–4570.

Szikriszt, B., Poti, A., Pipek, O., Krzystanek, M., Kanu, N., Molnar, J., ... Csabai, I. (2016). A comprehensive survey of the mutagenic impact of common cancer cytotoxics. *Genome Biology, 17*, 99.

Tate, J. G., Bamford, S., Jubb, H. C., Sondka, Z., Beare, D. M., Bindal, N., ... Dawson, E. (2019). COSMIC: The catalogue of somatic mutations in cancer. *Nucleic Acids Research, 47*, D941–D947.

Teng, B., Burant, C. F., & Davidson, N. O. (1993). Molecular cloning of an apolipoprotein B messenger RNA editing protein. *Science (New York, N. Y.), 260*, 1816–1819.

Vogelstein, B., Papadopoulos, N., Velculescu, V. E., Zhou, S., Diaz, L. A., Jr., & Kinzler, K. W. (2013). Cancer genome landscapes. *Science (New York, N. Y.), 339*, 1546–1558.

White, P. A., Luijten, M., Mishima, M., Cox, J. A., Hanna, J. N., Maertens, R. M., & Zwart, E. P. (2019). In vitro mammalian cell mutation assays based on transgenic reporters: A report of the International Workshop on Genotoxicity Testing (IWGT). *Mutation Research – Genetic Toxicology and Environmental Mutagenesis, 847*, 403039.

Xiao, W., Ren, L., Chen, Z., Fang, L. T., Zhao, Y., Lack, J., ... Kerrigan, L. (2021). Toward best practice in cancer mutation detection with whole-genome and whole-exome sequencing. *Nature Biotechnology, 39*, 1141–1150.

Yamamoto, Y., & Fujiwara, Y. (1990). Uracil-DNA glycosylase causes 5-bromodeoxyuridine photosensitization in Escherichia coli K-12. *Journal of Bacteriology, 172*, 5278–5285.

CHAPTER SIX

Defining APOBEC-induced mutation signatures and modifying activities in yeast

Tony M. Mertz[a], Zachary W. Kockler[b], Margo Coxon[c], Cameron Cordero[a], Atri K. Raval[a], Alexander J. Brown[c], Victoria Harcy[c], Dmitry A. Gordenin[b], and Steven A. Roberts[a,*]

[a]Department of Microbiology and Molecular Genetics, University of Vermont Cancer Center, University of Vermont, Burlington, VT, United States

[b]Genome Integrity & Structural Biology Laboratory, National Institute of Environmental Health Sciences, Durham, NC, United States

[c]School of Molecular Biosciences, Washington State University, Pullman, WA, United States

*Corresponding author. e-mail address: srober23@med.uvm.edu

Contents

Abstract

APOBEC cytidine deaminases guard cells in a variety of organisms from invading viruses and foreign nucleic acids. Recently, several human APOBECs have been implicated in mutating evolving cancer genomes. Expression of APOBEC3A and APOBEC3B in yeast allowed experimental derivation of the substitution patterns they cause in dividing cells, which provided critical links to these enzymes in the etiology

Methods in Enzymology, Volume 713
ISSN 0076-6879, https://doi.org/10.1016/bs.mie.2024.11.041
Copyright © 2025 Elsevier Inc. All rights are reserved, including those for text and data mining, AI training, and similar technologies.

of the COSMIC single base substitution (SBS) signatures 2 and 13 in human tumors. Additionally, the ability to scale yeast experiments to high-throughput screens allows use of this system to also investigate cellular pathways impacting the frequency of APOBEC-induced mutation. Here, we present validated methods utilizing yeast to determine APOBEC mutation signatures, genetic interactors, and chromosomal substrate preferences. These methods can be employed to assess the potential of other human APOBECs and APOBEC orthologs in different species to contribute to cancer genome evolution as well as define the pathways that protect the nuclear genome from inadvertent APOBEC activity during viral restriction.

Abbreviations

APOBEC Apolipoprotein B mRNA editing catalytic polypeptide-like.
AID Activation Induced cytidine Deaminase.
WGS whole genome sequencing.
ORF open reading frame.
DNA deoxyribonucleic acid.
RNA ribonucleic acid.
ss single strand.
ds double strand.
dU deoxyuridine.
COSMIC Catalog of Somatic Mutations in Cancer.
SBS Single base substitution.
SC Synthetic Complete.

1. Introduction

Apolipoprotein B mRNA editing catalytic polypeptide-like (APOBEC) cytidine deaminases are nucleic acid modifying enzymes with roles in RNA metabolism and adaptive and innate immune responses (Pecori, Di Giorgio, Paulo Lorenzo, & Nina Papavasiliou, 2022; Refsland & Harris, 2013). These enzymes are present in a broad range of organisms (Conticello, Thomas, Petersen-Mahrt, & Neuberger, 2005). However, the specific number of APOBEC genes varies significantly, particularly for APOBEC3 subfamily members (Conticello et al., 2005). 11 APOBEC genes are encoded in the human genome, with 9 members displaying documented catalytic activity of promoting sequence specific conversion of cytidine to uridine in single-stranded DNA or RNA (Mertz, Collins, Dennis, Coxon, & Roberts, 2022). When applied to foreign DNAs in a cell (such as viral DNA) or immunoglobulin genes undergoing somatic hypermutation, this activity produces C to T or C to G substitutions within the preferred sequence context of the APOBEC enzyme catalyzing the deamination reaction (Mertz et al., 2022; Pecori et al., 2022). AID and APOBEC3G prefer to deaminate cytidines within W*R*C (W = A or T; R=A or G) and CC sequences (target C is underlined), respectively. All the remaining catalytically active human APOBECs (i.e. APOBEC1, APOBEC3A, APOBEC3B, APOBEC3C, APOBEC3DE, APOBEC3F, and APOBEC3H) prefer TC dinucleotides. Because of this sequence specificity, APOBEC activity leaves a characteristic signature on its target substrates.

Sequencing of human cancer genomes has unexpectedly revealed an over-representation of the APOBEC-related mutation signatures, specifically COSMIC single base substitution signatures (SBS) 2 and 13 (for TC-specific APOBECs) and 84 and 85 (for AID) (Alexandrov et al., 2020; Burns, Temiz, & Harris, 2013; Nik-Zainal et al., 2012; Roberts et al., 2012; Roberts et al., 2013). While SBS84/85 are primarily limited to myeloid malignancies (Alexandrov et al., 2020), which is consistent with off-target AID activity during somatic hypermutation or class switch recombination processes occurring in these immune cell types (Pecori et al., 2022), SBS2/13 occur broadly across human cancer types and are one of the more abundant sources of mutation in tumors (Alexandrov et al., 2020; Burns et al., 2013; Roberts et al., 2013). This broad distribution of the SBS2/13 signatures initially obscured the causative association of APOBEC3 family members and these mutation signatures. Consequently, experimental characterization of the APOBEC3 family's mutagenic profile was required to provide strong support of these enzymes in signature generation. Expression of APOBEC3 enzymes in *S. cerevisiae* and *E. coli* produced data that strengthened the association of APOBEC3A and APOBEC3B with cancer mutagenesis.

First, whole genome sequencing (WGS) of yeast expressing APOBEC3A or APOBEC3B identified tetranucleotide target sequences specific to each enzyme (Chan et al., 2015). APOBEC3A preferred YTCA motifs, while APOBEC3B preferred RTCA motifs, which were subsequently confirmed to be enriched specifically among human cancer genomes containing the SBS2/13 signatures. This data provided strong support indicating that both enzymes cause cancer mutations and suggested, contrary to the paradigm at the time (Burns et al., 2013; Burns et al. 2013), that APOBEC3A induces more mutations in human cancers than APOBEC3B. Eventual CRISPR/Cas9 knockout of APOBEC3A and APOBEC3B in breast cancer cell lines confirmed this relationship (Petljak et al., 2022).

Secondary confirmation of APOBEC generation of SBS2/13 was later obtained through similar profiling of mutations identified from WGS of yeast and bacteria expressing APOBEC3A and APOBEC3B (Hoopes et al., 2016), or APOBEC3G (Bhagwat et al., 2016), respectively. Expression of any of these enzymes in the absence of deoxyuridine (dU) removal enzymes resulted in the accumulation of widely spread C to T substitutions across the organisms' genomes. Strand-specific profiling of these mutations revealed strong strand bias of APOBEC3-induced mutations flanking origins of replication. Each APOBEC3 enzyme primarily induced G to A substitutions upstream of origins, while inversely inducing C to T mutations downstream of origins. This bias is consistent with the APOBEC3 enzyme deaminating single-stranded portions of the lagging strand template during Okazaki fragment synthesis. In the yeast genome, which has bidirectional replication forks, the pattern gradually transitions between replication origins creating an inversion point at replication termination zones. Perturbation of replication factors also increases APOBEC3A and APOBEC3B activity (Hoopes et al., 2016; Sui et al., 2020). Importantly, SBS2/13 mutations in human cancers display similar strand biases when profiled against the direction of replication across the human genome (Haradhvala et al., 2016; Morganella et al., 2016; Seplyarskiy et al., 2016), indicating that these mutations are caused by an APOBEC damaging the lagging strand template during DNA replication. Thus, the high conservation between APOBEC3A and APOBEC3B mutation patterns in yeast (i.e., sequence motif, targeting the lagging strand template, and ability to induce kataegis (Taylor et al., 2013)) and SBS2/13 in human tumors provided convincing evidence for APOBEC3s causing SBS2/13 during human cancer cell proliferation.

In addition to characterizing APOBEC3A and APOBEC3B substrate preferences, expression of other human APOBEC family members or orthologs from other organisms in yeast is likely to be highly productive for evaluating their mutagenic nature and potential to contribute to cancer mutagenesis. This organism lacks a p53-mediated damage-induced apoptotic pathway (Wahl & Carr, 2001), allowing accumulation of larger amounts of DNA damage and mutation than other eukaryotic systems. Yeast also has a very compact genome (Goffeau et al., 1996) enabling whole genome sequencing (WGS) of many independent isolates, which increases the number of mutations that can be obtained for analysis and ensures reproducibility of the results. Moreover, yeast has low spontaneous mutation rates and lacks endogenous APOBEC enzymes, allowing the generation of highly accurate mutation signatures for each enzyme tested (Chan et al., 2012; Chan et al., 2015; Dennen et al., 2024; Dutko, Schafer,

Kenny, Cullen, & Curcio, 2005; Mayorov et al., 2005; Schumacher, Nissley, & Harris, 2005). Finally, knowledge of specific mutant alleles (e.g., *cdc13-1*) in yeast enables specific experimental generation of ssDNA (Booth, Griffith, Brady, & Lydall, 2001; Chan et al., 2012; Hoopes et al., 2017; Yang, Sterling, Storici, Resnick, & Gordenin, 2008), the required substrate for most APOBECs (Mertz et al., 2022).

The conservation of many of the processes involving DNA packaging (i.e., chromatinization and organization) and transactions (i.e., DNA replication, transcription, and DNA repair) between yeast and mammalian cells also allows the expression of APOBECs in yeast to be used to assess specificity for mesoscale DNA structures (e.g., hairpin forming sequences (Sanchez et al., 2024)) and other larger chromosomal features like genic regions. This has already been employed for APOBEC3B targeting of tRNA genes (Saini et al., 2017), and AID, APOBEC3G (Taylor, Wu, & Rada, 2014) and APOBEC3C (Brown, 2024) for transcription start sites. Similarly, the cellular processes that modulate the spectrum and abundance of APOBEC-induced mutations can be easily assessed with targeted and high-throughput screening methods (Mertz et al., 2023). For example, APOBEC3G activity on end-resected telomere DNA provided evidence for the roles for Rev1 and Pol ζ in creating APOBEC C to T and C to G substitutions during abasic site bypass (Chan et al., 2012) that underlie the SBS2 and SBS13 cancer signatures, respectively. Homology-directed template switching abasic site bypass mechanisms have been implicated in limiting APOBEC-induced mutation by both targeted studies (Hoopes et al., 2017; Rosenbaum et al., 2019; Saini et al., 2017) and screening approaches, along with base excision repair removal of APOBEC-induced deoxyuridine specifically in tRNA genes (Saini et al., 2017). Screening methodologies have also identified replication-associated factors like the Mrc1-Tof1-Csm3 complex, Ctf8-Ctf18 complex and histone deposition factors in limiting APOBEC3B mutagenesis (Mertz et al., 2023).

Similar methodologies can likewise be employed for any other DNA mutator to determine its mutagenic specificities, substrate preferences, and to determine what genetic factors alter their mutability. Here, we describe a suite of yeast expression vectors suitable for addressing questions of APOBEC activity in yeast. We also provide protocols to employ this system to screen for genetic modulators of APOBEC activity and assay APOBEC-induced mutation spectra through high-throughput amplicon sequencing or WGS.

2. A suite of vectors for APOBEC expression in yeast

The expression vectors that we have developed for APOBEC expression in yeast are based upon an original plasmid, pCM252-A3G, which has the C-terminal catalytic domain of APOBEC3G cloned downstream of a doxycycline-inducible promoter, an ampicillin resistance gene, and the yeast gene *TRP1* for auxotrophic selection of the plasmid in yeast (Chan et al., 2012). The A3G open reading frame (ORF) was removed from the vector and the *TRP1* gene was replaced with a hygromycin resistance marker through recombination in yeast to create a new empty vector construct, pySR419 (Chan et al., 2015; Hoopes et al., 2016). Initially, codon optimized ORFs for yeast expression of APOBEC3A and APOBEC3B were cloned into pySR419, resulting in two plasmids, pSR435, and pSR440, respectively (Chan et al., 2015; Hoopes et al., 2016). Transformation of pSR435 or pSR440 into yeast increases mutation frequency, even without doxycycline induction (Chan et al., 2015; Hoopes et al., 2016). However, pSR435 produced ~10-fold lower mutagenesis compared to pSR440 (Hoopes et al., 2016), indicating expression of APOBEC3A in yeast resulted in less active protein than expression of APOBEC3B. This occurred despite similar levels of transcription (Hoopes et al., 2016). Although addition of doxycycline to culture media containing yeast transformed with these vectors increases the transcription of the APOBEC genes, it results in a relatively small increase in mutagenesis. Consequently, these vectors are frequently used in uninduced conditions.

To eliminate concerns of potential auto-mutation of the *APOBEC3A* gene in pSR435 during propagation of the vector in *E.coli*, we modified pSR435 by cloning the *ACT1* intron into the N-terminal portion of the *APOBEC3A* gene, generating pVH-A3A (Mertz et al., 2023). Surprisingly, the presence of this intron resulted in a greater than 10-fold increase in mutation rate induced by the APOBEC3A expressed from pVH-A3A compared to APOBEC3A expressed from pSR435. Furthermore, APOBEC3A expressed from pVH-A3A induced mutations at rates equal to or greater than those of APOBEC3B expressed from pSR440. This occurred without any significant increase in transcript levels, suggesting that the splicing of the transcript may impact the translation or folding of the APOBEC3A protein. Careful determination of the growth rates of yeast carrying pySR419, pSR440, and pVH-A3A revealed that pVH-A3A transformed yeast grew slower specifically on hygromycin containing media than yeast transformed with the comparable empty vector or APOBEC3B-expression vector. This occurred even in diploid cells indicating that the growth inhibition was not likely to be the result of lethal mutagenesis. Sequencing of the entire pVH-A3A plasmid identified a non-synonymous substitution mutation in the hygromycin resistance gene. This was corrected by site-directed mutagenesis (producing pZAK031), which fully restored normal growth characteristics. For experiments requiring high levels of APOBEC3A-induced mutagenesis, we currently recommend using pZAK031.

Efforts to convert the intron-containing *APOBEC3A* and *APOBEC3B* into galactose-inducible vectors revealed that initial use of the lower tetracycline-inducible system was fortuitous. Cloning *APOBEC3A* or *APOBEC3B* into pBEVY-GL produced vectors that were lethal to yeast following addition of even very low levels of galactose (e.g., 0.1 %) to the culture media. Evidently, the Z1 cytidine deaminase domain-containing APOBECs are incompatible with high levels of expression driven by galactose-inducible promoters and require expression from weaker promoter systems.

We have since created additional variants of pySR419, pSR435, pVH-A3A, and pSR440 replacing the hygromycin resistance gene with additional selectable markers to increase the number of applications in which they can be utilized. Additional vectors containing the clonNAT or KanMX antibiotic cassettes were created to allow expression of these APOBECs in yeast already containing the hygromycin resistance gene from other genetic disruptions. Auxotrophic marked vectors *K. lactis LEU2* and *S. pombe HIS5*, which complement *S. cerevisiae leu2* and *his3* mutations respectively, were created to allow selection on lower cost media and enable larger scale experiments. All these vector systems produce similar levels of mutagenesis relative to their corresponding hygromycin vectors. A list of sequence-validated vectors is presented in Table 1. Full plasmid and sequences are provided in the Appendix.

3. Yeast transformation

All APOBEC-induced mutation analyses and screens presented in this chapter require transformation of the desired APOBEC expression vector into a yeast strain of choice. We suggest utilizing the protocol below for rapid and efficient uptake of these vectors.

Table 1 List of APOBEC expression plasmids.

Plasmid	APOBEC	Intron	Selection marker	Promoter
pCM-252	empty vector	None	*TRP1*	Doxycycline-inducible CYC1 promoter
pCM-252-A3G	APOBEC3G	None	*TRP1*	Doxycycline-inducible CYC1 promoter
pySR419	empty vector	None	Hygromycin resistance	Doxycycline-inducible CYC1 promoter
pSR433	APOBEC1	None	Hygromycin resistance	Doxycycline-inducible CYC1 promoter
pSR435	APOBEC3A	None	Hygromycin resistance	Doxycycline-inducible CYC1 promoter
pSR440	APOBEC3B	None	Hygromycin resistance	Doxycycline-inducible CYC1 promoter
pSR469	APOBEC3C	None	Hygromycin resistance	Doxycycline-inducible CYC1 promoter
pSR470	APOBEC3DE	None	Hygromycin resistance	Doxycycline-inducible CYC1 promoter
pSR471	APOBEC3F	None	Hygromycin resistance	Doxycycline-inducible CYC1 promoter
pSR472	APOBEC3H	None	Hygromycin resistance	Doxycycline-inducible CYC1 promoter
pVH-A3A	APOBEC3A	ACT1	Hygromycin resistance	Doxycycline-inducible CYC1 promoter
pZAK031	APOBEC3A	ACT1	Hygromycin resistance	Doxycycline-inducible CYC1 promoter
pVH-20	APOBEC3A	None	*LEU2*	Inducible GAL4 promoter
pTM-019	empty vector	None	*K.L. LEU2*	Doxycycline-inducible CYC1 promoter

pTM-020	APOBEC3A	None	*K.L. LEU2*	Doxycycline-inducible CYC1 promoter
pTM-021	APOBEC3B	None	*K.L. LEU2*	Doxycycline-inducible CYC1 promoter
pTM-485	empty vector	None	NAT resistance	Doxycycline-inducible CYC1 promoter
pTM-486	APOBEC3A	None	NAT resistance	Doxycycline-inducible CYC1 promoter
pTM-487	APOBEC3B	None	NAT resistance	Doxycycline-inducible CYC1 promoter
pMC-30	APOBEC3A	ACT1	NAT resistance	Doxycycline-inducible CYC1 promoter
pTM-488	empty vector	None	G418 resistance	Doxycycline-inducible CYC1 promoter
pTM-489	APOBEC3A	None	G418 resistance	Doxycycline-inducible CYC1 promoter
pTM-490	APOBEC3B	None	G418 resistance	Doxycycline-inducible CYC1 promoter
pTM-491	empty vector	None	*S.P. HIS5* (complements *S.C. his3*Δ)	Doxycycline-inducible CYC1 promoter
pTM-492	APOBEC3A	None	*S.P. HIS5* (complements *S.C. his3*Δ)	Doxycycline-inducible CYC1 promoter
pTM-493	APOBEC3B	None	*S.P. HIS5* (complements *S.C. his3*Δ)	Doxycycline-inducible CYC1 promoter
pMC-31	APOBEC3A	ACT1	*S.P. HIS5* (complements *S.C. his3*Δ)	Doxycycline-inducible CYC1 promoter

1. Grow a 5- or 10-mL overnight culture from a colony or 2 mm by 2 mm section of a patch:
   a. for YPDA: -start cultures later in the day (3 to 6 pm).
   b. Synthetic complete (SC) dropout: start cultures in the morning.
      - Note: growth takes 24 h to reach near saturation in SC dropout media.
2. The next morning, inoculate a "fresh" culture by diluting the overnight 1:10 in new media.
   a. 1.5 mL is needed per transformation, usually 500 μL culture into 4.5 mL fresh media.
   b. YPDA culture - grow 4 h.
   c. SC culture - grow 6 h.
   d. This puts yeast in early log phase.
3. Transfer 1.5 mL of culture to a microcentrifuge tube and pellet (max RCF for 2 min).
4. Pour off the supernatant and remove the remaining media with a pipette tip.
5. Add 10 μL of 10 mg/mL salmon sperm DNA (this should be boiled for 5 min and placed on ice before use).
6. Add DNA 40–100 ng of extrachromosomal plasmid.
7. Briefly vortex or mix by running tube along rack, to re-suspend the cells.
8. Add 500 μL of 1X LTE/PEG mix.
9. Add 50 μL DMSO.
10. Vortex for 10 s
11. Incubate at room temp or 30 °C with mild agitation for 30 min.
12. Heat shock for 15 to 30 min.
    a. The length of time for successful plasmid uptake is strain dependent.
13. Pellet yeast (max RFC, 2 min).
14. Remove supernatant and add 500 mL of sterile H_2O.
15. Invert several times, but do not vortex.
16. Spin again for 2 min at max RFC.
17. Remove Supernatant.
18. For extrachromosomal and integrative plasmids that complement and auxotrophic defect, plate directly on the appropriate SC dropout media. For plasmid utilizing antibiotic selection, plate on YPDA (without selections). Allow the yeast to grow overnight, then replica plate to YPDA with the appropriate selective agent. For plating, resuspend the cell pellet in 150 μL H_2O plate on two plates:
    1. 140 μL resuspended cells.

2. 10 μL resuspended cells + 140 μL H_2O on the plate.

19. The exact volumes are not critical; the goal is to plate a high concentration of cells on one plate and a lower concentration on the other.

4. Screening for genetic modulators of APOBEC mutagenesis

Genetic screens for gene deletions that influence the rate of APOBEC-induced mutagenesis are easiest to conduct utilizing a mating-sporulation method (Zhang et al., 2012) that crosses a wild-type yeast strain harboring the APOBEC expression plasmid with another yeast strain of opposite mating type that contains a genetic deletion to make a diploid yeast containing the APOBEC and one copy of the gene deletion (Mertz et al., 2023) (Fig. 1). Subsequently, sporulation of these diploid cells produces gamete cells that have the APOBEC plasmid and gene deletion allele randomly segregated. Plating of these haploid cells on selective media enables efficient selection of specifically mating type a yeast containing both the APOBEC plasmid and the genetic deletion. These isolates can then be assessed for altered mutation rate by *CAN1* mutation frequency. For screening, we transformed the *LEU2* marked pTM-21 vector into the yeast strain, yTM-02 (MATα, ura3-Δ, leu2-Δ, his3-Δ, lys2-Δ, rpl28-Q38K, mfa1Δ::MFA1pr-HIS3, V34205::LYS2, V29617::hphMX) (Mertz et al., 2023). Utilization of this vector-yeast combination allows compatible mating with the BY4741 deletion library (purchased from Open Biosystems; currently available from Transomic Technologies; Huntsville, AL), and subsequent selection of MATa haploid cells containing the APOBEC vector and a gene deletion by plating on synthetic complete (SC) media lacking leucine, histidine, and containing G418 and cycloheximide. Note that use of antibiotics in SC dropout media requires making low-salt media, which can be made by utilizing yeast nitrogen base without both amino acids and ammonium sulfate and using 1 g of monosodium glutamate in place of ammonium sulfate as the source of nitrogen.

4.1 Crossing strains

1. Set out and label 10 cm plates containing YPDA and 96 well microplates
2. Fill about 106 96-well plates with 100 μL of 50 % glycerol. This is for re-freezing the deletion library.

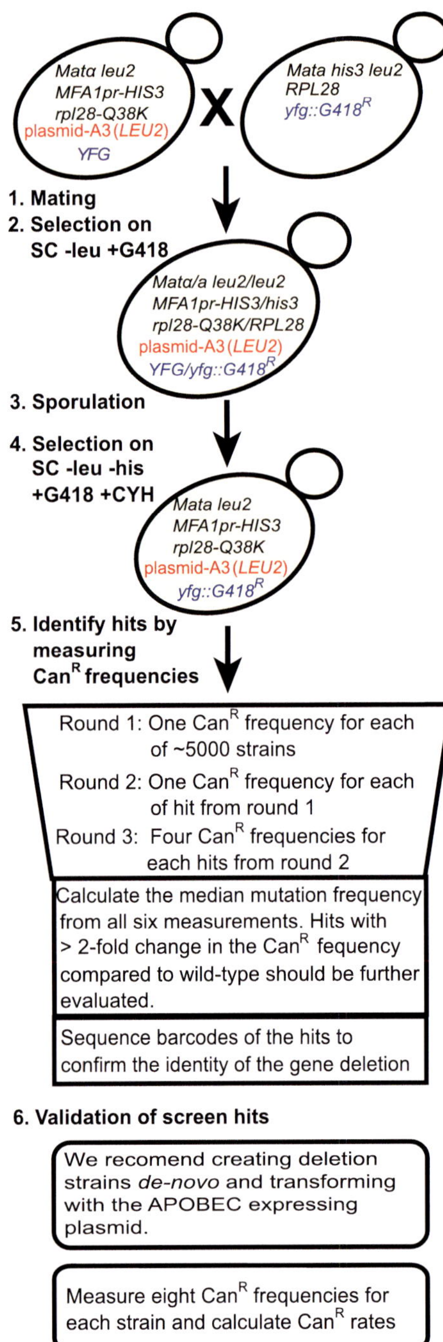

Fig. 1 Schematic of yeast screen used to identify genes that elevate APOBEC-induced mutation. Begin by mating a haploid MATα yeast carrying the APOBEC

3. Fill 53 96 well plates with 400 μL of YPDA. This is for growing library strains.
   a. this should take less than an hour
4. Thaw the yeast deletion library.
5. Inoculate YPDA-containing 96 well plates with 10 μL of yeast deletion library stock.
6. Cover each plate with a sterile breathable cover.
7. Plate 96 well plates in plastic baskets bolted to the shaker platform of an incubator. Shake at about 250 RPM at 30 °C overnight.
8. Inoculate 100 mL of SC-leu media with yTM02 +plasmid-A3 (*LEU2*) yeast and grow overnight at 30 °C in a shaking incubator.
9. Re-freeze the original library by resealing the plates and placing them directly in a −80 °C freezer.
10. The next day, dispense 100 μL of each culture from the deletion library grown overnight in 96 well plates into the labeled 96 well plates containing 50 % glycerol. Mix well, seal, and freeze at −80 °C to create fresh frozen stocks of the deletion library. This can be done efficiently using a Viaflo96 using 300 μL tips. We recommend the custom program of Mix 300 μL (4 cycles) and multiple dispensing 100 μL.
11. Set up plates for mating.
12. Pellet the overnight culture of yTM02 + plasmid-A3 (*LEU2*) and resuspend in 25 mL H_2O.

---

expression plasmid (plasmid-A3) with each MATa library strain carrying a yeast gene deletion (yfg::G418^R) to generate diploid strains. The selective markers used here for selection are an example. Other selective markers are usable (i.e., for screening different libraries), provided you maintain selection of the APOBEC expressing plasmid and can select diploids and later haploid cells. Post sporulation, to select haploid cells, we utilized the MFA1pr-HIS3 allele, which expresses *HIS3* only in MATa cells, and cycloheximide (CYH), which selects against cells with a wild-type *RPL28* gene. The resulting haploid cells were subjected to serial rounds of Can^R frequency measurements to identify gene deletions that likely augment APOBEC-induced mutagenesis. Note that round 1 and 2 of screening are each based on a single mutation frequency, which likely leads to loss of some modulators of APOBEC mutagenesis. However, this approach greatly reduces the effort required to complete the screening process. We recommend validating the factors that modulate APOBEC-induced mutagenesis using mutation rates measurements, which are more robust than mutation frequencies. Figure adapted from (Mertz et al., 2023) with permission under publication with a CC-BY 4.0 license.

13. Using sterile glass beads, plate 80 μL of the resuspended yTM02 + plasmid-A3 (*LEU2*) on 160 pre-labeled YPDA plates. This will make a lawn of the MATα haploid strain.
14. Vortex each 96 well library culture plate immediately before transferring yeast.
15. Using an 8-channel pipette, dot 2 μL of the deletion library strains onto the YPDA plates inoculated with a "lawn" of yTM02 +pTM-21.
16. Write the plate number and coordinates A1 to H4 of the strains used on the YPDA plates.
17. About 160 plates, with 32 strains each, will be needed.
18. Grow overnight in a 30 °C incubator.
19. Replica plate each plate to 2 SC-leu+G418 plates to select for diploid cells.
20. Grow overnight in a 30 °C incubator.
21. Replica one set of diploid strains to pre-spo media. Save the other plate at 4 °C as a backup.
22. Grow the pre-spo plates for 2 days in a 30 °C incubator.
23. Replica each pre-spo plate to Spo media.
24. Grow 7–12 days in a 30 °C incubator to generate yeast tetrads.
25. Replica plate each Spo plate to 2 plates of SC-his-leu +G418 +cycloheximide media to select for MATa haploids containing both plasmid-A3 (*LEU2*) and the gene deletion from the deletion library.
26. Grow for 2 days at 30 °C.
27. From each patch generated on the SC-his-leu+G418 +cycloheximide media, take a small number of cells with a sterile wooden applicator and streak to single colonies on new SC-his-leu+G418 +cycloheximide media to obtain clonal isolates of the desired haploid cells. This will require ~5000 single cell isolation streaks.
28. Individual colonies from these streaks will be utilized directly for measuring canavanine resistance (Can^R) frequencies in the next step.

4.2 Canavanine resistance frequencies

To determine the impact of each gene deletion on APOBEC-induced mutation, clonal isolates are tested for the frequency of resistance to the drug canavanine, which selects for mutations in the *CAN1* gene (Mertz et al., 2023). For the full screen, this will involve approximately 5000 mutation frequency measurements, comparing growth on canavanine containing media to SC control media. To reduce the number of plates needed, we recommend minimizing the cultures used and plating on media in quartered 10 cm petri plates. This allows 4 strains to be spread on each plate. We also made efforts to grow colonies to similar sizes. Strains that were noticeably slower growing were individually struck out and grown to approximately 1×10^7 cells. And done separately, but otherwise identically to the "normal growing" strains.

Protocol:

1. Make and label SC-arg+canavanine and SC plates in quartered petri dishes. We recommend keeping track strains identity by using plate and coordinates from original library. For example, labeling plate quarters with Stain 1-A-1 indicating the strain from plate 1, coordinate A-1 of the deletion library. When measuring frequencies, make a secondary key for your resuspension plates using a similar nomenclature: Plate 1, Well 1 (of the resuspended LEU+, G418^R colonies) – strain 1-A-1 abbreviated to P1-W1, 1-A-1 when labeling plates.
2. Add ~4 sterile glass beads to each quarter of each plate.
3. Pick a single LEU+, G418^R colony from each library cross that has grown to about 10,000,000 cells.
4. Resuspend the yeast colonies in 96 well plates containing 140 μL of H_2O.
5. Vortex.
6. Make two serial dilutions of each plate using either a multichannel pipette or a Viaflo96 (Integra). First take 5 μL of the resuspended culture and dispense it into 100 μL of H_2O in a 96 well plate.
7. Vortex.
8. Next remove 5 μL of the first dilution of culture and dispense it into another 100 μL of H_2O in a 96 well plate.
9. Vortex.
10. Repeat for all plates of LEU+ , G418^R colonies to be assessed.
11. From each of the cross cultures, plate 30 μL of the undiluted resuspension of SC-arg+canavanine media with 200 μL of H_2O. This should be about 2×10^6 cells.
12. Then plate 60 μL of the 1:40,000 dilution of each culture on SC with 100 μL H_2O. This should be about 100 cells.
13. Spread yeast by shaking the plates containing the beads.
14. Grow at 30 °C for 4 days.
15. For each strain, count the cells on the corresponding SC-arg+canavanine and SC quarters.
16. Multiply the number of colonies on the SC quarter by the dilution factor (i.e., 40,000) and divide by the volume plated 60 μL (to determine

the number of cells per μL in the undiluted resuspension). Multiply this number by the volume used to resuspend the colony (140 μL) to find the total number of cells.

17. Multiply the number of colonies by (140/30) to find the total number Can^R cells in the culture. Divide total Can^R cells by total cells to calculate the mutation frequency.

18. Compile the results of round 1 screen.

19. Due to the amount of work in measuring one Can^R frequency per crossed strain, we recommend making repeat measurements only for genetic deletions that appear to significantly alter APOBEC mutagenicity in the first round of screening.

20. Make 3 additional mutation frequency measurements on each hit and calculate a median Can^R frequency for these measurements.

21. Make genomic DNA preps on each bonafide hit. Amplify and sequence the deletion cassette barcode with oligos AJB097 (GATGT CCACGAGGTCTCT) and AJB098 (CGGTGTCGGTCTCGTAG) to confirm identity of mutation.

5. Generating *CAN1* mutation spectra

After completing a yeast screen, spectra of the mutations occurring in *CAN1* can be generated efficiently using long-read next generation sequencing methods (Guo et al., 2015; Hoopes et al., 2017; Mertz et al., 2023). These spectra can be useful in determining how specific genetic deletions alter APOBEC mutagenesis by providing specifics of strand bias for the mutations and the ratio of C to T to C to G substitutions (Mertz et al., 2023). This data can also reveal sequence preferences for different APOBEC enzymes. To generate this data, around 50 independent *CAN1* mutants for each strain should be assessed. This results in a very large number of isolates that need to be processed. Make sure to keep track of genotypes and sample identity by properly labeling plates. We recommend making some type of systematic labeling scheme as a key.

5.1 Isolate independent *CAN1* mutants

CAN1 mutants can be obtained by one of the two methods:

Method 1:

1. Post completing mutation rate measurements, pick one colony from each of the SC-arg+canavanine plates that are derived from independent clonal cultures and streak to single colonies on SC-arg+canavanine.

2. Once the streaked yeast isolates are grown, pick an individual colony from each streak/Can^R clone and make a patch on YPDA.
3. Grow for 1–2 days.
4. Isolate yeast genomic DNA from yeast patches (see next section for protocol).

Method 2:

1. Plate yeast strain at a density of 30 to 200 colonies per plate (media should maintain section for APOBEC expression plasmid).
2. Once colonies are grown, 2–4 days, colonies should be patched to the same selective media. Grow overnight.
3. Patches are then replica plated to SC-arg+canavanine.
4. Incubate plated 3–5 days at 30 °C, or suitable temperature.
5. Can^R colonies should be apparent in the shadow of the original patch.
6. Make one streak to single colonies from 1 papilla each patch.
7. Let the streak to single dishes grow for 2–4 days.
8. From each streak, pick a single colony and use it to make a single patch on YPDA. You will harvest genomic DNA from this patch.

5.2 High-throughput genomic DNA isolation

We recommend utilizing 96 well plate format for isolation of this large number of independent yeast isolates, although it can also be done in microcentrifuge tubes and using the following columns, http://omegabiotek.com/store/product/hibind-dna-mini-columns/. The set up and layout of the 96 well plates is important, and it is vital to have meticulous notes as to which well of what plate corresponds to each sample. Also make sure to have wells without yeast added during the genomic DNA isolation. These will serve as negative controls during the PCR amplification and serve as controls to ensure there is no cross contamination between wells during the genomic DNA purification.

Example plate layout.

	1	2	3	4	5	6	7	8	9	10	11	12
A	geno-type A, sample 1	geno-type A, sample 9	geno-type A, sample 17	geno-type A, sample 25	geno-type A, sample 33	geno-type A, sample 41	geno-type B, sample 1	geno-type B, sample 7	geno-type B, sample 15	geno-type B, sample 23	geno-type B, sample 31	geno-type B, sample 39
B	geno-type A, sample 2	geno-type A, sample 10	geno-type A, sample 18	geno-type A, sample 26	geno-type A, sample 34	geno-type A, sample 42	geno-type B, sample 2	geno-type B, sample 8	geno-type B, sample 16	geno-type B, sample 24	geno-type B, sample 32	geno-type B, sample 40

C	genotype A, sample 3	genotype A, sample 11	genotype A, sample 19	genotype A, sample 27	genotype A, sample 35	genotype A, sample 43	genotype B, sample 3	genotype B, sample 9	genotype B, sample 17	genotype B, sample 25	genotype B, sample 33	genotype B, sample 41
D	genotype A, sample 4	genotype A, sample 12	genotype A, sample 20	genotype A, sample 28	genotype A, sample 36	genotype A, sample 44	genotype B, sample 4	genotype B, sample 10	genotype B, sample 18	genotype B, sample 26	genotype B, sample 34	genotype B, sample 42
E	genotype A, sample 5	genotype A, sample 13	genotype A, sample 21	genotype A, sample 29	genotype A, sample 37	genotype A, sample 45	genotype B, sample 5	genotype B, sample 11	genotype B, sample 19	genotype B, sample 27	genotype B, sample 35	genotype B, sample 43
F	genotype A, sample 6	genotype A, sample 14	genotype A, sample 22	genotype A, sample 30	genotype A, sample 38	genotype A, sample 46	genotype B, sample 6	genotype B, sample 12	genotype B, sample 20	genotype B, sample 28	genotype B, sample 36	genotype B, sample 44
G	genotype A, sample 7	genotype A, sample 15	genotype A, sample 23	genotype A, sample 31	genotype A, sample 39	negative control	negative control	genotype B, sample 13	genotype B, sample 21	genotype B, sample 29	genotype B, sample 37	genotype B, sample 45
F	genotype A, sample 8	genotype A, sample 16	genotype A, sample 24	genotype A, sample 32	genotype A, sample 40	negative control	negative control	genotype B, sample 14	genotype B, sample 22	genotype B, sample 30	genotype B, sample 38	genotype B, sample 46

When making genomic DNA preps from yeast patches, using fresh yeast patches is key. The reason for this phenomenon is unclear, but genomic DNA harvested from yeast that have been sitting around for days, or stored at 4 °C for weeks, amplifies very poorly compared to genomic DNA from yeast patches freshly grown.

Protocol:

1. To each well of 1 mL deep 96 well plates, add 100 µL of Buffer-1/glass beads mixture using a repeater or electronic pipette.
   a. Cut the end of each pipette tip such that beads do not clog it.
2. Add 100 µL of Buffer 1 to each well.
3. Add a full patch of yeast cells (approximately 1 × 4 cm in size) to each well.
4. Cover tightly with adhesive foil seal
5. Vortex briefly for 2–3 s
6. Incubate at 70 °C for 20 min in a hybridization oven (indirect heating).
7. Vortex for 3 min using a microplate vortexer.

8. Incubate on ice for 5 min.
9. Remove foil and add 200 μL of cold Buffer 2 (i.e., 4 M NaCl). Note: If there is liquid at the top of the wells stuck to the seal (here or in subsequent steps), you should briefly centrifuge the plate to reduce the likelihood of cross contamination when removing the foil seals.
10. Replace the foil seal (use a new seal) and vortex briefly, place on ice for 5 min
11. Centrifuge 20 min max (2000–4000) RCF at 4 °C.
12. Transfer 140 μL of supernatant to a new 2 mL microplate.
    a. This leaves a lot of liquid behind but decreases the likelihood of picking up debris from the bottom.
    b. We complete this step with a 96-channel pipette (Viaflow96) that allows for precise placement of pipette tips at the same height during aspiration. It is possible to complete this with a multi-channel pipette but it can be difficult.
    c. This dirty supernatant can be frozen, and the following completed later.
13. Add 750 μL of binding buffer and mix by pipetting up and down a volume of 800 μL 6X (or vortex 2–3 s with foil seal in place).
14. Transfer sample to DNA binding plate
    a. We use Omega Bio-tek 96 well plates (BD96-01: http://omegabiotek.com/store/product/hibind-dna-mini-columns-copy/)
15. Place DNA binding plate on a 96 well plate vacuum manifold.
16. A apply gentle vacuum, the sample will pass through individual columns on the plate and into the waste reservoir. DNA will be bound to the silica matrix within each column.
17. Once all sample is bound to silica matrix, transfer the 96 well DNA binding plate to a 96 well deep well plate. Centrifuge for 5 min at 1000 RCF in a low-speed swing bucket microplate rotor. This step ensures all the lysate passes through the column before the wash step, which can decrease guanidinium thiocyanate, a component of the lysis buffer, carryover into the purified DNA.
18. Discard the flow through.
19. Add 1200 μL of wash buffer (WB).
20. Return the 96 well DNA binding plate to the vacuum manifold and apply vacuum.
21. Discard the flow through.
22. Repeat steps 19–22 twice (for a total of 3 washes).

23. Dot bottom of binding plate on paper towels.
24. Repeat several times until liquid is not visible.
25. Transfer DNA binding plate to a clean collection plate
    a. an empty tip box works for this
26. Centrifuge 15 min at max RCF (2000 to 4000) and again blot on clean paper towels.
27. If any liquid is visible in the tube, invert the plate and centrifuge for 5 s
28. Set plate on a paper towel upright in the hybridization oven set at 70 °C for 30 min to dry off residual ethanol.
29. Set binding plate on a clean collection plate
    a. use a PCR plate or shallow microplate
30. Add 100 μL of elution buffer warmed to 60 °C directly to column. The binding matrix must be fully "wetted." Gentle tapping can be used to make sure the buffer reaches the matrix.
31. Incubate 7 min at 37 °C.
32. Centrifuge for 10 min at 1000 RCF.
33. Cover the plate with adhesive foil and store at −20 °C.
34. 1–10 ng/μL of DNA by qubit is a good yield.
35. Use these DNAs as the templates for PCR amplification of *CAN1* in the next section.

5.3 High-throughput amplification *CAN1*

To utilize long-read next-generation sequencing technologies (Marx, 2023) (i.e., PacBio or Nanopore) for sequencing *CAN1* mutants, each PCR will have a unique pair of barcoded primers to allow demultiplexing of the sequencing data (Guo et al., 2015). These barcodes will become unique identifiers for all the reads stemming from a specific sample. Both the forward and reverse primers contain barcode sequences and can be used in demultiplexing (Guo et al., 2015; Hoopes et al., 2017; Mertz et al., 2023). Each forward-reverse primer pair can only be used once for each pooled PacBio or Nanopore run. We generally include the reverse primer in separate PCR master mixes and the forward primers are pipetted to individual PCR reactions from a 96 well plate of forward primers. The unique barcode on the end of each primer is not optimized for PCR, and can lead to primer dimers, hairpins, and other phenomenon that can decrease the efficiency or cause failure of the PCR. So far, we have not experienced major issues with this. However, we recommend a hot-start polymerase for PCR, setting up reactions on ice, and adding the PCR plate to a thermocycler pre-heated to 95 °C. Since this process also involves parallel amplification of the same target, cross-contamination can be a problem. Therefore, use recently cleaned pipettes and filter tips for assembling the PCR. Also never use the same pipettes for assembling these PCRs that have been previously used for pipetting *CAN1* PCR products. Wells from the genomic DNA isolation (see prior section) that lack yeast will serve as a control to ensure cross-contamination is not an issue.

Protocol:

1. Prior to doing the PCR prepare a 96 well plate containing Forward primers (10 μM).

Example:

	1	2	3	4	5	6	7	8	9	10	11	12
A	canT-M_F1	canT-M_F9	canT-M_F17	canT-M_F25	canT-M_F1	canT-M_F9	canT-M_F17	canT-M_F25	canT-M_F1	canT-M_F9	canT-M_F17	canT-M_F25
B	canT-M_F2	canT-M_F10	canT-M_F18	canT-M_F26	canT-M_F2	canT-M_F10	canT-M_F18	canT-M_F26	canT-M_F2	canT-M_F10	canT-M_F18	canT-M_F26
C	canT-M_F3	canT-M_F11	canT-M_F19	canT-M_F27	canT-M_F3	canT-M_F11	canT-M_F19	canT-M_F27	canT-M_F3	canT-M_F11	canT-M_F19	canT-M_F27
D	canT-M_F4	canT-M_F12	canT-M_F20	canT-M_F28	canT-M_F4	canT-M_F12	canT-M_F20	canT-M_F28	canT-M_F4	canT-M_F12	canT-M_F20	canT-M_F28
E	canT-M_F5	canT-M_F13	canT-M_F21	canT-M_F29	canT-M_F5	canT-M_F13	canT-M_F21	canT-M_F29	canT-M_F5	canT-M_F13	canT-M_F21	canT-M_F29
F	canT-M_F6	canT-M_F14	canT-M_F22	canT-M_F30	canT-M_F6	canT-M_F14	canT-M_F22	canT-M_F30	canT-M_F6	canT-M_F14	canT-M_F22	canT-M_F30
G	canT-M_F7	canT-M_F15	canT-M_F23	canT-M_F31	canT-M_F7	canT-M_F15	canT-M_F23	canT-M_F31	canT-M_F7	canT-M_F15	canT-M_F23	canT-M_F31
F	canT-M_F8	canT-M_F16	canT-M_F24	canT-M_F32	canT-M_F8	canT-M_F16	canT-M_F24	canT-M_F32	canT-M_F8	canT-M_F16	canT-M_F24	canT-M_F32

2. Thaw reagents on ice.
3. Vortex reagents and make sure no precipitate is in the 10X buffer.
4. Centrifuge both the primer plate and template plates (i.e., genomic DNAs from the prior section) to collect all liquid at the bottom of the plate before opening.
5. Hold on ice.
6. For completing 96 well plates, make 3 master mixes as follows (each with a different reverse primer):

		mix for
PCR mix for confirmation PCR	**1 rxn 25ul**	**35**
H_20	13.4	**475.3**
Buffer 10X	2.5	**87.5**
dNTPs 2 mM	2.5	**87.5**
MgSO4 (25 mM)	1.9	**66.5**
primer reverse barcoded (100 uM)	0.2	**0.7**
KOD Hot Start polymerase	*0.5*	**17.5**
		mix/rxn
total rxn volume	**25**	**21**
Added separately		
primer Forward barcoded (10 uM)	2	
DNA template	2	

   a. The 10x Buffer, 2 mM dNTPs, and $MgSO_4$ are all included with the KOD Hot Start DNA Polymerase (Millipore-Sigma).

7. Once everything is combined, mix very well, but do not vortex.

8. Centrifuge master mixes to get everything at the bottom (just a few seconds).

9. Fill the 96 well PCR plate with mixes as indicated below:
   a. Reverse primer 1 mix - rows 1–4
   b. Reverse primer 2 mix - rows 5–8
   c. Reverse primer 3 mix - rows 9–12

10. Next use a 96 well pipette (or multichannel) to transfer 2 μL from the Forward primer plate to the 96 well PCR plate containing the master mix.
   a. Make sure plates have the same orientation.

11. Transfer 2 μL from the 96 well plate containing the genomic DNA samples.
   a. make sure plates have the same orientation.
   b. recording what primers go with which genomic DNA is crucial. Failure to do this will prevent you from determining what mutations correspond to what samples.

12. Cover the 96 well plates. We prefer using strip tube caps for a very secure seal. Good adhesive seals specific for thermocycling should also work well.

13. Centrifuge the 96 well plate briefly at 1000 RCF for 2 min.

14. Place PCR plate in the thermocycler for amplification.
  a. Recommended conditions:

Step	Temperature	Time
1	95 °C	2:00
2	95 °C	0:20
3	65.5 °C	0:10
4	70 °C	0:45
5	Go to #2	36X
6	70 °C	0:20
7	8 °C	forever

15. Pool successful amplifications and submit to sequencing core for either PacBio or Nanopore sequencing.

Post PCR Notes:

Avoid using column cleanup kits on pooled amplicons. DNA amplicons are denatured and reannealed during these cleanup methods. This results in strands from different amplicons annealing and being ligated to PacBio adapters. The resulting circular consensus sequences will contain multiple barcodes and mutations from different samples. These methods are fine on individual samples but result in extremely poor quality mutation calling for pooled amplicons. The individual PCR products (most often) do not need to be purified. Most core labs can purify and size select pooled amplicons without denaturing the DNA. We recommend consulting your core lab, or PacBio sequencing provider prior to starting.

It is advisable to pool a similar amount (ng of PCR product) prior to submitting the samples. Because one cannot accurately quantify DNA in the presence of primers and dNTPs, we recommend running the PCR samples on a gel along with a fragment of DNA of similar size and known amount (we use 100 ng of linearized pUC19). Then use ImageQuant (Cytiva), ImageLab (BioRad), imageJ, or a similar software to quantify the DNA bands. We normalize via the adjusted volume/signal by dividing each PCR bands signal by that of pUC19, then multiply by 100 (to calculate the ng/band), then divide by the volume of the PCR product loaded, which give the (ng specific PCR product/μL PCR product). We then use this value to determine how much of each sample to pool.

5.4 Variant calling from high throughput amplicon sequencing

The sequencing of hundreds to thousands of *CAN1* mutants can be efficiently and cost effectively (as low as $0.0005 per base) completed by utilization of long-read next generation sequencing platforms like Pacific Biosciences (PacBio) or Oxford Nanopore sequencing. In either case, pooled barcoded PCR products are submitted to the sequencing facility as a single sample and run on either a PacBio Sequel or MinION sequencer. Generally, circular consensus reads from PacBio have significantly better sequencing quality compared to Nanopore. However, the sequencing capacity for the PacBio Sequel is greater than necessary for most amplicon sequencing applications, therefore generating more data than is necessary to call mutations for each *CAN1* mutant (usually greater than 5 reads per barcode set is sufficient) and increasing the cost per experiment. By contrast, the lower sequencing quality of the Nanopore systems requires that more reads per clone (>10) be obtained to increase confidence in any mutation call. We generally recommend obtaining 100-fold more reads than samples to ensure that most barcode sets will have at least 10 reads associated with them.

Both sequencing platforms will produce a singular fastq output containing sequencing reads for all PCR products pooled in the sample. To obtain mutation calls from this data, the fastq file must be split into multiple files specific to the barcode sets included in the experiment. These individual files specific to a single PCR product are then individually aligned to a reference sequence and the sequences compared to identify variants consistently occurring among the majority of the sequencing reads for the amplicon. Multiple aligners (e.g., Geneious (https://www.geneious.com/), BWA-SW (Li & Durbin, 2010), or LRA (Ren & Chaisson, 2021)) and variant calling (e.g., Geneious, VarScan2 (Koboldt et al., 2012), Strelka2 (Kim et al., 2018), etc.) software can be used in these downstream steps. Below we present the steps to obtain mutation calls from this data using the commercially available software suite, Geneious (Geneious Prime® 2023.2.1, Biomatters Ltd).

Protocol:

1. Demultiplex long-read fastq file using 5' and 3' barcodes. Most demultiplexing softwares require sequential parsing of the 5' and 3' barcodes, requiring the software to be run multiple times. We therefore recommend utilizing our code, Sort_Reads_by_Barcode_MP.py (available at S-RobertsLab GitHub https://github.com/S-RobertsLab/)

(Brown et al., 2018; Hanscom et al., 2022), for this step. This script, written in python3 can read either fasta or fastq formats and searches the forward and complementary sequences of each read for specified forward-reverse barcode pairs, creating dictionaries of the reads that have each barcode set. It then subsequently writes either fasta or fastq output files for each barcode pair, containing only the reads containing each barcode. Any reads not having exact matches to any barcode par are written to a Skipped.fasta or Skipped.fastq file. Due to errors in primer synthesis, PCR amplification, and sequencing, upwards of 50% of the long-read sequencing output is likely to be skipped as the script requires an exact match within the barcode sequence. Sort_Reads_by_Barcode_MP.py requires an input fasta or fastq file, and two files containing your forward and reverse barcodes.

2. Create your barcode files. Forward and reverse barcodes need to be listed in two separate tab-delimited txt files. Each row of the files corresponds to a forward-reverse barcode pair. Therefore, the number of rows in each file needs to be equal and the same barcode will be repeated in either file if you used different combinations of the same 5' and 3' barcodes in your dataset. Barcodes can simply be the entire forward and reverse primer sequences. However, this increases the length of the barcode, increasing the chance for a sequencing error to occur in that region of your sequencing read, and increasing the number of reads skipped by the code. Barcode sequences can both be in the 5' to 3' direction of either DNA strand since the code checks for reverse complement sequences. All barcode sequences must be in uppercase to match the output convention of the fastq file.
3. Execute Sort_Reads_by_Barcode_MP.py. Sort_Reads_by_Barcode_MP.py can be executed in a terminal of a linux shell from the directory containing the script using the command: **python3 Sort_Reads_by_Barcode_MP.py -i path to input.fastq -f path to forward barcode.txt -r path to reverse barcode.txt -p number of threads**
4. This will create a series of new fastq files, one for every barcode pair, within the directory of the input.fastq file. Each new file will be named by the barcode sequences of each pair.
5. Import fastq files into Geneious. Once the individual fastq files are made, open up Geneious and drag the parsed fastq files into a Geneious folder for analysis. You will be questioned if you would like to separate all the sequences or "**Create sequence list**." Chose "**Create sequence list**" to keep all of the sequences with the same barcodes in a single document.
6. Align to a reference sequence. Select all imported sequence files in your Geneious directory. Click "**Align/Assemble**" from the toolbar. Chose

"**Map to Reference**." Select a reference sequence. We recommend using fasta file of your unmutated amplicon sequence. Geneious annotates this sequence with detected variants. Therefore, starting with an unannotated sequence is preferred. Check "**Assemble each sequence list separately**." Use the Geneious Mapper on Medium Sensitivity/Fast. Check "**Find structural variants, short insertions, and deletions of any size**." Do not trim your reads and only save contigs. We recommend naming the assembly with the Reads Name only. Click "**OK.**"

7. Find Variants. Geneious finds variants as part of the Annotate & Predict function. Select all alignments generated from the previous step. Select "**Find Variants/SNPs**" from the "**Annotate & Predict**" dropdown menu. You will be warned that these annotations cannot be undone. Click "**Continue**." Select a minimum coverage of 4 reads and a minimum variant frequency of 0.5. We recommend calculating Variant P-values, Merging adjacent variants, and using separate annotations for each variant at a position. Click "**OK**."
8. Export a table of variants. Find your original reference sequence file that has now been annotated with all the variants identified in step 7. Open the reference sequence file in Geneious and click on the annotations tab. Click on "**Type**" and select all. Click on "**Track**" and select all. Next, click on columns to select what information about the variants you would like in your table. We recommend including columns: Track Name, Type, Change, Coverage, Polymorphism Type, Variant Frequency, Variant P-vale, Minimum, Maximum, Length, Variant Nucleotides, and Reference Nucleotides. Click "**Export table**" and save the resulting file to your computer.
9. The resulting list of mutations can then be further filtered for accuracy and reformatted for downstream sequence analysis.

6. Whole genome sequencing of yeast expressing APOBEC enzymes

6.1 Mutation accumulation

Passaging yeast (Fig. 2) allows cells to accumulate mutations over time and generations. It lets the cells go through multiple cycles of replication, increasing likelihood of generating these mutations (Hoopes et al., 2016).

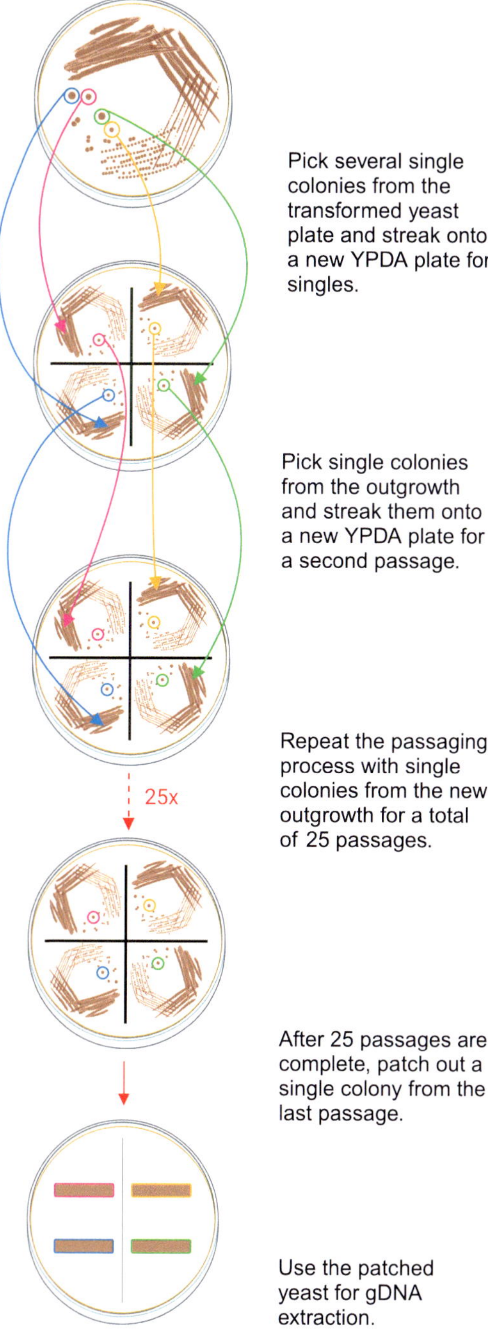

Fig. 2 Passaging of yeast strains expressing APOBEC cytidine deaminases for mutation accumulation during replication. From a selective plate with yeast

(Continued)

6.1.1 Outgrowth strains

1. First transform yeast with an APOBEC expression vector using the previously described method. We typically transform Ung1-deficient yeast when assessing APOBEC mutation signatures or determining cellular targets of APOBEC activity as no dU repair occurs in these strains (Hoopes et al., 2017), allowing all APOBEC activity to be evaluated. For assessing modulators of APOBEC mutagenesis, wild-type yeast should be used (Mertz et al., 2023).
2. Pick a single isolated yeast colony from the selective media of the yeast transformation using a sterile applicator. We typically use autoclaved wooden applicators or toothpicks.
3. In a quarter section of a new selective media plate (to maintain the APOBEC expression vector in the yeast), apply the yeast to the corner.
4. Perform quadrant streaks to streak to single colonies, creating a bottleneck effect, and ensuring generation of a clonal population, as represented in Fig. 2.
   a. Note: Some yeast colonies post transformation may have inactivated APOBEC activity. We recommend using a small amount of the streaked yeast to assess the mutator phenotype of each isolate prior to serial passaging. Do this by patching the yeast on a new selective plate, allowing it to grow for 2 days at 30 °C, and subsequently replica plating to SC-arg+canavanine media to evaluate the number of Can^R papillae.
5. Fill additional plates with quadrant streaks from other individual yeast colonies from the transformation step to generate independent outgrowth lines.
6. Incubate at 30 °C for 2–3 nights.
7. When individual colonies appeared after incubation, pick a single colony from each quadrant and re-streak it on a new selective media plate.
8. Repeat this process for at least 25 passages.
9. Once the desired numbers of passages are completed, patch out a single colony from the last passage for each isolate, as represented in Fig. 2.

Fig. 2—Cont'd transformed with an APOBEC expression plasmid, select single colonies and streak them in quadrants to obtain clonal populations. For each outgrowth, select a single isolated colony and streak onto another selective plate to maintain the plasmid through each passage. Repeat for at least 25 passages. After the number of passages are complete, from the last passage, patch out a single colony for gDNA extraction. Image created with BioRender.com; Agreement number RA279GTDF2.

10. Incubate the plates for 3 nights at 30 °C until a thick patch is formed. This patch can be used to generate frozen stocks (−80 °C) of the accumulation strains by resuspending each patch in separate freezer tubes containing 50 % glycerol. Additionally, yeast from the patch can be used to inoculate YPDA cultures for genomic DNA isolation using the protocol below (section 6.2).

6.1.2 Telomeric ssDNA accumulation strains

Yeast strains used in this protocol should occur in a background with a *cdc13*-1 allele (Chan et al., 2012; Chan et al., 2015), which will produce single stranded DNA at the telomere ends (Fig. 3).

1. Add 5 mL YPDA media with 30 μL 50 mg/mL hygromycin B to a culture tube for each yeast strain to be tested.
2. For each yeast strain to be tested, pick a single yeast colony with a wooden applicator and mix into the YPDA + hygromycin B media in the respective culture tube.
3. Incubate cultures at 23 °C for 72 h.
4. Dilute each yeast culture 1:10 to create a new 5 mL YPDA + hygromycin B culture for each original yeast strain.

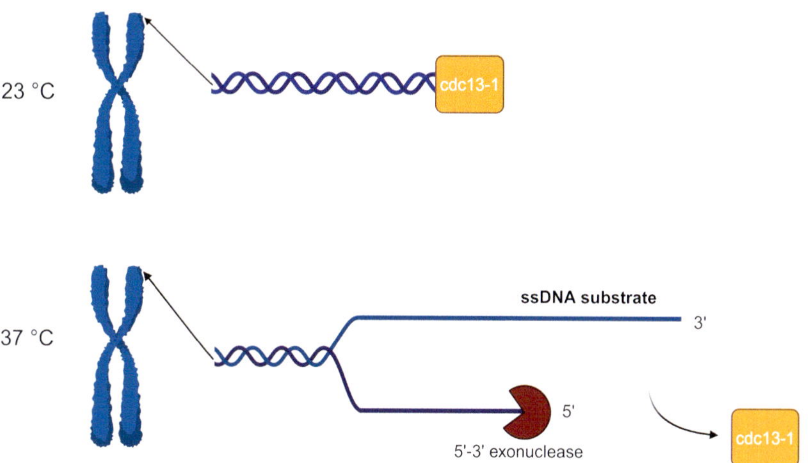

Fig. 3 Method for mutation accumulation on end-resected chromosomal ssDNA. Yeast grown at 23 °C have a functional cdc13–1 telomere cap. When shifted to 37 °C, cdc13–1 detaches from chromosomal DNA, leaving telomeric dsDNA uncapped and susceptible to resection by 5′ to 3′ exonucleases and subsequent generation of telomeric ssDNA for mutagens that favor a ssDNA substrate (such as APOBECs) to act on. Image created with BioRender.com; Agreement number HT278CRMVX.

5. Incubate diluted yeast cultures at 37 °C for 6 h.
6. Pellet yeast by centrifuging at 4000 RCF for 10 min.
7. Remove supernatant by pouring off or pipetting.
8. Wash yeast by resuspending yeast pellet in 5 mL PBS (phosphate buffered saline).
9. Centrifuge yeast in PBS mixture at 4000 RCF for 10 min.
10. Remove supernatant.
11. Resuspend the pellet in 5 mL PBS.
12. Incubate yeast in PBS at 37 °C for 42 h.
13. Pipette a small aliquot (~100 μL) from each culture into a 0.6 mL tube and dilute the aliquot 1:10 in sterile water.
14. Count yeast cells on hemocytometer using the 1:10 diluted sample to obtain the number of yeast cells/mL of each culture (accounting for the tenfold dilution factor).
15. Plate 1×10^7 cells onto 2 plates containing canavanine media and 1 plate containing synthetic complete medium per yeast strain tested. Note: the number of cells plated onto canavanine plates may need to be adjusted for each strain to obtain sufficient colonies.
16. Incubate plates at 23 °C for 5 days or until there are 50–100 colonies present per plate.
17. Pick the desired number of colonies off canavanine plates for each yeast strain and streak them out individually on YPDA plates. Grow for 5 days at 23 °C.
18. Make patches from streaks.
19. Let grow at 23 °C.

6.2 Preparation of yeast genomic DNA for whole genome sequencing

Reagents needed for this method:

- YPDA media (liquid & containing agar)
- Hygromycin B (GoldBio, H-270-1)
- 1X S-Buffer, 2X S-buffer
- Canavanine media
- Synthetic complete media
- 1X TE
- Zymolyase (USBiological, Z1004)
- Tris pH 8
- 3 M sodium acetate pH 5.3
- 100 % and 70 % ethanol

- Phenol-chloroform
- RNAseA
- Low EDTA TE (10 mM Tris-HCl pH 8, 0.1 mM EDTA)
- 5 M potassium acetate
- 20 % SDS
- Beta-mercaptoethanol

Protocol:

1. Start 7 mL YPDA culture in a 15 mL plastic culture tube from each streaked-out yeast colony.
2. Incubate cultures overnight (about 20–23 h) at 23 °C on a rotator.
3. Pellet the yeast in each culture at 3200 RCF for 15 min.
4. Pour off the supernatant.
5. To each yeast pellet add 196 μL TE Buffer, 210 μL molecular biology grade water and 2.1 μL beta-mercaptoethanol.
6. Completely resuspend cell pellet. This can be completed by running the tube along microcentrifuge rack and vortex mixing.
7. Transfer entire volume to a 1.5 mL tube.
8. Incubate at 30 °C for 45 min.
9. Centrifuge 1.5 mL tube for 30 s at top speed in a microcentrifuge.
10. Remove and discard supernatant.
11. Resuspend pellet in 350 μL 1X S-Buffer.
12. Centrifuge for 30 s at top speed. Remove and discard supernatant.
13. Resuspend cell pellet in 350 μL 1X S-Buffer containing 7 units of zymolyase.
14. Incubate at 35 °C for 1 h to allow for degradation of yeast cell wall.
15. Centrifuge at 4000 RCF for 2 min.
16. Resuspend the yeast spheroplast pellet in 245 μL TE buffer containing 2 % SDS by volume. Be gentle while mixing spheroplasts, as they are fragile. Tip: resuspend pellet in an aliquot of TE buffer first to allow for easier resuspension then add the remaining TE buffer with SDS added (e.g., add 200 μL TE buffer, mix, add TE Buffer with enough SDS added to create a final 2 % V/V solution).
17. Incubate at 65 °C for 25 min.
18. Invert tube and put it on ice.
19. Add 245 μL of 5 M potassium acetate, mix by inverting.
20. Incubate on ice for 30 min.
21. Centrifuge at max speed at 4 °C for 15 min.
22. Transfer the supernatant to a new 1.5 mL tube.

23. Add 2 volumes (~980 μL) of 100 % ethanol to the supernatant and mix by inverting.
24. Pellet DNA by centrifuging samples at max speed at 4 °C for 25 min.
25. Remove and discard supernatant.
26. Wash pellet with 70 % ethanol: add 1000 μL to pellet, spin at max speed for 5 min, then remove and discard supernatant.
27. Centrifuge at max speed for 30 s, then remove any residual liquid by pipetting.
28. Dry pellet for 10–15 min (necessary time may vary) at 37 °C with tube lid open.
29. Resuspend dried pellet in 170 μL low EDTA TE buffer. Running tube along tube rack and vortex can help resuspension.
30. Add 14 μL 10 mg/mL RNAseA, pipette to mix.
31. Incubate at 37 °C for 1 h.
32. Transfer supernatant for a heavy phase lock tube.
33. Add an equal volume of phenol chloroform (pH 8) to the volume of TE Buffer + RNAseA in each tube (in this case, 184 μL).
34. Mix on an inverter for 5 min (30 RPM). Pulse vortex for 30 s to mix phases.
35. Centrifuge at 12,000 RCF for 5 min to separate the phases.
36. Transfer upper aqueous phase to a new 1.5 mL tube and note the average volume transferred across samples.
37. Add 0.1 vol 3 M sodium acetate pH 5.3 and 2 volumes of 100 % ethanol to the 1.5 mL tube.
38. Mix by inverting, place tubes on ice for 20 min to allow DNA to precipitate.
39. Centrifuge at 15,000 RCF for 30 min at 4 °C.
40. Remove and discard supernatant.
41. Add 1 mL of 70 % ethanol to the pellet.
42. Centrifuge at 15,000 RCF for 5 min at 4 °C, remove and discard supernatant.
43. Repeat steps 39 and 40 for a total of 2 washes.
44. Centrifuge at 15,000 RCF for 30 s
45. Remove any residual ethanol by pipetting.
46. Dry DNA pellet at 37 °C for 10 min (times may vary) with tube lid open. Check each tube to assess whether the pellet is dry (i.e., no residual ethanol) and increase the drying time if necessary. Care should be taken to not over dry the pellet, which can make the DNA difficult to resuspend.

47. Once dry, add 40 μL of 10 mM Tris pH 8 to the pellet.
48. Incubate at 37 °C for 5 min.
49. Pipette up and down to break up the pellet.
50. Incubate at 37 °C for 10 min.
51. Vortex in pulses for 30 s
52. Continue incubation at 37 °C as necessary to resuspend the DNA pellet.
53. Measure DNA concentration. Samples can be run on a 1 % agarose gel to perform a quality check for the presence of RNA contamination and any degradation of DNA in samples.

6.3 Whole genome sequencing analysis

Illumina 150 bp sequencing continues to be the dominant method of whole genome sequencing (WGS) in the field of biology (Satam et al., 2023). This technique, known as short-read sequencing, differs from more recent methods like Nanopore or PacBio long-read sequencing. Although long-read sequencing has its advantages, particularly for humans and other complex organisms, short-read sequencing is the optimal choice for yeast. The yeast genome is simple, compact, and well-defined, allowing for 100x coverage in a single lane on an Illumina sequencer. Despite being lower-order eukaryotes, yeast systems often mirror those of higher-order eukaryotes, making sequencing and mutagenesis experiments straightforward, cost-effective, and highly informative.

Processing sequencing data can require proficiency in Linux/Unix operating systems, command line tools, and an understanding of the computational methods necessary for accurate interpretation and troubleshooting. This guide aims to provide one example pipeline from obtaining raw sequencing data to obtaining variant calls (Fig. 4) and to help navigate the challenges associated with processing large datasets.

General Tips:

Before delving into the intricacies of each step, the following items should be considered before deploying any whole genome sequencing pipeline.

1. **Minimize Disk I/O and File Size:**

   Whole genome sequencing FASTQ and alignment files are large. Based on storage considerations alone, we highly recommend file compression, as the compressed FASTQ and SAM files are typically 3 to 5 times smaller than their uncompressed counterparts. This increases the available space to write new intermediate files during variant calling,

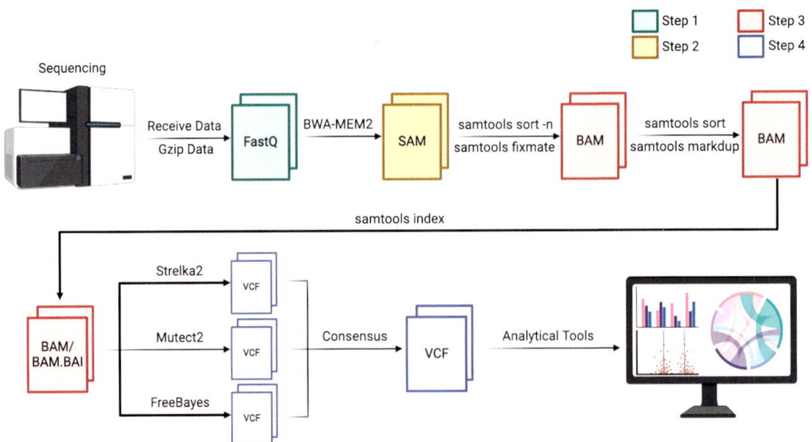

Fig. 4 Overview of yeast next generation sequencing data processing workflow. The workflow outlines the steps needed to process raw yeast sequencing data to obtain variant call (VCF) files. Step 1: compress FastQ files using Gzip. Step 2: Use BWA-MEM2 to align reads. Step 3: Use samtools for processing SAM files into sorted and indexed BAM files. Step 4: Use Strelka2, Mutect2, and FreeBayes to generate VCF files for downstream analysis. Image created with BioRender.com; Agreement number TP278BO2CB.

which would fail if the available space were exceeded. Furthermore, Disk I/O, also known as Disk Input/Output, refers to the read and write operations performed by a computer's storage system, such as a hard drive or SSD. Disk I/O measures the speed and efficiency of data transfer between the disk, memory, and CPU which is one of the biggest bottlenecks in computational biology because of the large file sizes. For example, one pair of yeast FASTQ files contains 50 million lines of data, the pairwise aligned reads contain over 10 million lines of data. The equivalent pair of human FASTQ files can be upwards of 8 billion lines and the aligned reads can be around 1.5 billion lines. When the size of files or the quantity of files being read grows, there will be a significant increase in the time required to complete the analysis and compressing your input files will decrease that.

2. **Parallelization/Utilization is Key:**

   Implementing parallelization in your tasks can significantly improve performance of intensive and time-consuming tasks. As an illustration, consider the scenario where sequencing results need to be compressed to conserve storage and minimize disk I/O. Assume there are 24 files and a CPU with 24 cores. One could utilize **gzip** for this task and have the option to compress all 24 files using the command **gzip** * . This command compresses the files in series, beginning with the first file and continuing until the last file has been compressed. Compressing files is typically limited to utilizing only one CPU core at a time, resulting in the remaining 23 files and cores remaining idle. By utilizing a tool such as GNU's **parallel**, all 24 Gunzip commands can be executed simultaneously, making use of all 24 cores. Although one could open a new terminal instance and enter each command individually, utilizing the parallel framework is highly recommended due to its efficient built-in queue system. Users can specify the desired number of tasks to be executed concurrently, along with the list of tasks, and the system will ensure that the maximum number of tasks is being executed simultaneously. In this scenario, the recommended approach would be to use the command **ls fastq_dir | parallel -j 24 gzip**. This command utilizes the **ls** command to retrieve all files in the **fastq_dir**, and then passes them into the parallel command using the **|** symbol.

By utilizing the **-j < jobs >** option, the parallel command enables users to specify the number of jobs, and potentially aligning it with the cores of the system, and subsequently using it as an argument for the **parallel** command. Assuming each compression step required 10 min, the sequential processing would have taken 4 h, while parallel processing would have taken only 10 min. Other tools, such as BWA (Li, 2013), STAR (Dobin et al., 2013), and Samtools (Li et al., 2009) possess built-in parallelization capabilities, enabling the concurrent utilization of multiple cores to accomplish the same task. When possible, we strongly advise opting for intra-process parallelization over inter-process parallelization as it is more efficient. However, we strongly advise fully utilizing your resources and making every effort to maximize CPU usage.

3. **Do not Overload Your System:**

   In accordance with the previous tip, there is a specific threshold where resource utilization starts to yield diminishing returns. Consider a scenario where a system has the same 24 cores but needs to allocate them to 36 tasks. If you were to launch all 36 tasks simultaneously, you would overload your system. Since there are more tasks than cores, the tasks either must share a core or wait for a core to become available. This situation can be compared to a traffic jam, where the number of cars (tasks) exceeds the available number of lanes (cores). As the traffic builds up, some cars must wait or slow down, causing congestion.

Similarly, the CPU's resources are occupied not only by task execution, but also by the management and coordination of these tasks. This management overhead acts like a traffic cop directing cars at a busy intersection, which itself consumes resources. The more tasks there are beyond the capacity of the system, the more time and resources are spent on this management, leading to inefficiencies and delays. Similar to how increasing the number of vehicles on a congested road can result in slower traffic flow, introducing more tasks than the system can effectively manage can lead to delays and diminished overall performance. Therefore, balancing the number of tasks with the system's capacity is crucial to maintaining optimal performance.

4. **Artificial Intelligence (AI) Is Great, Not Perfect:**

   AI models like ChatGPT (OpenAI. (2024). *ChatGPT* (Version 4.0) [Large language model]. https://chat.openai.com/) are becoming common for code troubleshooting, generating quick bash scripts, and seeking information on existing tools for one's analysis. AI models are highly effective tools, but you cannot depend entirely on them, considering that certain information they generate in responses may be incorrect. Learning how to most effectively prompt AI tools to help with analysis is an acquired skill that comes with experience. We suggest consistently prompting AI tools with questions regarding specific details to reduce the possibility of incorrect information. Utilize them, but do not solely depend on them. Always verify results and consistently check your AI's sources. Many newer models report the websites used in the response, so it is recommended to verify the credibility and accuracy directly from the source.

5. **Informative Titles and Organization are Key:**

   With each successive experiment and increase in sequencing, the task of data management becomes increasingly challenging. When organizing your data, it is recommended to assign meaningful titles to both file names and folder names. By employing this method, researchers can ensure that there is no uncertainty about the experiment, treatment, or genotype corresponding to the sample that was analyzed. When files are copied or moved, the date stamps may also be lost. Therefore, it is important to include the date in the file or folder name as needed. We recommend prioritizing organization and use informative titles, even if they become lengthy, as the long-term advantages outweigh any inconvenience. If you encounter file path length limitations, consider creating a metadata file to accompany your data, which will help you keep track of this information.

Alignment:

Many tools have been developed to handle sequence alignment. Two highly effective alignment tools, BWA and Bowtie2 (Langmead & Salzberg, 2012), are widely utilized for whole-genome sequencing alignments. Traditionally, BWA has been perceived to be the more accurate aligner while Bowtie2 is significantly faster. We recommend utilizing BWA-MEM2 (Vasimuddin, Misra, Li, & Aluru, 2019) because it is 2–5 times faster than the original BWA-MEM and can be faster than Bowtie2 in large datasets or when multithreading. It is also more accurate than Bowtie2, especially when the sequencing data contains insertion/deletion mutations (INDELs) or complex variation events. BWA-MEM2 is also more accurate at aligning highly repetitive DNA; overall BWA-MEM2 has a slight advantage when aligning longer reads and handling INDELs, leading to an increase in initial accuracy. BWA-MEM2 can also be used for aligning sequencing from higher eukaryotes, allowing users the application of the same pipelines developed for processing yeast data. Alignments can be processed in multiple ways. Below is the general workflow we have employed.

1. **Preparing for Alignment:**

   Sequencing results are often in a FASTQ format with a **.fastq** or **.fq** file extension. Additionally, these files will likely be in a GNU Gunzip (Free Software Foundation, Inc. (2024). *gunzip* (version 1.10) [Software]. Available at: https://www.gnu.org/software/gzip/) compressed format (to conserve storage space and enhance file processing speed), giving it the **.gz** file extension. Gunzip (**gzip**) is the most common compression tool on Linux and is used frequently in genomics because it has faster compression and decompression speeds compared to the standard zip compression method. Most aligners, like BWA, implement on-the-fly decompression, which comes with some additional computational burden, but the resulting reduction in disk I/O makes it worth it. If you need to Gunzip your files, use the command **gzip [OPTIONS]... [FILE]...** (See General Tip 2).

   We advise the creation of a project folder where the FASTQ files can be stored in a clearly labeled dedicated folder. Take the time to rename uninformative filenames with informative titles including information about genotype, treatment, or time point. Bear in mind that making significant renaming modifications downstream can complicate the process of retracing your steps. Prioritize the initial organization and renaming of your files, ensuring that a comprehensive log is kept to help with error tracing (See General Tip 5).

2. **Aligning Your Sequencing:**

   bwa-mem2 mem -M -t [cores] [reference_genome] [read file (s)] > output.sam

   Once data files are organized and named well, begin with the alignment tool that best suits your analysis. We run BWA-MEM2 with the command above where the -M option marks shorter split reads as secondary alignments, which can facilitate downstream identification of structural variants. The **-t** lets users select the number of threads used for the alignment. We recommend processing samples serially but maximizing multithreading within each alignment instead of parallelizing multiple single threaded alignments at once. Opting for the former approach yields a significant performance advantage. Last, users redirect the standard output to a SAM file using the > symbol.

3. **Processing Steps:**

   Once the reads have been correctly aligned, a few steps must be performed to clean up the information and make sure duplicate reads from PCR amplification have been removed and that the reads are prepared for variant callers. If the data was generated from single-end reads then proceed to Step c, Sort2.

a. **Sort 1:**

samtools sort -n -m 4 G -@ [cores – 1] -O bam output1.sam > output2.bam.

First, **sort** the SAM file and, at the same time, compress the output into a BAM using **-O bam**. The first sort is by read name using **-n** in preparation for the next step. Depending on the user's system memory, adjust the memory usage to something above the default of 768 Mb using **-m INT** with a **K/M/G** suffix specifying the units of memory per thread. Note that with **samtools**, the **-@** specifies additional threads, not total threads like BWA.

b. **Fixing Mate Information:**

samtools fixmate -@ [cores – 1] -m output2.bam output3.bam.

For paired end reads, fix any potential inconsistencies in mate information generated during alignment. This is important for downstream steps of removing duplicates for more accurate variant calling. The **-m** option allows for the marking of mate pairs.

c. **Sort 2:**

samtools sort -m 4G -@ [cores – 1] -O bam output3.bam > output4.bam.

This sorting step uses the default sorting method, which sorts by genomic coordinates in preparation for removing duplicates.

d. Mark Duplicates:

samtools markdup -@ [cores – 1] -r output4.bam output5.bam.

This step marks duplicate reads likely resulting from PCR amplification artifacts and removes them with the **-r** option.

e. Index BAM Files:

samtools index -@ [cores – 1] output5.bam.

This step creates an index file for the BAM, and creates a secondary file called **output5.bam.bai** in this process. This indexed BAM is used by programs like IGV and mutation callers like Strelka2.

f. Generate MPILEUP (Optional):

parallel -j [cores] 'samtools mpileup -q 30 -f reference_genome {} > mpileup_dir/{/.}.mpileup'::: $(ls bam_dir/*.bam).

Some mutation callers, like VarScan2, use MPILEUP files for variant calling. MPILEUP files can be generated with Samtools. While multi-sample MPILEUPS are ideal, the management of those and resources can be difficult, so we suggest making a single sample mpileup for each BAM file. The above command takes the desired MPILEUP directory, the BAM directory, and reference genome to make the MPILEUPs in parallel. The {} acts as a placeholder for the input argument into the GNU **parallel** (Tange, 2018) command, and it will output the file to the desired directory **mpileup_dir/** using the file stem (the file name without the extension) {/.} of the input file and will then add the **mpileup** extension to the file. Then it uses command substitution with **$(...)** and takes the output of the **ls** command to list all BAM files ***.bam** from the given directory **bam_dir/** and command substitution captures the output and passes them into parallel as arguments.

Additional tools can be introduced into this pipeline to potentially improve accuracy and quality of mutation calls. For example, GATK (McKenna et al., 2010) has a BaseRecalibrator tool designed to correct systematic errors in base-calling errors during sequencing. It can adjust the quality scores and allow for more accurate mutation calling. This is just one example and depending on your sequencing method or the sensitivity of your calls, additional steps like this may be appropriate. In the scenario of diploid or haploid yeast, due to their clonality and limited genomic complexity, many of these steps are deemed unnecessary. However, they could only benefit the quality of your mutation calls.

4. Variant Calling:

There is a wide range of variant callers that are accessible at no cost, each with distinct applications. The mutation callers commonly regarded as the "industry standard" include Mutect2 (Cibulskis et al., 2013) developed by the Broad Institute, Strelka2 (Kim et al., 2018) by Illumina, VarScan2 (Koboldt et al., 2012) by the Genome Institute at Washington University, and FreeBayes (Garrison & Marth, 2012) by Erik Garrison and colleagues. These four callers are commonly utilized across a broad spectrum of applications. The caller that performs the best depends on the anticipated types of mutations in the samples. Choose a caller that is more sensitive to the mutation types of interest in the sample, but that avoids over-filtering your data. For example, certain callers, like VarScan2, struggle to identify complex mutation events whereas a caller like DeteX (Cui et al., 2022) may be better suited. DeteX is a recently developed caller that aims to enhance overall accuracy with the added benefit of improving the calling of complex events.

Choosing the appropriate caller requires conducting some research. However, if you are unsure, take a general approach using any of the four callers previously mentioned. Additionally, we recommend gathering a consensus from multiple callers (e.g., mutations consistently identified by Strelka2, Mutect2, and FreeBayes). The statistical methods used by each caller for identifying mutations vary, and so do the situations in which false positives can arise. By aggregating the consensus of multiple callers, the false positive rate of the entire set is reduced in comparison to that of an individual caller. As a result, a set of variants is generated with greater confidence than any individual set of calls.

The consensus for any mutation can be obtained through different approaches, such as the presence in a fraction or all the callers, or by generating a weighted confidence score from each tool and then filtering based on some value. Regardless of the method, compare the chromosome, position, reference base, and alternate base of each mutation between datasets to identify the mutations shared among the different calling methods. An easy way to do this would be with a Python script or an R script. To approach creating a consensus in a binary manner, one simple method is to generate a hash map of the data and include the frequency of this mutation across the three callers for each sample. One could iterate through the VCF files in the directory, extracting the chromosome, position, alternate base, and reference base from each line in a sequential manner, keeping track of which programs called each specific mutation. If one wants mutations that occur in at least 2 of the 3 sets of calls, the hash map can be selectively filtered to keep variants that have values that are 2 or greater. Alternatively, **vcftools** using the **vcf-isec** command can be employed. An example usage to take the consensus of two or more files would be:

vcf-isec -f -n + 2 mutect2.vcf strelka2.vcf freebayes.vcf> consensus.vcf.

To find unique and common variants could be:

vcf-isec -p output_prefix mutect2.vcf strelka2.vcf freebayes.vcf.

Additional filtering steps can be applied to variant call data before or after generating consensus calls, for example, filtering by allelic fraction or removing recurrent mutations. We typically filter for mutations that have an allelic fraction of 0.4 for haploids and an allelic fraction of 0.2 for diploid yeast to allow for rare mutations that occur in amplified regions of the genome. Even with this low stringency filter, most variants from clonal yeast samples display allelic fractions greater than 0.9 for haploid yeast and 0.4 for diploids. For identifying mutations acquired during outgrowth or short-term exposure of APOBECs to ssDNA in resected telomeres, the frequency of mutation is unlikely to be high enough to generate recurrent hotspot mutations. Therefore, most recurrent mutations identified are likely preexisting in the yeast and were not a result of the change in time, genotype, or treatment. If the data has expected sites of recurrent mutation, such as selected mutations in a gene like *CAN1*, we recommend skipping this step. We also suggest evaluating the mutation spectra before and after filtering steps to see if non-specific mutagenesis is reduced, or if overall mutations are reduced. Statistical power comes with numbers, but it also comes with high signal to noise, finding the balance between reducing your non-specific mutations and keeping statistical power should be determined on a case-by-case basis, but remember to apply all filters generally so the data is not biased toward specific certain mutation types.

Strelka2:

Strelka2 can be tedious to manage for variant calling on multiple samples due to its directory structure and job scheduling mechanisms. The provided Python script addresses these challenges by automating directory creation, job execution, and dynamic resource management.

Script Overview:

Strelka2's intra-process parallelization strategy is unique compared to other bioinformatics tools. While it allocates multiple threads to each sample, it runs different tasks on these cores instead of splitting a single job, which can cause some cores to become idle as tasks finish at different times. This uneven distribution can lead to suboptimal CPU utilization. To address this, we incorporated a dynamic multiprocessing approach. The script monitors the system's CPU usage every 10 s. When it detects that at least 8 cores are free, it launches the next Strelka2 job. This method allows for continuous job execution without overwhelming the system, effectively optimizing CPU utilization while avoiding system overload. On a 24-core system, for instance, three jobs might start initially, and as each job progresses and threads become available, new jobs are initiated attempting to maximize CPU usage without overload.

The script also emphasizes efficient file organization, which is critical when handling multiple samples and similarly named output VCF files. It creates a structured directory system where each sample's outputs are stored in a dedicated folder, making it easier to manage and retrieve results. The script labels the result VCF files systematically using the name of the BAM file, which is particularly beneficial for downstream analysis and ensures that results from different samples or runs are not confused.

Running the Script:

To run the script, users need to adjust the path to the **configureStrelkaSomaticWorkflow.py** script provided by Strelka2. The script takes several arguments:

1. **pairs_file**: A tab-delimited file with two columns. The first column contains the paths to the treated sample BAM files, and the second column contains the paths to the untreated (normal) sample BAM files. This file lists all the sample pairs for which variants need to be called.
2. **reference_fasta**: The reference genome in FASTA format, which must match the reference used during the BAM file alignment.
3. **output_dir**: The directory where Strelka2 configuration files and results will be saved. The script automatically creates subdirectories for each sample pair to organize outputs.
4. **jobs**: An optional argument specifying the number of jobs to run per sample. Although Strelka2 suggests using 8–16 jobs, we found that using 8 jobs per sample provided the most efficient balance between speed and resource utilization.

This runs Strelka2's **configureStrelkaSomaticWorkflow.py** and then runs the **runWorkflow.py** from that directory with the mode set to local with **-m local**, the number of jobs set to the argument passed in and then runs it on **–quiet** so it only writes to log files and not the terminal as well. All parameters for the variant calling are set to default.

The described script enhances the efficiency of running Strelka2 by automating critical tasks such as directory creation, BAM file indexing, and job scheduling. By dynamically managing system resources, it ensures that the available CPU cores are effectively utilized, minimizing idle time while avoiding system overload. This approach not only speeds up the variant calling process but also simplifies the management of multiple samples, making use of general tips provided above.

7. Summary and conclusion

Yeast has traditionally been an excellent model system for studying mutagenic processes in an eukaryotic genome due to the organism's small genome size, rapid cell division, easy cultivation at large scale, and lack of DNA damage-induced programmed cell death pathways. These characteristics combine to allow the generation of large numbers of mutagenic events within experimentally controlled settings at a low cost and rapid speed that far surpasses the capacity of other model systems. The application of next-generation sequencing technologies to yeast mutagenesis assays has only further advanced the capabilities of this model organism for studying mutation processes. Here, we provide methods to utilize these sequencing platforms to produce large scale mutation data sets from both selected mutations by long-read amplicon sequencing and for unselected mutations with short-read whole genome sequencing. Both methodologies provided critical insights into the mechanisms leading to APOBEC-induced mutations in nuclear DNA. The mutation patterns established for APOBEC3A and APOBEC3B in yeast have enhanced the understanding of how these enzymes contribute to human cancer genome evolution. Moreover, the conservation of DNA metabolic processes between yeast and human cells allows yeast to be further utilized to identify key aspects of nuclear organization and DNA replication and transcription dynamics that limit aberrant APOBEC cytidine deamination. Future application of these methods to other human APOBECs and APOBEC orthologs promises to provide additional information and their potential to enzymatically contribute to ssDNA editing, including within cancer. Whether orthologous APOBECs have the capacity to significantly mutagenize the yeast genome could indicate how likely APOBECs contribute to cancer in non-primates. Additionally, the APOBEC expression vectors described in this chapter have the potential to be utilized in other yeast applications to test the impacts of dU or abasic sites in modulating DNA transactions.

Appendix A. Supporting information

Supplementary data associated with this article can be found in the online version at https://doi.org/10.1016/bs.mie.2024.11.041.

References

Alexandrov, L. B., Kim, J., Haradhvala, N. J., Huang, M. N., Tian Ng, A. W., Wu, Y., ... Consortium, P. (2020). The repertoire of mutational signatures in human cancer. *Nature, 578*(7793), 94–101. https://doi.org/10.1038/s41586-020-1943-3.

Bhagwat, A. S., Hao, W., Townes, J. P., Lee, H., Tang, H., & Foster, P. L. (2016). Strand-biased cytosine deamination at the replication fork causes cytosine to thymine mutations in Escherichia coli. *Proceedings of the National Academy of Sciences of the United States of America, 113*(8), 2176–2181. https://doi.org/10.1073/pnas.1522325113.

Booth, C., Griffith, E., Brady, G., & Lydall, D. (2001). Quantitative amplification of single-stranded DNA (QAOS) demonstrates that cdc13-1 mutants generate ssDNA in a telomere to centromere direction. *Nucleic Acids Research, 29*(21), 4414–4422. https://doi.org/10.1093/nar/29.21.4414.

Brown, A. J., Al-Soodani, A. T., Saul, M., Her, S., Garcia, J. C., Ramsden, D. A., ... Roberts, S. A. (2018). High-throughput analysis of DNA break-induced chromosome rearrangements by amplicon sequencing. *Methods in Enzymology, 601*, 111–144. https://doi.org/10.1016/bs.mie.2017.11.028.

Brown, G. W. (2024). The cytidine deaminase APOBEC3C has unique sequence and genome feature preferences. *Genetics, 227*(4), https://doi.org/10.1093/genetics/iyae092.

Burns, M. B., Lackey, L., Carpenter, M. A., Rathore, A., Land, A. M., Leonard, B., ... Harris, R. S. (2013). APOBEC3B is an enzymatic source of mutation in breast cancer. *Nature, 494*(7437), 366–370. https://doi.org/10.1038/nature11881.

Burns, M. B., Temiz, N. A., & Harris, R. S. (2013). Evidence for APOBEC3B mutagenesis in multiple human cancers. *Nature Genetics, 45*(9), 977–983. https://doi.org/10.1038/ng.2701.

Chan, K., Roberts, S. A., Klimczak, L. J., Sterling, J. F., Saini, N., Malc, E. P., ... Gordenin, D. A. (2015). An APOBEC3A hypermutation signature is distinguishable from the signature of background mutagenesis by APOBEC3B in human cancers. *Nature Genetics, 47*(9), 1067–1072. https://doi.org/10.1038/ng.3378.

Chan, K., Sterling, J. F., Roberts, S. A., Bhagwat, A. S., Resnick, M. A., & Gordenin, D. A. (2012). Base damage within single-strand DNA underlies in vivo hypermutability induced by a ubiquitous environmental agent. *PLoS Genetics, 8*(12), e1003149. https://doi.org/10.1371/journal.pgen.1003149.

Cibulskis, K., Lawrence, M. S., Carter, S. L., Sivachenko, A., Jaffe, D., Sougnez, C., ... Getz, G. (2013). Sensitive detection of somatic point mutations in impure and heterogeneous cancer samples. *Nature Biotechnology, 31*(3), 213–219. https://doi.org/10.1038/nbt.2514.

Conticello, S. G., Thomas, C. J., Petersen-Mahrt, S. K., & Neuberger, M. S. (2005). Evolution of the AID/APOBEC family of polynucleotide (deoxy)cytidine deaminases. *Molecular Biology and Evolution, 22*(2), 367–377. https://doi.org/10.1093/molbev/msi026.

Cui, Y., Li, H., Liu, P., Wang, H., Zhang, Z., Qu, H., ... Fang, X. (2022). DeteX: A highly accurate software for detecting SNV and InDel in single and paired NGS data in cancer research. *Frontiers in Genetics, 13*, 1118183. https://doi.org/10.3389/fgene.2022.1118183.

Dennen, M. S., Kockler, Z. W., Roberts, S. A., Burkholder, A. B., Klimczak, L. J., & Gordenin, D. A. (2024). Hypomorphic mutation in the large subunit of replication protein A affects mutagenesis by human APOBEC cytidine deaminases in yeast. *G3 (Bethesda)*. https://doi.org/10.1093/g3journal/jkae196.

Dobin, A., Davis, C. A., Schlesinger, F., Drenkow, J., Zaleski, C., Jha, S., ... Gingeras, T. R. (2013). STAR: Ultrafast universal RNA-seq aligner. *Bioinformatics (Oxford, England), 29*(1), 15–21. https://doi.org/10.1093/bioinformatics/bts635.

Dutko, J. A., Schafer, A., Kenny, A. E., Cullen, B. R., & Curcio, M. J. (2005). Inhibition of a yeast LTR retrotransposon by human APOBEC3 cytidine deaminases. *Current Biology: CB, 15*(7), 661–666. https://doi.org/10.1016/j.cub.2005.02.051.

Garrison, E. P., & Marth, G. T. (2012). Haplotype-based variant detection from short-read sequencing. *arXiv: Genomics*.

Goffeau, A., Barrell, B. G., Bussey, H., Davis, R. W., Dujon, B., Feldmann, H., ... Oliver, S. G. (1996). Life with 6000 genes. *Science (New York, N. Y.), 274*(5287), 546, 547–563. https://doi.org/10.1126/science.274.5287.546.

Guo, X., Lehner, K., O'Connell, K., Zhang, J., Dave, S. S., & Jinks-Robertson, S. (2015). SMRT sequencing for parallel analysis of multiple targets and accurate SNP phasing. *G3 (Bethesda), 5*(12), 2801–2808. https://doi.org/10.1534/g3.115.023317.

Hanscom, T., Woodward, N., Batorsky, R., Brown, A. J., Roberts, S. A., & McVey, M. (2022). Characterization of sequence contexts that favor alternative end joining at Cas9-induced double-strand breaks. *Nucleic Acids Research, 50*(13), 7465–7478. https://doi.org/10.1093/nar/gkac575.

Haradhvala, N., Polak, P., Stojanov, P., Covington, K. R., Shinbrot, E., Hess, J., ... Getz, G. (2016). Mutational strand asymmetries across cancer reveal mechanisms of DNA damage and repair. *Cell*.

Hoopes, J. I., Cortez, L. M., Mertz, T. M., Malc, E. P., Mieczkowski, P. A., & Roberts, S. A. (2016). APOBEC3A and APOBEC3B preferentially deaminate the lagging strand template during DNA replication. *Cell Reports, 14*(6), 1273–1282. https://doi.org/10.1016/j.celrep.2016.01.021.

Hoopes, J. I., Hughes, A. L., Hobson, L. A., Cortez, L. M., Brown, A. J., & Roberts, S. A. (2017). Avoidance of APOBEC3B-induced mutation by error-free lesion bypass. *Nucleic Acids Research, 45*(9), 5243–5254. https://doi.org/10.1093/nar/gkx169.

Kim, S., Scheffler, K., Halpern, A. L., Bekritsky, M. A., Noh, E., Kallberg, M., ... Saunders, C. T. (2018). Strelka2: Fast and accurate calling of germline and somatic variants. *Nature Methods, 15*(8), 591–594. https://doi.org/10.1038/s41592-018-0051-x.

Koboldt, D. C., Zhang, Q., Larson, D. E., Shen, D., McLellan, M. D., Lin, L., ... Wilson, R. K. (2012). VarScan 2: Somatic mutation and copy number alteration discovery in cancer by exome sequencing. *Genome Research, 22*(3), 568–576. https://doi.org/10.1101/gr.129684.111.

Langmead, B., & Salzberg, S. L. (2012). Fast gapped-read alignment with Bowtie 2. *Nature Methods, 9*(4), 357–359. https://doi.org/10.1038/nmeth.1923.

Li, H. (2013). Aligning sequence reads, clone sequences and assembly contigs with BWA-MEM. *arXiv: Genomics*.

Li, H., & Durbin, R. (2010). Fast and accurate long-read alignment with Burrows-Wheeler transform. *Bioinformatics (Oxford, England), 26*(5), 589–595. https://doi.org/10.1093/bioinformatics/btp698.

Li, H., Handsaker, B., Wysoker, A., Fennell, T., Ruan, J., Homer, N., ... Genome Project Data Processing, S. (2009). The sequence alignment/map format and SAMtools. *Bioinformatics (Oxford, England), 25*(16), 2078–2079. https://doi.org/10.1093/bioinformatics/btp352.

Marx, V. (2023). Method of the year: Long-read sequencing. *Nature Methods, 20*(1), 6–11. https://doi.org/10.1038/s41592-022-01730-w.

Mayorov, V. I., Rogozin, I. B., Adkison, L. R., Frahm, C., Kunkel, T. A., & Pavlov, Y. I. (2005). Expression of human AID in yeast induces mutations in context similar to the context of somatic hypermutation at G-C pairs in immunoglobulin genes. *BMC Immunology, 6*, 10. https://doi.org/10.1186/1471-2172-6-10.

McKenna, A., Hanna, M., Banks, E., Sivachenko, A., Cibulskis, K., Kernytsky, A., ... DePristo, M. A. (2010). The Genome Analysis Toolkit: A MapReduce framework for analyzing next-generation DNA sequencing data. *Genome Research, 20*(9), 1297–1303. https://doi.org/10.1101/gr.107524.110.

Mertz, T. M., Collins, C. D., Dennis, M., Coxon, M., & Roberts, S. A. (2022). APOBEC-induced mutagenesis in cancer. *Annual Review of Genetics, 56*, 229–252. https://doi.org/10.1146/annurev-genet-072920-035840.

Mertz, T. M., Rice-Reynolds, E., Nguyen, L., Wood, A., Cordero, C., Bray, N., ... Roberts, S. A. (2023). Genetic inhibitors of APOBEC3B-induced mutagenesis. *Genome Research, 33*(9), 1568–1581. https://doi.org/10.1101/gr.277430.122.

Morganella, S., Alexandrov, L. B., Glodzik, D., Zou, X., Davies, H., Staaf, J., ... Nik-Zainal, S. (2016). The topography of mutational processes in breast cancer genomes. *Nature Communications, 7*, 11383. https://doi.org/10.1038/ncomms11383.

Nik-Zainal, S., Van Loo, P., Wedge, D. C., Alexandrov, L. B., Greenman, C. D., Lau, K. W., ... Breast Cancer Working Group of the International Cancer Genome, C. (2012). The life history of 21 breast cancers. *Cell, 149*(5), 994–1007. https://doi.org/10.1016/j.cell.2012.04.023.

Pecori, R., Di Giorgio, S., Paulo Lorenzo, J., & Nina Papavasiliou, F. (2022). Functions and consequences of AID/APOBEC-mediated DNA and RNA deamination. *Nature Reviews. Genetics, 23*(8), 505–518. https://doi.org/10.1038/s41576-022-00459-8.

Petljak, M., Dananberg, A., Chu, K., Bergstrom, E. N., Striepen, J., von Morgen, P., ... Maciejowski, J. (2022). Mechanisms of APOBEC3 mutagenesis in human cancer cells. *Nature, 607*(7920), 799–807. https://doi.org/10.1038/s41586-022-04972-y.

Refsland, E. W., & Harris, R. S. (2013). The APOBEC3 family of retroelement restriction factors. *Current Topics in Microbiology and Immunology, 371*, 1–27. https://doi.org/10.1007/978-3-642-37765-5_1.

Ren, J., & Chaisson, M. J. P. (2021). lra: A long read aligner for sequences and contigs. *PLoS Computational Biology, 17*(6), e1009078. https://doi.org/10.1371/journal.pcbi.1009078.

Roberts, S. A., Lawrence, M. S., Klimczak, L. J., Grimm, S. A., Fargo, D., Stojanov, P., ... Gordenin, D. A. (2013). An APOBEC cytidine deaminase mutagenesis pattern is widespread in human cancers. *Nature Genetics, 45*(9), 970–976. https://doi.org/10.1038/ng.2702.

Roberts, S. A., Sterling, J., Thompson, C., Harris, S., Mav, D., Shah, R., ... Gordenin, D. A. (2012). Clustered mutations in yeast and in human cancers can arise from damaged long single-strand DNA regions. *Molecular Cell, 46*(4), 424–435. https://doi.org/10.1016/j.molcel.2012.03.030.

Rosenbaum, J. C., Bonilla, B., Hengel, S. R., Mertz, T. M., Herken, B. W., Kazemier, H. G., ... Bernstein, K. A. (2019). The Rad51 paralogs facilitate a novel DNA strand specific damage tolerance pathway. *Nature Communications, 10*(1), 3515. https://doi.org/10.1038/s41467-019-11374-8.

Saini, N., Roberts, S. A., Sterling, J. F., Malc, E. P., Mieczkowski, P. A., & Gordenin, D. A. (2017). APOBEC3B cytidine deaminase targets the non-transcribed strand of tRNA genes in yeast. *DNA Repair (Amst), 53*, 4–14. https://doi.org/10.1016/j.dnarep.2017.03.003.

Sanchez, A., Ortega, P., Sakhtemani, R., Manjunath, L., Oh, S., Bournique, E., ... Buisson, R. (2024). Mesoscale DNA features impact APOBEC3A and APOBEC3B deaminase activity and shape tumor mutational landscapes. *Nature Communications, 15*(1), 2370. https://doi.org/10.1038/s41467-024-45909-5.

Satam, H., Joshi, K., Mangrolia, U., Waghoo, S., Zaidi, G., Rawool, S., ... Malonia, S. K. (2023). Next-generation sequencing technology: Current trends and advancements. *Biology (Basel), 12*(7), https://doi.org/10.3390/biology12070997.

Schumacher, A. J., Nissley, D. V., & Harris, R. S. (2005). APOBEC3G hypermutates genomic DNA and inhibits Ty1 retrotransposition in yeast. *Proceedings of the National Academy of Sciences of the United States of America, 102*(28), 9854–9859. https://doi.org/10.1073/pnas.0501694102.

Seplyarskiy, V. B., Soldatov, R. A., Popadin, K. Y., Antonarakis, S. E., Bazykin, G. A., & Nikolaev, S. I. (2016). APOBEC-induced mutations in human cancers are strongly enriched on the lagging DNA strand during replication. *Genome Research, 26*(2), 174–182. https://doi.org/10.1101/gr.197046.115.

Sui, Y., Qi, L., Zhang, K., Saini, N., Klimczak, L. J., Sakofsky, C. J., ... Zheng, D. Q. (2020). Analysis of APOBEC-induced mutations in yeast strains with low levels of replicative DNA polymerases. *Proceedings of the National Academy of Sciences of the United States of America, 117*(17), 9440–9450. https://doi.org/10.1073/pnas.1922472117.

Tange, O. (2018). *GNU Parallel 2018*. https://doi.org/10.5281/zenodo.1146013.

Taylor, B. J., Nik-Zainal, S., Wu, Y. L., Stebbings, L. A., Raine, K., Campbell, P. J., ... Neuberger, M. S. (2013). DNA deaminases induce break-associated mutation showers with implication of APOBEC3B and 3A in breast cancer kataegis. *Elife, 2*, e00534. https://doi.org/10.7554/eLife.00534.

Taylor, B. J., Wu, Y. L., & Rada, C. (2014). Active RNAP pre-initiation sites are highly mutated by cytidine deaminases in yeast, with AID targeting small RNA genes. *Elife, 3*, e03553. https://doi.org/10.7554/eLife.03553.

Vasimuddin, M., Misra, S., Li, H., & Aluru, S. (2019). *Efficient Architecture-Aware Acceleration of BWA-MEM for Multicore Systems*.

Wahl, G. M., & Carr, A. M. (2001). The evolution of diverse biological responses to DNA damage: Insights from yeast and p53. *Nature Cell Biology, 3*(12), E277–E286. https://doi.org/10.1038/ncb1201-e277.

Yang, Y., Sterling, J., Storici, F., Resnick, M. A., & Gordenin, D. A. (2008). Hypermutability of damaged single-strand DNA formed at double-strand breaks and uncapped telomeres in yeast Saccharomyces cerevisiae. *PLoS Genetics, 4*(11), e1000264. https://doi.org/10.1371/journal.pgen.1000264.

Zhang, Y., Shishkin, A. A., Nishida, Y., Marcinkowski-Desmond, D., Saini, N., Volkov, K. V., ... Lobachev, K. S. (2012). Genome-wide screen identifies pathways that govern GAA/TTC repeat fragility and expansions in dividing and nondividing yeast cells. *Molecular Cell, 48*(2), 254–265. https://doi.org/10.1016/j.molcel.2012.08.002.

CHAPTER SEVEN

Biochemical assays for AID/APOBECs and the identification of AID/APOBEC inhibitors

Priyanka Govindarajan, Ying Zeng, and Mani Larijani*
Simon Fraser University, Burnaby, BC, Canada
*Corresponding author. e-mail address: mani_larijani@sfu.ca

Contents

Abstract

Activation-induced cytidine deaminase (AID) and apolipoprotein B-mRNA editing catalytic polypeptide 3 (APOBEC3 or A3) proteins belong to the AID/APOBEC family of cytidine deaminases. While AID mediates somatic hypermutation and class-switch recombination in adaptive immunity, A3s restrict viruses and retroelements by hypermutation. Mis-regulated expression and off-target activity of AID/A3 can cause genome-wide mutations promoting oncogenesis, immune evasion, and therapeutic resistance due to tumor and viral evolution. In these contexts, inhibition of AID/A3 represents a promising therapeutic approach. Competitive inhibition

Methods in Enzymology, Volume 713
ISSN 0076-6879, https://doi.org/10.1016/bs.mie.2024.12.001
Copyright © 2025 Elsevier Inc. All rights are reserved, including those for text and data mining, AI training, and similar technologies.

could be achieved with different strategies: one class would be small molecules that bind in the catalytic pocket (active site) and block access for the substrate cytidine. Another type of larger molecule inhibitor would bind the enzymes' surface more broadly and compete with the binding of the polynucleotide substrates prior to deamination catalysis. Several biochemical assays developed to assess AID/A3 activity can be employed to screen for potential inhibitors. These include *in cellulo* and in vitro activity-based as well as binding-based assays. In this chapter, we discuss the key considerations for designing robust enzyme assays and provide an overview of assays that we and others have established or modified for specific applications in AID/A3 enzymology, including measurement of inhibition. We provide detailed protocols for the two most widely used in vitro enzyme assays that directly measure the activities of purified AID/A3s on DNA and/or RNA substrates, namely, the gel-based alkaline cleavage assay and multiple variations of PCR/sequencing-based assays.

1. Introduction

1.1 AID/APOBECs and their multi-faceted roles in immunity, viral restriction and cancer

Tumor evolution, drug resistance and immune escape in viral or cancer diseases are common hindrances to therapeutic efficacy (Grillo et al., 2022). Major drivers of therapeutic resistance are mutations. In addition to factors like oxidative stress, errors during DNA replication, DNA damaging alkylation events, many such disease-promoting mutations in viral infections and cancer are caused by endogenous enzyme-catalyzed DNA cytidine to uridine deamination mostly mediated by various members of the activation induced cytidine deaminase/apolipoprotein B-mRNA editing catalytic polypeptide (AID/APOBEC) family. This family includes the single-deaminase domain AID and four types of APOBECs, including 7 members of the APOBEC3 branch (A3s), which include both single-deaminase domain members (e.g., A3A, A3H) and double-domain members (e.g., A3B, A3F, A3G), with the former and latter being ~ 200 and 400 aa in length (Swanton et al., 2015). While the primary role of AID is to mediate secondary antibody diversification processes by mutating the immunoglobulins loci, and that of A3s is the restriction of exogenous viruses and endogenous retroelements by hypermutation of viral genomes (Borzooee et al., 2018; Conticello, 2008; Dickerson et al., 2003; Feng et al., 2020; Krishnan et al., 2018; Larijani et al., 2007; Rogozin et al., 2007; Salter & Smith, 2018), their misregulated expression and rampant mutational activities are often associated with mistargeted genome-wide mutagenesis which drives oncogenesis. AID-mediated off-target mutations often lead to double-stranded breaks (DSBs) and subsequent immunoglobulin locus-oncogene chromosomal translocations driving a multitude of leukemia and lymphomas (Borzooee et al., 2018; King et al., 2021; Mechtcheriakova et al., 2012; Pasqualucci et al., 2008).

A3s were first identified as viral restricting factors that hypermutate viral genomes, incapacitating them and leading to non-infectious progeny (for example, A3G and Human Immunodeficiency Virus – I, HIV-I) (Liddament et al., 2004; Mangeat et al., 2003; Sheehy et al., 2002; Zhang et al., 2003). On the other hand, viruses like HIV either express factors that counteract A3s, like the virion infectivity factor (Vif), or use A3s to be selectively mutated to produce variants that evade therapy or host immune responses (Borzooee et al., 2018; Fourati et al., 2010; Grillo et al., 2022; Lecossier et al., 2003; Monajemi et al., 2014; Sheehy et al., 2002; Uriu et al., 2021; Xu et al., 2020; Yebra & Holguin, 2011; Yu et al., 2004). When misregulated or over-expressed, A3s turn against the host genome leading to oncogenesis (Burns et al., 2013; Rebhandl et al., 2015; Roberts et al., 2013; Swanton et al., 2015); however, in some cases, A3-mediated mutations of cancer genomes could also lead to anti-tumor effects, for instance, by producing tumor neo-antigens (DiMarco et al., 2022; Driscoll et al., 2020; Leonard et al., 2016). Therefore, A3s can either cause lethal mutational loads in viruses and tumor cells or lead to the "just right" amount of genetic diversity that is selectively propagated during viral and tumor evolution, ensuring their ability to evade immune barriers, resist therapy, and persist (Cahill et al., 1999; Swanton et al., 2015).

While both scenarios are possible, most evidence points to pro-tumorigenic activities of AID/A3s, wherein their off-target activities instigate tumorigenesis and accelerate tumor clonal evolution and drug resistance (Caswell et al., 2024; Durfee et al., 2023; Isozaki et al., 2023; Klemm et al., 2009; Law et al., 2020; Law et al., 2016). Thus, there has been an intense effort to identify small molecule inhibitors for AID and A3s (Alvarez-Gonzalez et al., 2021; Barzak et al., 2019; Cen et al., 2010; Grillo et al., 2022; Harjes et al., 2023; King et al., 2021; Kurup et al., 2022; Kvach et al., 2019; Kvach et al., 2020; Li et al., 2012; Serrano et al., 2022). This calls for high-throughput methods for screening large libraries of potential inhibitors (small molecules or other types of agents) in in vitro biochemical assays using purified preparations of AID and APOBECs as a first step to identify and refine candidate inhibitors.

1.2 The first challenge for enzyme assays: purification of AID/APOBECs

Prokaryotic systems such as bacteria, and eukaryotic systems including human, insect, mouse, and yeast cells have been employed for the expression and purification of AID/APOBECs. Different expression hosts pose advantages and disadvantages. Although bacterial system allows rapid, scalable and cost-effective production of these enzymes, the expressed proteins may not be folded correctly or contain essential post-translational modifications, hence being enzymatically inactive which is particularly the case for the larger double-domain members like A3G, A3B (Borzooee & Larijani, 2019; Iwatani et al., 2006). On the other hand, eukaryotic systems effectively produce properly folded and (near) native post-translationally decorated active enzymes but are expensive, time-consuming and require more specialized infrastructure to scale up. No matter the expression system or the specific host cell type, the major challenge in the purification of AID/APOBECs is inherent to their structures; these enzymes have highly charged surfaces (Borzooee & Larijani, 2019; King et al., 2015). While these facilitate substrate DNA or RNA capture, they pose a challenge to purification due to extensive non-specific protein-protein and protein-nucleic acid interactions that are a hallmark of AID/APOBECs, making purification of highly soluble forms of these enzymes with sufficient purity difficult and necessitating additional technical steps (Chelico et al., 2006; Chelico et al., 2008; Iwatani et al., 2006; King et al., 2015; Larijani & Martin, 2012).

While it is essential to use different forms of a purified enzyme in an enzyme assay – expressed using different host systems and using different kinds of N- or C-terminal fusion tags – given the charged surface and consequent non-specific interactions of AID/APOBECs, not all tags work suitably well with these enzymes and indeed some may interact with the enzyme surface affecting folding and/or activity. For instance, in our extensive experience and trials, and that of colleagues working in the field, such is the case for His-tag, which is a staple and highly used tag for many other enzymes/proteins. Rather, more bulky tags (e.g., GST, MBP) yield enzyme preparations that are easier to purify, more soluble and enzymatically active, presumably due to the bulky tag contributing to fusion protein stabilization and interfering with detrimental polydisperse oligomerization and precipitation. Hence, careful optimization of the purification methods is imperative, and also an important factor in the choice for enzyme assay, as it affects yield and specific activity. For a detailed protocol about APOBEC purification, see Chapter 6.

1.3 Key reasons for developing effective enzyme assays

An optimal enzyme assay calls for high sensitivity and reproducibility and ought to be able to effectively reveal differences in catalytic activity with respect to the subtlest of changes to substrate/inhibitor/enzyme concentrations, temperature and pH differences, substrate sequence/structure differences (given that the members of this family have unique preferences for certain NNC motifs and secondary structures of target nucleic acids over others (Diamond et al., 2019; Ghorbani et al., 2022; Larijani & Martin, 2007; Sharma et al., 2015; Shi et al., 2017)) (Fig. 1A). Due to the challenges mentioned above in purifying active, properly folded AID/APOBECs, it is equally important to have an assay that has the least number of steps possible and can be carried out in a quick and efficient manner.

AID/A3s possess certain innate biochemical and structural properties that act as in-built regulatory mechanisms – properties like substrate sequence/structure preference, enzyme processivity, structural features that regulate substrate selectivity, binding and catalysis, optimal temperature/pH preference, interactions with other proteins and cofactors. *In vitro* biochemical assays have been instrumental in identifying and assessing a host of these inherent enzymatic properties which have proven to be key regulators and determinants of the in vivo activities of these enzymes, across the realms of antibody diversification, virus restriction and oncogenesis. Specifically, in vitro enzyme assays under controlled conditions have revealed precise substrate sequence specificity of AID/A3s. AID shows a preference for WRC in ssDNA regions of 7 nt bubbles or other branched substrates like G-quadruplexes. A3A and A3B favor TpC (thymidine followed by a cytidine) motifs in stem-loops (Larijani & Martin, 2007; Qiao et al., 2017; Sanchez et al., 2024). These specificities directly correlate with their in vivo functions like somatic hypermutation, class switch recombination in the case of AID, or viral restriction by the A3s, where these motifs in such structural contexts are frequently targeted for cytidine deamination by these enzymes. The relationship between AID/APOBEC activity and processes like transcription and replication are well-established (Canugovi et al., 2009; Hu et al., 2015; Meng et al., 2014; Nambu et al., 2003; Pavri et al., 2010; Qian et al., 2014; Seplyarskiy et al., 2016; Sohail et al., 2003). In in vitro cell-free assays using purified AID and supercoiled dsDNA substrate coupled with transcription, transient "breathing" ssDNA and R-loops are generated, making the DNA more accessible for AID to target. This behavior, observed in in vitro settings, mirrors AID's behavior in vivo (Branton et al., 2020; Canugovi et al., 2009; Shen & Storb, 2004), which

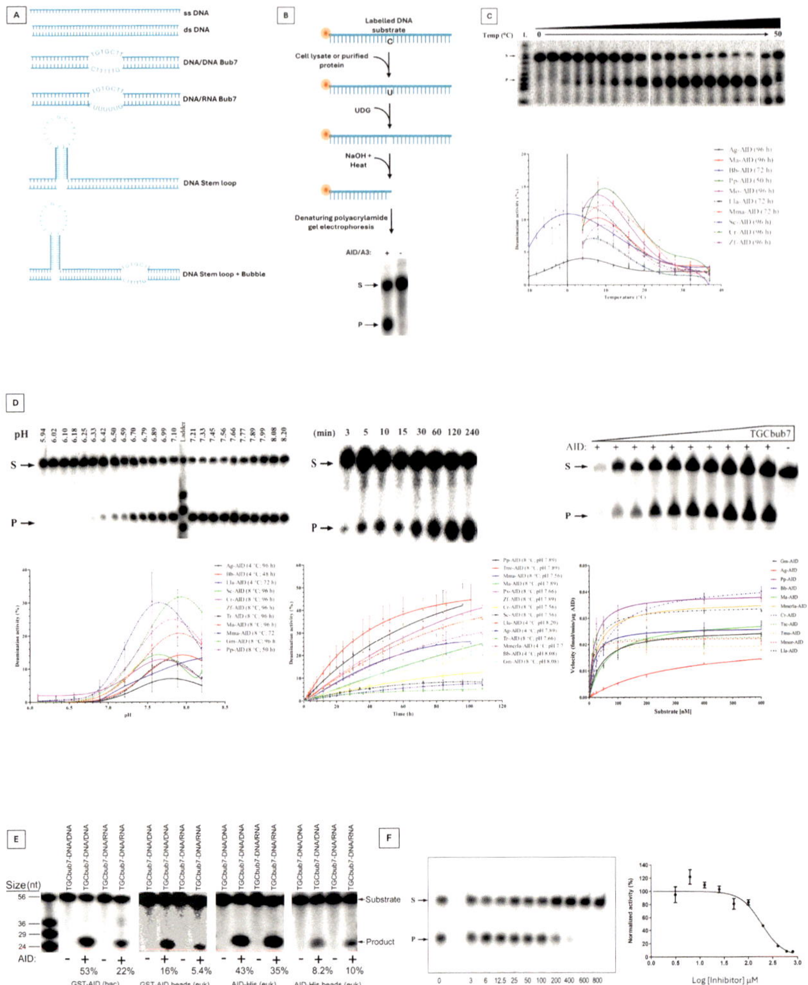

Fig. 1 Gel-based alkaline cleavage assay is a sensitive and reliable assay to test various DNA substrate structures across different reaction conditions: (A) Substrates with different secondary structures can be used to test AID/A3 activity by alkaline cleavage assay. Some of them shown here are single-stranded (ss), double-stranded (ds), DNA/DNA bubble with a 7nt bubble, DNA/RNA bubble of size 7, stem loop, substrate with stem loop and bubble. The sequences shown here at the bubble or stem are for representation. (B) Schematic showing the alkaline cleavage assay principle using a ssDNA substrate with the target C at the center. The ssDNA can be labeled with either a fluorophore or radioactive P-32 (orange sphere) at the 5′-end. Purified AID/A3 or cell extracts expressing the enzymes is incubated with the labeled substrate. After uracil excision by UDG, NaOH and heat are used to induce cleavage at the abasic site of uracil and denaturation. This is followed by electrophoresis on a denaturing polyacrylamide gel. A representative gel shows the full-length substrate and the cleaved product on a denaturing polyacrylamide gel with and without AID/A3. Alkaline

further enables a detailed mechanistic understanding of substrate interactions and preferences of these enzymes in vivo.

AID and A3G exhibit processive deamination of the target DNA strand upon binding (Chelico et al., 2006; Chelico et al., 2009) which has been studied in vitro and corresponds to the clustered mutations termed "kataegis" mediated by AID/A3s commonly observed in cancer cells and viruses (Alexandrov et al., 2013; Lada et al., 2012). Furthermore, structural and functional studies using enzyme kinetics and mutational assays have been pivotal in identifying residues near the active site and those that make up the substrate-binding grooves that regulate substrate access, binding, and catalysis

cleavage assay is a highly sensitive and reliable assay that can detect differences in activity due to even subtle changes in reaction conditions like pH, temperature, and incubation time. (C) (Top) A representative gel depicting alkaline cleavage assay performed across a temperature gradient ranging from 0 to 50 degrees Celsius. (Bottom) Graph representing temperature sensitivity profiles of many AID orthologs tested in parallel. (D) Representative gels (top) and graphs (bottom) denoting the efficient use of alkaline cleavage assay for the determination of optimal pH (left - tested in 0.1 increments of pH), appropriate time of incubation by performing time course experiments (middle - substrates incubated with AID/A3 enzymes at different time points represented in minutes). The preliminary experiments to determine optimal temperature, pH and appropriate incubation time are used to determine the Michaelis Menten parameters (right). Representative gel showing alkaline cleavage assay across different substrate concentrations (TGC bubble substrate of size 7 in this case) and the corresponding graph of velocity of the enzyme (fmol product/μg of enzyme/min of incubation) vs. substrate concentration. The graphs in panels C and D represent a subset of nearly 80 different AID orthologs (represented in different colors in the graphs) that were tested from three to six individual experiments using two to four independently purified preparations expressed from at least two different host systems. The graph and the gels in C-D were adapted from different papers to highlight the scalability, sensitivity and efficiency of alkaline cleavage assay. The gels do not correspond to the data on the graphs. (E) Performing enzyme assays across different expression systems, using different fusion tags ensures that the enzyme behavior observed is "bonafide" activity and is not due to expression or purification artifacts. This figure depicts representative alkaline cleavage gels testing AID activity on DNA/DNA bubble and DNA/RNA bubble of size 7 containing TGC motif performed using AID expressed in bacteria using N-Terminal glutathione-S-transferase (GST) tag (first and third gels) and AID expressed in mammalian cells (HEK293T) with a C-Terminal histidine tag (second and fourth gels). The enzyme was tested either as purified protein or as protein bound to the Sepharose or Ni beads. (F) Representative gel (left) depicting the use of alkaline cleavage assay to test the inhibition of AID/A3 enzyme activity by a potential inhibitor tested along a range of inhibitor concentrations. Normalized percentage activity taking 100 % activity in the absence of inhibitor is plotted against log[inhibitor] (right) to determine the IC50 concentration. (S - Substrate, P - Product). *Adapted from Ghorbani et al. (2022), Ghorbani et al. (2022), Borzooee & Larijani (2019), Abdouni et al. (2018).*

of AID/A3s (King & Larijani, 2017; King et al., 2015; Qiao et al., 2017; Shi et al., 2017). This information is, in turn, valuable during inhibitor design against AID/A3s. Additionally, pH and temperature sensitivities of AID/A3s measured in cell-free biochemical assays (Ghorbani et al., 2022; Ito et al., 2017) reflect the physiological conditions where they operate in vivo, such as A3s' ability to inhibit viral replication in acidic cellular compartments like endosomes (Pham et al., 2013). Reconstitution experiments also provided a better understanding of complex protein-protein interactions of AID/A3s, including the association of AID with RNA polymerase machinery or other transcription-associated protein co-factors (Besmer et al., 2006; Canugovi et al., 2009; Chaudhuri & Alt, 2004; Conticello et al., 2008; Pavri et al., 2010; Shen et al., 2005; Shen & Storb, 2004; Zheng et al., 2015).

Thus, in vitro enzyme assays have proved to be essential in dissecting the fundamental properties of AID and APOBEC3 enzymes. They allow us to precisely control experimental variables, yielding a clear understanding of how these enzymes' in vitro behaviors translate into their regulatory roles in vivo.

1.4 Enzyme assays for measuring cytidine deamination by AID/A3s

Inhibitors screened for AID/A3s are either small-molecule compounds that bind into or near the catalytic pocket of the enzyme blocking the cytidine from accessing the pocket (Alvarez-Gonzalez et al., 2021; King et al., 2021; Li et al., 2012) or transition state analogues of cytidine incorporated into longer DNA strands folded into the enzymes' preferred substrate secondary structures (e.g., stem loops preferred by A3A) that in turn competes with the substrate for the active site thereby affecting enzyme activity (Harjes et al., 2023; Serrano et al., 2022). When screening such inhibitors against AID/A3s various types of assays can be employed.

Broadly, they can be divided into two categories, *in cellulo* or in vitro assays.

- **In cellulo assays:**
  - o Bacterial/yeast antibiotic reversion assay: One of the earliest and most commonly used assays for cytidine deaminase activity is the *E. coli* or yeast antibiotic reversion assay that measures mutations that confer resistance to antibiotics like rifampicin. This assay can be performed *in cellulo* or in vitro. In the *in cellulo* version, AID/A3s are ectopically expressed in bacteria or yeast, where their mutagenic activity induces reversion mutations in antibiotic resistance genes. The mutation levels are determined by

comparing the number of resistant colonies formed on a selective media as opposed to cells not expressing AID/APOBEC. Additionally, the target genes are also sequenced to detect C/G→T/A mutation levels (Harris et al., 2002; Lada et al., 2011), assuming that additional mutations do not destroy resistance, a general limitation of this assay.

In the in vitro version, a plasmid containing an antibiotic resistance gene (e.g., *rpoB*, *amp*, or *kan*) interrupted by a stop codon with an AID/APOBEC hotspot motif is used. The plasmid is incubated with purified AID/APOBECs and transformed into uracil repair-deficient *E. coli* or yeast. Plasmids are isolated from colonies that have gained antibiotic resistance through reversion mutations mediated by AID/APOBEC activity and sequenced by Sanger method (Branton et al., 2020; Sohail et al., 2003). Though this assay provided many early insights on AID/APOBEC activity (Canugovi et al., 2009; Shen et al., 2005; Shen & Storb, 2004), an inherent bias and limitation is that the only observable AID/APOBEC-mediated mutations are those occurring on plasmids that also carry the antibiotic resistance reversion mutation.

- **Yeast kataegis assay:** This assay is used to model the clustered mutations mediated by AID/A3s in cancers called "kataegis", in yeast cell system. AID/A3s expressed in *Saccharomyces cerevisiae* deaminate cytidine residues in the yeast genome. Genome sequencing reveals mutational signatures of the expressed proteins (Taylor et al., 2013).
- **Viral restriction assay:** The activity of A3s like A3G and A3B can be tested for their virus restricting ability. 293 T cells expressing viral packaging plasmids encoding a fluorescent reporter are co-transfected with or without vectors expressing A3s (Grillo et al., 2022; Harris et al., 2003). The reduction in fluorescence is taken as a measure of viral restriction by the A3s tested.
- **Fluorescent reporter-based assay:** A real-time assay that measures activity of APOBECs *in cellulo* uses a reporter system called AMBER (APOBEC-mediated base-editing reporter). This system is constructed with a constitutively expressing mCherry and eGFP reporters with a single mutation that ablates GFP fluorescence. Cytidine deaminase activity at this position reverses the mutation that restores fluorescence. The ratio of eGFP to mCherry is used as a measure of APOBEC activity(Grillo et al., 2022; Martin et al., 2019).

Though these myriad cell-based assays have provided immense insights into the activities of AID/A3s, they are not discussed further here, as they cannot be considered true "enzyme assays" since readouts are impacted by several other cellular factors, processes and conditions, upstream/regulating/downstream of AID/A3 activity (e.g. DNA repair and/or biological selection for the function imposed or lost due to specific selections). Henceforth, we devote the remainder of this article to true (i.e., in vitro) enzyme assays that directly measure the enzymatic activities of highly purified AID/A3s on DNA and/or RNA substrates.

- **In vitro assays:** These assays utilize purified enzymes or cell lysates to directly measure cytidine deamination activity on DNA or RNA substrates.
  - **Alkaline-cleavage assay (Gel-, plate- or HPLC based):** This assay could be gel-, plate- or HPLC based and all three are based on the principle that post cytidine to uridine deamination by AID/A3s, the uracil base is removed using Uracil DNA glycosylase (UDG) leaving an abasic site that is cleaved upon treatment with sodium hydroxide (NaOH) (Bransteitter et al., 2003; Larijani & Martin, 2007) and heat (Fig. 1B). The gel-based method, which will be discussed in more detail in the subsequent sections, uses DNA substrates labeled at one end (most commonly 5′, but 3′ also possible) either with radioactive ^{32}P or fluorescent tags. The labeled substrates, after incubation with AID/APOBECs, UDG, and NaOH treatments, are analyzed using denaturing polyacrylamide gel electrophoresis that separates the longer original substrate from the shorter cleaved deamination product (Fig. 1B). The larger (substrate) and shorter cleaved (product) bands are then quantified using band densitometry and their relative proportion is a direct measure of cytidine deaminase activity (Iwatani et al., 2006; King et al., 2015; Larijani & Martin, 2007). The plate-based assay uses DNA substrates labeled at either ends with a fluorophore-quencher pair and the emission signals from the fluorophore due to separation from the quencher is measured as a function of cytidine deaminase activity (Grillo et al., 2022; Li et al., 2012; Olson et al., 2013; Wang et al., 2021). The HPLC-based assay uses liquid chromatography to separate and detect the fluorescent-labeled oligonucleotide and cleaved product (Serrano et al., 2022).
  - **PCR/sequencing-based assays:** This assay uses supercoiled or linear dsDNA, ssDNA or RNA substrates containing a target sequence that will be PCR amplified after incubation with AID/A3. The amplicons are then sequenced, and cytidine deamination is measured as C/G→T/A mutations (Barka et al., 2022; Branton et al., 2020; Larijani, Frieder, Basit, et al., 2005).

- o **NMR-based activity assays:** NMR spectroscopy could also be used to assess cytidine deaminase activity where the chemical shifts occurring due to deamination of cytidine to uridine on DNA substrates are measured (Barzak et al., 2019; Kvach et al., 2019).

Additionally, inhibitor screening against AID/A3s is also done using binding-based assays that measure the effect of candidate inhibitors on the enzyme:substrate binding. They include electrophoretic mobility shift assays (Hellman & Fried, 2007; Larijani et al., 2007), NMR-based binding assays, fluorescence anisotropy/polarization, isothermal calorimetry (Barzak et al., 2019; Harjes et al., 2017; Kvach et al., 2019; Sasaki et al., 2018), surface plasmon resonance (Cen et al., 2010; Grillo et al., 2022), and biolayer interferometry (Ma et al., 2018). While binding assays are essential and informative in the process of inhibitor design, ultimate inhibition must be measured using enzyme activity assays. In the following sections, we focus on the gel-based alkaline cleavage assay and several variations of PCR/sequencing-based assays that are utilized to this end.

1.5 Gel-based biochemical assays: the alkaline cleavage assay for cytidine deaminase activity

One of the earliest developed in vitro assays to test the catalytic activity of AID/A3s has been extensively applied to AID/APOBECs present in cell lysates and in purified form. The gel-based alkaline cleavage assay typically uses an ssDNA substrate containing the target cytosine labeled with either fluorescent tags such as FAM, Cy5, AlexaFluor, or radioactive γ-^{32}P. Following incubation with the enzyme, the reaction is treated with UDG and NaOH. The resulting cleaved products and the full-length substrate are resolved using denaturing polyacrylamide gel electrophoresis, visualized by imaging and quantified by band densitometry (Fig. 1A and B) (Larijani & Martin, 2007; Shi et al., 2017). This assay is highly sensitive to small changes in reaction conditions such as temperature and pH (Fig. 1C and D). It is suitable for testing various substrate structures (ss, ds, bubble, stem loops) and different sequences effectively across a wide range of substrate and enzyme concentrations and hence, ideal for inhibition and inhibitor half-maximal inhibitor concentration or IC50 dosage screening as well as general Michelis Menten (MM) derivation of enzyme kinetic parameters (Fig. 1A–F). For this reason, this assay has proven a cornerstone for various types of studies: first, large scale enzyme-comparative studies that have compared the enzymatic parameters of many AID/APOBECs and/or orthologous enzymes (Ghorbani, 2022); second, substrate-comparative studies that have evaluated substrate preferences for each AID/APOBECs by testing a range of substrates that vary in secondary structure and/or sequence (Abdouni et al., 2018; Diamond et al., 2019); and third, optimal condition studies that have examined optimal temperatures, buffers, and pH conditions of activity for each AID/APOBEC (Ghorbani et al., 2022; Quinlan et al., 2017). To ensure reliable derivation of MM kinetics parameters, it is critical to perform preliminary experiments to establish the appropriate reaction conditions (optimal temperature, pH, reaction duration, [enzyme], [substrate] to ensure that the reactions adhere to the MM regime), including occurrence in the linear initial velocity phase of enzyme activity. The high sensitivity of the alkaline cleavage assay to conditions, time, [enzyme] and [substrate] and its consistency makes it the ideal assay for these types of inquiries.

1.6 PCR/sequencing-based assays

PCR/sequencing-based assays rely on Sanger or Next Generation sequencing (NGS) of PCR amplicons of long (few hundred bases to kilobases) DNA substrates after incubation with the AID/APOBEC enzyme (Barka et al., 2022; Branton et al., 2020; Larijani, Frieder, Basit, et al., 2005). The principle of the assay is that it is performed using DNA or RNA substrates where active cytidine deamination results in uracil, which is detected as thymine. During amplification of DNA substrate by PCR or reverse-transcription of RNA followed by PCR on cDNA using low fidelity polymerases such as Taq, the uracil is read as thymine and Taq incorporates adenine opposite uracil in the newly synthesized DNA resulting in C/G- > T/A mutations (Branton et al., 2020; Canugovi et al., 2009; Shen et al., 2005; Shen & Storb, 2004; Sohail et al., 2003). We established two other PCR/sequencing-based assays to screen for cytidine deamination on long stretches of target DNA, (1) Deam-PCR assay (Fig. 2A) (Branton et al., 2020; Larijani, Frieder, Basit, et al., 2005; Larijani, Frieder, Sonbuchner, et al., 2005; Quinlan et al., 2017) – a nested PCR that uses deamination-specific primers that binds and amplifies only if the cytosines in the primer region have been deaminated to uracils and, (2) Degen-PCR (Fig. 2B) which employs degenerate primers allowing for unbiased amplification of both wildtype and mutated sequences (Branton et al., 2020). Thus, advocate for the usage of degen-PCR whenever possible, as it does not carry the bias and limitation of observing only mutated sequences and/or only mutated sequences carrying specific mutations (e.g., the antibiotic reversion mutation, in the aforementioned bacterial antibiotic resistance reversion assay) (Branton et al., 2020). In addition, these assays are amenable to using target sequences derived from genes or sequences tailored to the specific

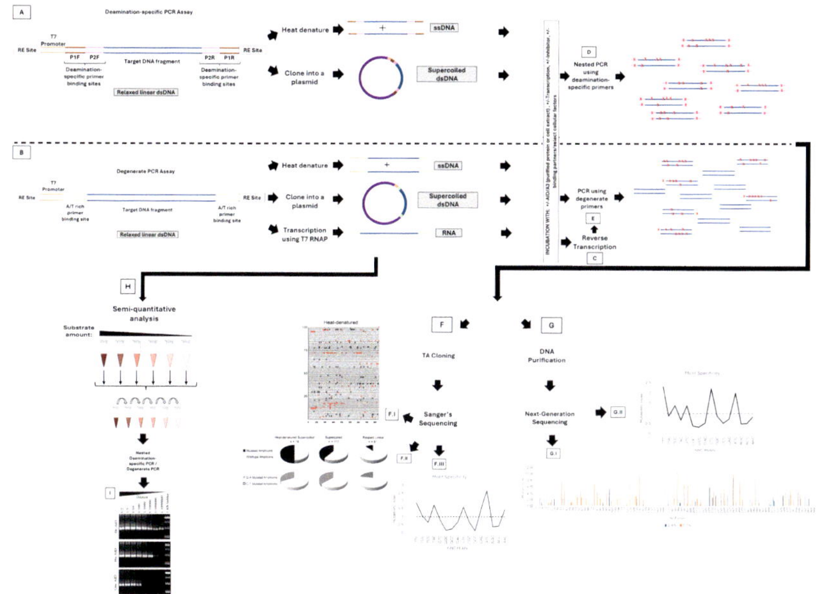

Fig. 2 Schematic of PCR/Sequencing-based Degenerate and Deamination-specific Assays: (A) Representation of the deamination-specific and (B) the degenerate PCR assays.The target DNA is designed with flanking deamination specific inner (P2F, P2R) and outer (P1F, P1R) primer-binding regions represented in pink and brown, respectively (A) or A/T-rich degenerate primer-binding regions, represented in green (B). Additionally, the target sequence contains a T7 promoter at the 5′-end (orange) and restriction sites. These assays allow the use of different kinds of substrates as shown here (gray boxes). The target DNA can be used as a relaxed linear dsDNA or cloned into a vector to be used as a supercoiled dsDNA or heat denatured ssDNA. In case of Degen-PCR assay, in vitro transcription of the target DNA can be performed, RNA purified and used as a substrate. The substrates are incubated either with purified AID/A3 or extracts from cells expressing the enzymes in the presence or absence of processes like transcription, factors like inhibitors, or other binding protein partners. Deamination-specific assay is a nested PCR assay using a pair of outer primers (P1F, P1R) and inner primers (P2F, P2R) (A & D), while degenerate PCR is a one-step PCR (B and E). If using RNA as a substrate, post enzyme incubation, the substrate should be reverse transcribed (C) and then subjected to degen-PCR (E). Amplification using A/T-rich degenerate primers enables an unbiased amplification of both mutated (mutations are depicted as red X's) and wildtype DNA while Deam-PCR is highly sensitive in that it specifically amplifies only deaminated DNA. Following PCR, the amplicons could be TA-cloned into a vector and the target sequence can be sequenced using Sanger's method (F). This allows visualizing mutations along single DNA molecules and assessing the mutational pattern (representative graph in F.I), or evaluating the relative amounts of mutated and wildtype clones, the preference and mutational behavior across different substrates as well as the amount of deamination between the two DNA strands in the form of G-A or C-T mutations (representative pie charts in F.II). Additionally, the preferred NNC motifs of each enzyme tested can also be

(Continued)

requirements of the experiment. They can be applied to test different DNA topologies, including supercoiled or relaxed dsDNA or denatured ssDNA wherein different types of ss regions (due to DNA "breathing") are available to be acted upon by AID/APOBECs (Fig. 2A and B) (Branton et al., 2020). In addition to DNA, sequencing-based activity assays could be used for RNA substrates. In case of RNA as a substrate, after incubation with the enzyme, the substrate is reverse transcribed and then PCR amplified. Both DNA- and RNA-derived amplicons are then sequenced using Sanger or next generation sequencing (NGS) (Fig. 2F and G) (Barka et al., 2022). By conducting reactions under serial [enzyme] or [substrate], both deam-PCR and degen-PCR could also be used as semi-quantitative methods for measuring relative rates of deamination between different reaction conditions, substrate topology/sequences, and even to reveal enzyme velocity differences between different deaminase enzymes (Fig. 2H) (further discussed in Methods section) (Branton et al., 2020; Ghorbani, 2022). These assays were traditionally designed using Sanger sequencing. However, to improve them into more efficient and high-throughput assays, deep sequencing using NGS can be applied. Moreover, the substrates can be subject to transcription in the case of DNA, and reverse transcription in the case of RNA before/after/during AID/APOBEC treatment to study the relationship between AID/APOBEC targeting and the processes / co-factors involved in generating ssDNA in the cell (Besmer et al., 2006; Branton et al., 2020; Dorner et al., 1998; Pham et al., 2019). In the

Fig. 2—Cont'd determined (an example graph comparing 16 NNC motifs, in F.III - y axis shows the mutability index, where 1 = average rate of mutation (dotted line) for all 16 NNC motifs). The relative preference for each individual NNC sequence is obtained by dividing its mutation rate by the average value for all 16 NNCs. The amplicons can be run on agarose gel, purified and sequenced by next-generation sequencing (NGS) methods (G) which can also be processed and analyzed for mutational pattern across the two strands (representative graph of mutation % vs sequence position as in G.I) and determine motif specificity (as in G.II). These assays can be optimized to be used as a semi-quantitative method (H). This is achieved by setting up the activity assay reactions using different dilutions of the DNA substrate (for instance, ranging from 100 ng to 0.2 ×10–3 ng) to cover a range of enzyme:substrate ratios. After incubation with the enzyme, 10-fold serial dilutions are prepared from each reaction from the previous step and 1–2 μL of the dilutions are used as template for deam-/degen-PCR. In case of deam-PCR assay, a higher level of cytidine deamination in a given reaction condition will result in successful amplification from a higher dilution. Therefore, the more extensive the deamination, the higher the dilution that will still yield a visible PCR product (Representative gel pictures comparing the activity of AID orthologs from zebrafish (Dr-AID), human (Hs-AID), Atlantic cod (Gm-AID)). The gel here depicts deamination-specific PCR. *Adapted from Branton et al. (2020), Ghorbani et al. (2022).*

sections below, we will discuss in detail the protocols of gel-based alkaline cleavage assay and PCR/sequencing-based assay using synthetic DNA or RNA substrates with purified AID/A3s.

2. Materials

2.1 Labeling substrates for gel-based assay (p32 labeling)

- DNA oligo substrates (IDT)
- T4 Polynucleotide Kinase (PNK) enzyme and 10X PNK Buffer (NEB, Catalog# M0201S)
- ATP- [γ-32P]−3000Ci/mmol, 5 mCi/mL, 250 μCi
- Mini quick-spin columns for purification of labeled oligos (Sigma, Catalog# 11-814-397-001)
- Tris-EDTA (TE) buffer, pH 8.0 (10 mM Tris + 1 mM EDTA)
- Thermocycler PCR system
- 0.2 mL PCR tubes/PCR strip tubes
- Table-top centrifuge

2.2 Alkaline-cleavage assay

- ^{32}P-labeled DNA substrates
- APOBEC Buffer – 100 mM Phosphate buffer (for AID) or HEPES buffer (for A3s) or other appropriate buffers at preferred pH as applicable
- Uracil DNA Glycosylase (UDG) enzyme with 10X UDG Buffer (NEB, Catalog# M0280L)
- Autoclaved MilliQ water
- 2 M Sodium Hydroxide
- Formamide
- Tris
- EDTA
- Boric acid
- 40 % Acrylamide:Bisacrylamide
- Ammonium persulphate (Sigma Aldrich, Catalog# A3678)
- N-N-N′-N′-Tetramethylethylenediamine (TEMED – Bio-Rad, Catalog# 1610800)
- Bromophenol blue (Sigma, Catalog# B0126)
- Vertical electrophoresis system (Biorad mini-PROTEAN vertical electrophoresis cell)

- Phosphorimaging screen
- Phosphorimager

2.3 PCR/sequencing-based assays

- Target DNA sequence of desired length (amenable to PCR amplification), flanked by primer binding regions, T7 promoter at the 5′-end
- Appropriate APOBEC enzyme buffer (see above)
- T7 RNA synthesis kit (NEB, Catalog# E2040S/L)
- TOPO-TA cloning kit (ThermoFisher Scientific, Catalog# 450641)
- RNase free DNase I kit (Qiagen, Catalog# 79254)
- cDNA Synthesis kit (Onescript plus, Catalog# G236)
- Taq Polymerase and 10X PCR buffer (Invitrogen, Catalog# 10342020)
- Deoxyribonucleotide triphosphates (dNTPs) (ThermoFisher Scientific, Catalog# 10297018)
- Magnesium chloride ($MgCl_2$)
- Forward and reverse primers
- Agarose
- Thermocycler and 0.2 mL PCR tubes
- DNA Gel electrophoresis system
- DNA Gel Extraction kit

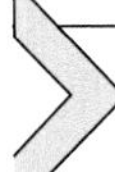

3. Methods

3.1 Alkaline-cleavage assay

3.1.1 Substrate labeling with p32

- To prepare 50 fmol/μL of labeled ssDNA substrate, mix the following in a 0.2 mL PCR tube in the order mentioned in Table 1.

Table 1 Reaction setup for substrate labeling with p32.

Sterile MilliQ H_2O5 μL	
10X PNK Buffer	1 μL (1X Final)
DNA Oligo for labelling (2.5 pmol/μL)	1 μL (yields 50 μL of 50 fmol/μL labelled substrate)
ATP[γ-^{32}P]	2 μL
PNK Enzyme (NEB – 10U/μL)	1 μL
Total reaction volume10 μL	

- Mix by pipetting and incubate the reaction mix at 37 °C for 1 h and heat inactivate PNK at 65 °C for 10 min.
- Any free ATP-[γ-^{32}P] is removed and the labeled oligo is purified using mini quick spin labeled oligo purification columns (Roche). Prepare the column as per the manufacturer's instructions and place the prepared column in a clean 1.5 mL Eppendorf tube.
- Dilute the above reaction mix with 15–20 μL of TE buffer and add to the center of the prepared column. Centrifuge at 800 g for 2 min. The labeled substrate will be collected in the Eppendorf tube.
- Adjust the final volume to 50 μL to get a final concentration of 50 fmol/μL. If ATP-[γ-^{32}P] used for labeling is fresh, this can be diluted to lower concentrations depending on the nature of the experiment. If higher substrate concentrations are to be labeled, proportionally higher amounts of ATP-[γ-^{32}P] should be used.

3.1.2 Preparation of 10X Tris-Borate EDTA running buffer

- To prepare 1000 mL of 10X TBE buffer, mix 108 g of Tris, 55 g Boric acid, 40 mL of 0.5 M Sodium EDTA (pH 8) and adjust the volume to 1000 mL with MilliQ water.

3.1.3 Preparation of urea/formamide/acrylamide gel mix (14 %)

- To prepare 1000 mL of 14 % gel mix that is appropriate for separating 30 – 60 nucleotide oligos (refer to notes section for more details) – add 100 mL of 10X TBE, 250 mL of formamide, 350 mL of 40 % Acrylamide:Bisacrylamide and 450 g of Urea.
- Completely dissolve the components using a magnetic stirrer overnight and store in a dark bottle at room temperature.
- To prepare one gel, use 6 mL of gel mix, 60 μL of 10 % Ammonium persulfate (APS) and 6 μL of TEMED. This is for preparing small polyacrylamide gels to be run using Bio-rad mini protean vertical system, which is quick, easy to handle and can be used in a high throughput manner that allows running multiple gels at a time and the length of the run is also ~1-1.5 h as opposed to 2-4 h for large gels.
- Alternatively, large format vertical electrophoresis systems from Bio-rad can also be used. To prepare 2 large format working gels, use 80 mL of gel mix, 800 μL of 30 % APS and 80 μL of TEMED.

3.1.4 Preparation of loading dye

- To 5 mL of formamide, add a pinch of bromophenol blue.

3.1.5 Enzyme activity assay

- There are three steps to this assay. The first step is enzyme incubation, second step uses UDG to remove uracil bases on the DNA substrate leaving an abasic site (UDG step) and the third step involves treating with NaOH and heating the sample at high temperature (95–98 °C) that cleaves the substrate at the abasic site to give a shorter product (alkaline cleavage step).
- Enzyme incubation step:
  - o In a 0.2 mL PCR tube, add 50–60 mM of APOBEC buffer of the desired pH, 1 μL labeled DNA oligo substrate and 3 μL of purified APOBEC of chosen concentrations to a total reaction volume of 10 μL. All components must be kept on ice at all times.
  - o Mix by pipetting and incubate at 37 °C for the required duration followed by heat inactivation at 85 °C for 10 min in a thermocycler. The incubation temperature mentioned here is for human AID/A3s. However, the temperature should be adjusted to the intended temperature according to the enzyme being tested (Refer Notes section for more details).
- UDG step:
  - o Prepare the following mix (per reaction) in Table 2.
  - o Prepare a master mix of the above components and add 10 μL to each enzyme reaction tube.
  - o Incubate at 37 °C for 30 min, heat inactivate UDG enzyme at 65 °C for 10 min.
  - o At this point, the reaction can be frozen at −20 °C until ready to run the gel.
- Alkaline Cleavage step:

Table 2 Reaction setup for UDG Step.

Sterile MilliQ H_2O7.9 μL	
10X UDG Buffer	**2 μL (1X Final)**
UDG Enzyme (5U/μL stock)	**0.1 μL**
Total volume10 μL	

- o Prepare the urea-formamide-acrylamide gel as mentioned above. The protocol here is described for use with the Bio-rad mini protean system. Pre-run the gel at 50 V for 30 min.
- o To each AID+UDG reaction mix, add 3 μL of 2 M NaOH and 7 μL of loading dye. Mix by pipetting.
- o Incubate at 96 °C for 10 min in a thermocycler.
- o After pre-run, clean the wells in the gel with a syringe using 1X TBE buffer before loading to obtain a clean gel. Load 10–15 μL of the sample and run the gel at 300 V for 1–1.5 h or until the dye reaches the bottom of the gel.
- o Wrap the gel with plastic wrap and expose to phosphorimaging screen for 4 h or overnight. Image the screen using a phosphorimager (GE Cytiva Amersham Typhoon IP Phosphor Imager). Make sure the image is clear with no overexposure.

3.1.6 Notes

- The gel mix protocol given above is for preparing a 14% denaturing polyacrylamide gel. The percentage of the gel depends on the size of the substrate/product that are being separated. Additionally, bromophenol blue is mentioned above as the tracking dye. Xylene cyanol can also be used. These dyes migrate to different sizes depending on the percentage of the gel. Bromophenol blue typically migrates to a size of 15 nucleotides on a 14% gel. Information on appropriate gel percentage according to oligo size and details on dye migration can be found online (for example here: https://www.thermofisher.com/ca/en/home/references/ambion-tech-support/rna-electrophoresis-markers/general-articles/gel-electrophoresis-tables.html).
- When working with AID/APOBEC enzymes, especially cold-adapted orthologs or enzymes with unknown optimal temperatures, tight temperature control is crucial to minimize variations and achieve reproducibility when comparing substrate or condition preferences. This can be accomplished by preparing master mixes in advance, pre-equilibrating thermocycler blocks to the desired incubation temperature, and retrieving the enzyme from storage (typically at −80 °C) just before adding it to the reaction mix and additionally, completing the heat inactivation step promptly. These steps are critical to avoid exposing the enzyme to non-optimal conditions (Ghorbani, 2022; Ghorbani et al., 2021; Quinlan et al., 2017).

- If using DNA substrates labeled with a fluorescent tag, the labeling step should be skipped and in place of bromophenol loading dye, Orange-G dye should be used, as bromophenol blue can self-fluoresce causing strong background interference. To prepare 50 mL of Orange-G loading dye, dissolve 20 g sucrose in 40 mL MilliQ water followed by 100 mg Orange-G and make up the volume to 50 mL with water. After running the gel, the gel can be directly imaged using any imaging system (ChemiDoc Imager from Bio-rad for example) with a filter appropriate to the fluorescent tag used.

3.1.7 Data quantification and analysis

- Use Image Lab software (Bio-rad) or other similar programs such as ImageJ, ChemDraw to load the gel image. Once the image is loaded, adjust the brightness and contrast to enhance clarity.
- Locate the substrate and product bands in each lane of the gel. Use the volume analysis tools to select a band shape (rectangle, square, freehand or circle, for example). Click and drag to draw a boundary around the substrate. Do the same for the product band as well. Ensure that the entire band is captured, avoiding adjacent bands or background noise. Repeat for each lane. In ImageLab, each band is labeled as U for Unknown / Std for Standard / B for Background. For the substrate and product bands, select "Unknown". By default, they will be labeled as U1, U2 and so on. This can be renamed as needed for clarity.
- To correct for background noise, draw a band in an empty area of the gel where there are no bands and set it as "B" for Background. This will adjust for background interference in the intensity measurements.
- Make sure all the lanes have been analyzed, and the corresponding intensities for substrate and product have been recorded and labeled correctly.
- After drawing the bands, the software will automatically calculate the integrated density for each band, which will be displayed in the results table.
- To calculate the percentage of deamination, calculate the ratio of the product band intensity to the sum of product and substrate band intensity (adjusted intensities after background subtraction) multiplied by 100.

3.2 PCR/sequencing-based assays

3.2.1 Synthesis of DNA and RNA substrates

- The target DNA is flanked by the primer binding regions, a T7 promoter region at the 5′ end, and appropriate restriction sites on either side to clone into a vector of choice.

- The plasmid with the target DNA fragment cloned, is used as the supercoiled dsDNA substrate for enzyme incubation and the target sequence will be PCR amplified after incubation with AID/APOBEC enzymes, as detailed in the next section (Section 3.2.2).
- For relaxed linear dsDNA substrate, linearize the plasmid containing the target DNA using a blunt-end restriction digestion enzyme with a single cut-site in the plasmid. Alternatively, a PCR-amplified fragment can be used after gel purification.
- For heat-denatured DNA substrate, incubate the target DNA fragment at 99 °C for 5 mins, snap cool in ice for 1 min and then immediately proceed to enzyme incubation by adding the remaining components of the assay reaction (detailed in Section 3.2.3).
- To synthesize the RNA substrate for this assay, set up in vitro transcription of the target DNA using T7 RNA polymerase (NEB) as per manufacturer's protocol using the primers designed for the assay.
- Treat the newly synthesized RNA with RNase-free DNase I (Thermo Fisher Scientific) and purify using RNA purification columns (Qiagen RNeasy mini kit). Repeat this step twice and use the purified RNA as substrate for the assay.

3.2.2 Primer design

- **Deam-PCR:** The primers for this assay are designed to specifically recognize and bind to regions flanking the target sequence where the cytosines are deaminated to uracils by AID/A3s. The primers are constructed such that the region of the primer corresponding to the deaminated cytosine, pairs with adenine in the complementary strand during PCR. This ensures selective amplification of highly deaminated sequences. This assay is performed as a nested PCR, which uses a pair of outer and a pair of inner deamination-specific primers (Fig. 2A).
- **Degen-PCR:** The forward and reverse primers could be designed to contain a Y (Y = C/T) in the position of dC keeping the primer region A/T-rich containing the minimum required G/C content for PCR amplification thereby adding more flexibility allowing amplification of both wildtype and mutated sequences (Branton et al., 2020). As an alternative, we designed purely A/T-rich primer binding regions, wherein the forward primer sequence is 5′- AAAATTTAAATTAAA AATATTTTAATA – 3′ and the reverse primer sequence is 5′ – ATATTAATATTAAATATTAATTATTTT – 3′. The A/T-rich region eliminates any possibility of bias towards the amplification of

mutated or wild-type sequences. These primer binding sites flank a synthetic DNA substrate containing all 16 NNC combinations (N is A/T/G/C, target cytosine underlined) repeated several times across a 400 bp sequence in different secondary structure contexts. We have performed several hundred sequencing-based PCR assay reactions for assessing cytidine deaminase activity using this design (unpublished work from Dr. Larijani's lab).

3.2.3 Assay protocols

- **Enzyme reaction set up:**
  - o All three PCR/Sequencing-based assays – deam-, degen-PCR using Sangers and NGS-based– follow the same reaction setup as below:
  - o To set up reaction with DNA as the substrate, prepare the enzyme reaction mix (on ice) with 150–300 ng of plasmid/relaxed linear/heat denatured ssDNA substrate, reaction buffer (50–60 mM final concentration), 1×10^{-3} U of Uracil DNA glycosylase inhibitor (UGI) and APOBEC enzyme and make up the final volume to 20 μL with RNase and DNase-free water or the storage buffer of purified APOBEC.
  - o To set up enzyme reaction using RNA as substrate, prepare the enzyme reaction using 150–300 ng of purified RNA substrate, reaction buffer (50–60 mM final concentration), 40-80 U RNase Inhibitor (AID/APOBEC purification often requires RNase application during purification, and hence the protein preparations may contain trace RNase), 5 mM DTT, APOBEC enzyme and make up the final volume to 20 μL with RNase and DNase-free water or the storage buffer of purified APOBEC.
  - o Incubate at 37 °C for the desired length of time. The reaction using DNA can be used as template for PCR after heat inactivation at 65 °C for 10 min.
  - o Heat inactivate the reaction with RNA as substrate following enzyme incubation at 95 °C for 10 min. Use 12 μL or 150–200 ng of RNA from this reaction as template to set up reverse transcription using in vitro reverse transcription kit (Onescript cDNA synthesis kit) as per manufacturer's protocol.
- **Deam-PCR protocol:**
  - o This assay is set up as a nested PCR reaction where 2 μL of the reaction mix is used as template to set up the first round of PCR with outer deam-specific primers followed by a second round of PCR with

the nested inner deam-specific primers using the first PCR reaction product as template.

- Each PCR reaction is set up using 2 μL of the reverse transcription mix or 2 μL of reaction mix that used DNA as substrate. Each 25 μL PCR reaction includes 10X PCR Buffer (1X final), 400 μM dNTPs, 2–4 mM $MgCl_2$, 0.5 μM each of forward and reverse primers, 2.5 U Taq polymerase (Fraggobio) and 2 μL of template made up to 25 μL total reaction volume with water. Set this up in a thermocycler for 35–37 cycles using appropriate annealing temperatures.
- The PCR products can be cloned into a vector using TOPO-TA cloning kit.
- Following TA cloning, transform the vectors into chemically competent *E. coli* and select the transformed clones by plating the transformed cells on a LB plate containing 20 mg/mL X-gal and 50 μg/mL of kanamycin. Successful cloning and transformation will disrupt the lacZα gene of the TOPO vector resulting in white colonies resistant to kanamycin. Extract the plasmids from the positive clones by miniprep (QIAGEN or Geneaid) and sequence by Sanger's sequencing method.
- We further optimized this assay to be used as a semi-quantitative assay to determine the relative amount of deaminated DNA between different reaction conditions (Branton et al., 2020). To achieve this, set up the activity assay reactions using different dilutions of the DNA substrate (for instance, ranging from 100 ng to 0.2×10^{-3} ng) to cover a range of enzyme:substrate ratios.
- After incubation with the enzyme, prepare 10-fold serial dilutions from each reaction from the above step using ultra-pure water and use 1-2 μL of each of the dilutions as template for the nested PCR reactions.
- Run the PCR products on an agarose gel. A higher level of cytidine deamination in a given reaction condition will result in successful amplification from a higher dilution. Therefore, the more extensive the deamination, the higher the dilution that will still yield a visible PCR product.

- **Degen-PCR Protocol:**
  - Degen-PCR is a one-step PCR reaction using degenerate primers as described earlier.
  - The PCR reaction is set up using 2 μL of the reverse transcription mix or 2 μL of reaction mix that used DNA as substrate. Each 25 μL reaction includes 10X PCR Buffer (1X final), 400 μM dNTPs,

2–4 mM $MgCl_2$, 0.5 μM each of forward and reverse primers, 2.5 U Taq polymerase (Fraggobio) and 2 μL of template made up to 25 μL total reaction volume with water. Set this up in a thermocycler for 35–37 cycles using appropriate annealing temperatures.

- o The PCR products can be cloned into a vector using TOPO-TA cloning kit.
- o Following TA cloning, positive clones can be selected using X-gal and kanamycin, as mentioned in the previous section, and plasmids extracted by miniprep and sequenced using the Sangers method.
- o Alternatively, instead of degenerate primers that have Y in place cytidines where Y can be C/T, purely A/T-rich primers can also be used for unbiased amplification of both mutated and wild-type sequences and the PCR reaction needs to be optimized for the appropriate annealing temperatures.

- **NGS-based assay:**
  - o The protocol for setting up the PCR for the NGS-based assay is the same as degen-PCR protocol except that after PCR cycles, the amplified products are run on an agarose gel and the target amplicons purified using gel purification columns (QIAGEN). The purified amplicons are then sequenced using NGS.
  - o After the nested PCR steps of deam-PCR as opposed to cloning into a vector and sequencing by Sangers, the amplicons can be purified and sequenced by NGS as well.

3.2.4 Notes

- For Deam-PCR, the annealing temperatures could be dialed to reduce the stringency allowing primers to anneal and amplify both deaminated and non-deaminated sequences. On the other hand, higher annealing temperatures could be used that would ensure stringent amplification of strictly deaminated sequences.
- For NGS-based sequencing, there are short/medium/long-read amplicon sequencing depending on the length of the target DNA to be analyzed. Appropriate pipelines should be chosen depending on the target sequence length.
- To ensure there is no DNA contamination in the purified RNA, before incubation with the AID/APOBEC, it is important to perform a direct No-RT PCR control using the purified RNA as template. No target amplicon band upon running on agarose gel will indicate a clean DNA-free RNA substrate for the assay.

- Amplicons from a no enzyme negative control where DNA or RNA substrate processed in parallel without APOBEC should be sequenced to compute and subtract the background mutation levels.
- The heat inactivation step for enzyme reaction with RNA substrate is crucial as APOBECs are predominantly DNA editors and any trace amounts of active APOBEC can act on the cDNA produced during the reverse transcription step and lead to false positive results.

3.2.5 Data processing and analysis

- Analysis of Sangers sequencing results can be done using commercially available chromatogram analysis software such as Seqman (DNASTAR) or Snapgene. The chromatogram should be visually analyzed to make sure the mutations observed are real mutations and a good quality read represented by a clear bonafide peak.
- There are different sequencing platforms in NGS depending on the desired read length and read depth. For a 400–600 bp target sequence, MiSeq sequencing platform could be used (Ravi et al., 2018) and amplicon-EZ service of Genewiz provides a pre-made workflow that includes library preparation and sequencing. The raw data is delivered as.fastq files which need to be processed for low quality reads, adapter and primer contamination, if any.
- Pre-processing, alignment, sorting, variant calling and annotating detailed in the following steps can be done using the Galaxy webtool (https://usegalaxy.org/) which is an open-source web-based interactive platform that hosts a library of data analysis tools. On the other hand, the data can be analyzed on a Linux system as well. The steps described below are using Galaxy.
- *Pre-Processing:* The quality of the reads is evaluated using FastQC (https://www.bioinformatics.babraham.ac.uk/projects/fastqc/) (Andrews), an easy to use quality control tool. This returns a report that gives overall quality statistics across the read length, per base sequence quality, and sequence length distribution and indicates if there is any contamination with any adapter sequences that were used for library preparation. This step only checks the quality of the raw data and provides quality control statistics across different categories, which are considered depending on the nature of the experiment and the samples used. This information can then be used to remove contaminating sequences and any low-quality reads using various data trimming and quality filtering tools, for example, trimmomatic (Bolger et al., 2014) or Cutadapt (Martin, 2011).

- *Read alignment and mapping:* The trimmed reads can then be aligned against the reference sequence using commonly used sequence alignment tools like Burrows-Wheeler Aligner (BWA) (Li & Durbin, 2009), Bowtie2 (Langmead & Salzberg, 2012). This step requires creating an index for the reference sequence which will be automatically created by Galaxy server using a fasta file of the reference.
- *Variant calling and annotation:* The last step is to identify all variants or single nucleotide polymorphisms (SNPs) at each position. The samples sequenced are PCR amplicons that contain the same sequence with variations at a few positions ranging from a few individual reads to a fraction of the total reads depending on the APOBEC being tested. Many variant calling algorithms filter variants depending on the read coverage (% of reads containing the variant), mapping and base quality parameters (assigned during read alignment and mapping). In order to avoid losing possible deamination spots, especially in the case of slow acting members like AID (King et al., 2015; Larijani et al., 2007) or enzymes like A3A that exhibits a highly selective but lower activity on RNA compared to DNA (Barka et al., 2022), it is recommended to use a tool that calls all possible variants at all positions which can then be normalized using no-enzyme controls processed in parallel to filter out background mutational levels and false-positives. Two python-based programs, Naïve Variant Caller (NVC) and Variant Annotator (Blankenberg et al., 2014) are available on Galaxy that can be used for the purpose.

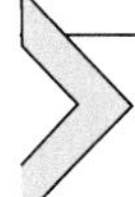

4. Choosing between gel-based vs. Sanger sequencing-based PCR vs. NGS-based PCR assays

The gel-based alkaline cleavage assay has been successfully used for years as a sensitive, reproducible, and efficient test for cytidine deamination. It is exquisitely sensitive to experimental conditions making it the ideal assay for measuring kinetic parameters of AID/A3s, both for comparative enzymology (comparing various human and AID/A3 orthologs) (Ghorbani, 2022), substrate preferences (comparing the activities of AID/A3s on a range of substrates with very well defined sequences and structures) (Diamond et al., 2019), and for testing inhibition and inhibitors IC50 (King et al., 2021). Additionally, the assay can be adapted to test DNA substrates with different secondary structures, such as single-stranded, double-stranded, bubble, and stem-loop configurations (Abdouni et al., 2018; Diamond et al., 2019; Quinlan et al., 2017). However, the gel-based alkaline cleavage assay has limitations due to the length and type of substrates it can accommodate. It is ideal for shorter DNA oligonucleotide substrates (< 100 nt, typical range 30–60 nt) with one or two target cytidines, as the interpretation of multi-cleaved products can become progressively more complex. In addition, working with RNA is more challenging and ^{32}P labeling, or fluorescent modifications add significant costs to the assay.

On the other hand, the gel-based alkaline cleavage assay, especially when used with a simple single-cytidine substrate, is highly quantitative and can provide precise Michaelis-Menten (MM) kinetic parameters and reveal even the slightest differences in enzyme:substrate kinetics and enzyme rates in a comparative manner. Though not inherently a "high throughput" assay, we have demonstrated its adaptability for large-scale comparative enzymology. In our hands, this assay has been applied to decipher optimal temperatures, pH, and MM kinetic parameters on nearly 100 orthologous AID enzymes, characterized in parallel (Ghorbani, 2022). We have successfully tested vertebrate AID orthologs at sub-zero temperatures (as low as −10 °C) and identified cold-adapted AID orthologs that retain nearly 50 % of their deaminase activity at near-zero celcius temperatures (Fig. 1C) (Ghorbani, 2022). For parallel enzymology at such large-scale, reactions can be frozen and electrophoresed sequentially in batches at convenience provided that – (1) they are expressed and purified simultaneously (when conducting comparative enzymology across different deaminases), (2) substrate master mixes are prepared in parallel (when conducting comparative substrate/condition preferences for a more limited set of enzymes) and, (3) reactions are carried out concurrently, in duplicates or triplicates with multiple independent enzyme preparations (a reproducible set up in our hands would include at least two independently purified preparations of an enzyme and duplicate reactions for each). In this manner, our largest scale comparative enzymology has included up to several thousand parallel reactions (Fig. 1C and D) (Abdouni et al., 2018; Diamond et al., 2019; Ghorbani, 2022; Quinlan et al., 2017).

Building on these strengths, the alkaline cleavage assay can be extended for high-throughput screening of potential AID/A3 inhibitors. By systematically testing multiple candidate inhibitors in parallel reaction setups, this assay can then be used to quantitatively measure enzyme activity across varying inhibitor concentrations, enabling the precise determination of

IC50 and hence, evaluate efficacy of the potential inhibitors (King et al., 2021). Thus, alkaline cleavage assay serves not only as a powerful tool for enzymology but also has immense potential to be used in screening of AID/A3 inhibitors (Fig. 1F).

In contrast, PCR/sequencing-based assays, while not as sensitive as the gel-based assay, present the advantages of (1) providing in the order of 10^5 nucleotides worth of sequencing data from a single enzyme reaction, (2) working with longer DNA substrates, which can be comprised of actual gene sequences, including well-characterized target sequences of AID/A3s (such as the variable and switch regions for AID or viral genome sequences for A3s). Additionally, these assays can be applied to RNA substrates, expanding the range of enzyme activities that can be tested for inhibitor effects by select drugs (Barka et al., 2022). The Deam-PCR assay on one hand provides two modes of verification of deamination. First, it only amplifies if the target region has been deaminated, so a direct visualization of the amplified product on an agarose gel confirms deamination (Fig. 2A). Second, the semi-quantitative version of the assay (Fig. 2H) can estimate the extent of deamination at different [inhibitor] or enzyme:substrate ratios, with results further validated by Sanger sequencing. The Degen-PCR, on the other hand, provides an unbiased analysis of the entire reaction mixture, including non-mutated sequences (Fig. 2B). While Sanger sequencing is highly accurate with a low error rate in base calling compared to Illumina NGS, it is time-consuming and requires screening tens-hundreds of clones to identify unique sequences for meaningful analysis. This increases the cost and labor involved in cloning, plasmid extraction, and sequencing. However, NGS overcomes this limitation by providing a large number of reads, allowing for a more confident and high-throughput analysis. NGS is also efficient, supports screening across a wide range of inhibitor concentrations, and enhances the overall productivity of the screening process. The downside of NGS is significantly higher background error rates, which can necessitate multiple repetitive no-enzyme control reactions, especially when studying "real" enzyme-mediated mutations for an enzyme that either supports a low mutation frequency, or whose optimal deamination conditions are still being worked out. Another advantage of the PCR/sequencing-based assays is that they provide an insight into larger scale patterns of mutations supported by an enzyme, from which substrate preference and processivity properties can be gleaned from a handful of reactions; the same insights could require exponentially more work using the gel-based alkaline cleavage assay, through testing hundreds of reactions under different conditions and substrates. However, on the upside, the alkaline cleavage assay provides the platform to precisely quantify inhibitor efficiency by measuring IC50 values which could prove to be challenging in terms of cost and sensitivity of the PCR/sequencing-based assay.

Overall, both gel-based and PCR/sequencing-based assays offer distinct advantages in screening for AID/APOBEC activity and inhibition. The latter is inherently significantly high throughput, especially when used in conjunction with NGS, in that one reaction can provide in the order of 10^5 nucleotides worth of data; on the other hand, the gel-based alkaline cleavage assay can provide exquisitely sensitive MM kinetics data and can be made to be semi high-throughput allowing for thousands of reactions to be analyzed in parallel, though this requires significant effort. In our hands, for the specific application of inhibitor design/screening, the gel-based assays offer an initial screening method, and the PCR/sequencing assays offer a more confirmatory role.

5. Conclusion and additional notes

A major challenge to designing effective in vitro enzyme assays for screening AID/APOBEC inhibitors is the issue of protein expression and purification in stable, active forms in sufficient quantities. The purity and stability of these enzymes are often affected by non-specific interactions with proteins and nucleic acids (King et al., 2015), improper folding when expressed in prokaryotic hosts (Iwatani et al., 2006), and oligomerization, which leads to protein aggregation (Polevoda et al., 2016; Polevoda et al., 2015). To ensure that the enzyme behavior observed is bona fide activity of the purified protein and is not due to factors contributed by the expression system, purification method, or tags used, our standard practice has been to purify the proteins from various expression systems, utilizing different tags fused at either the N- or C-terminus, and ensure that results obtained in enzyme assays are consistent across at least two different versions of the purified enzyme, and across multiple independent preparations (at least 2 preparations, but preferably 3 or 4) of each enzyme (Abdouni et al., 2018; Borzooee & Larijani, 2019; Diamond et al., 2019; Holland et al., 2018) (Fig. 1).

Enzyme assays that are high-throughput and highly sensitive to testing conditions like minute changes in pH, temperature, enzyme and substrate concentrations as well as substrate types, sequence and structures are desirable. The antibiotic reversion assay has been one of the most commonly used to test APOBEC activity. However, typical substrates use a small subset of motifs targeted by AID/APOBECs and mutation detection depends on specific mutations that could make the cells antibiotic resistant while all other mutations would go unnoticed. The gel-based assay commonly involves three steps – enzyme incubation, UDG and finally NaOH and heat steps. For high-throughput inhibitor screening, assays with as few steps as possible are often desirable. Some studies have attempted to reduce the number of steps after enzyme incubation by adding an excess of UDG during the enzyme incubation step (Petersen-Mahrt & Neuberger, 2003). The 2-step reaction could be further reduced to a 1-step assay to measure APOBEC activity real time by using ssDNA oligo substrates with fluorophores like FAM at the 5′ end and quenchers like TAMRA or Iowa Black FQ (IAB) at the 3′ end with the target C at the center. Additionally, instead of UDG and treatment with NaOH and heat steps, APE or EndoQ enzyme from *Pyrococcus furiosus* can be added in excess to the enzyme reaction which would cleave the substrate adjacent to the uracil thus formed by deamination of cytidine. The activity is measured from the FAM signal emitted upon substrate cleavage that happens as the cytidine is deaminated thereby reducing it to a single-step assay (Belica et al., 2024). Alternatively, sequencing-based assays provide flexibility in terms of sequence and length of substrates with the added advantage of deep sequencing which provides a wealth of data from a single enzyme:substrate reaction. However, they are not as sensitive as gel-based or plate-based assays for precise quantitative enzymology (i.e., MM parameters). Given the diverse range of assays currently in use, each with its own advantages and disadvantages and use cases, it will be crucial to develop new assays with the goal of overcoming the limitations while bridging the strengths of the current methods.

References

Abdouni, H. S., King, J. J., Ghorbani, A., Fifield, H., Berghuis, L., & Larijani, M. (2018). DNA/RNA hybrid substrates modulate the catalytic activity of purified AID. *Molecular Immunology, 93*, 94–106. https://doi.org/10.1016/j.molimm.2017.11.012.

Alexandrov, L. B., Nik-Zainal, S., Wedge, D. C., Aparicio, S. A., Behjati, S., Biankin, A. V., et al. (2013). Signatures of mutational processes in human cancer. *Nature, 500*(7463), 415–421. https://doi.org/10.1038/nature12477.

Alvarez-Gonzalez, J., Yasgar, A., Maul, R. W., Rieffer, A. E., Crawford, D. J., Salamango, D. J., et al. (2021). Small molecule inhibitors of activation-induced deaminase decrease class switch recombination in B cells. *ACS Pharmacology & Translational Science, 4*(3), 1214–1226. https://doi.org/10.1021/acsptsci.1c00064.

Andrews, S. *FastQC A Quality Control tool for High Throughput Sequence Data*. http://www.bioinformatics.babraham.ac.uk/projects/fastqc/.

Barka, A., Berrios, K. N., Bailer, P., Schutsky, E. K., Wang, T., & Kohli, R. M. (2022). The base-editing enzyme APOBEC3A catalyzes cytosine deamination in RNA with low proficiency and high selectivity. *ACS Chemical Biology, 17*(3), 629–636. https://doi.org/10.1021/acschembio.1c00919.

Barzak, F. M., Harjes, S., Kvach, M. V., Kurup, H. M., Jameson, G. B., Filichev, V. V., et al. (2019). Selective inhibition of APOBEC3 enzymes by single-stranded DNAs containing 2′-deoxyzebularine. *Organic & Biomolecular Chemistry, 17*(43), 9435–9441. https://doi.org/10.1039/c9ob01781j.

Belica, C. A., Carpenter, M. A., Chen, Y., Brown, W. L., Moeller, N. H., Boylan, I. T., et al. (2024). A real-time biochemical assay for quantitative analyses of APOBEC-catalyzed DNA deamination. *The Journal of Biological Chemistry, 300*(6), 107410. https://doi.org/10.1016/j.jbc.2024.107410.

Besmer, E., Market, E., & Papavasiliou, F. N. (2006). The transcription elongation complex directs activation-induced cytidine deaminase-mediated DNA deamination. *Molecular and Cellular Biology, 26*(11), 4378–4385. https://doi.org/10.1128/MCB.02375-05.

Blankenberg, D., Von Kuster, G., Bouvier, E., Baker, D., Afgan, E., Stoler, N., et al. (2014). Dissemination of scientific software with Galaxy ToolShed. *Genome Biology, 15*(2), 403. https://doi.org/10.1186/gb4161.

Bolger, A. M., Lohse, M., & Usadel, B. (2014). Trimmomatic: A flexible trimmer for Illumina sequence data. *Bioinformatics (Oxford, England), 30*(15), 2114–2120. https://doi.org/10.1093/bioinformatics/btu170.

Borzooee, F., Asgharpour, M., Quinlan, E., Grant, M. D., & Larijani, M. (2018). Viral subversion of APOBEC3s: Lessons for anti-tumor immunity and tumor immunotherapy. *International Reviews of Immunology, 37*(3), 151–164. https://doi.org/10.1080/08830185.2017.1403596.

Borzooee, F., & Larijani, M. (2019). Pichia pastoris as a host for production and isolation of mutagenic AID/APOBEC enzymes involved in cancer and immunity. *New Biotechnology, 51*, 67–79. https://doi.org/10.1016/j.nbt.2019.02.006.

Bransteitter, R., Pham, P., Scharff, M. D., & Goodman, M. F. (2003). Activation-induced cytidine deaminase deaminates deoxycytidine on single-stranded DNA but requires the action of RNase. *Proceedings of the National Academy of Sciences of the United States of America, 100*(7), 4102–4107. https://doi.org/10.1073/pnas.0730835100.

Branton, S. A., Ghorbani, A., Bolt, B. N., Fifield, H., Berghuis, L. M., & Larijani, M. (2020). Activation-induced cytidine deaminase can target multiple topologies of double-stranded DNA in a transcription-independent manner. *The FASEB Journal, 34*(7), 9245–9268. https://doi.org/10.1096/fj.201903036RR.

Burns, M. B., Temiz, N. A., & Harris, R. S. (2013). Evidence for APOBEC3B mutagenesis in multiple human cancers. *Nature Genetics, 45*(9), 977–983. https://doi.org/10.1038/ng.2701.

Cahill, D. P., Kinzler, K. W., Vogelstein, B., & Lengauer, C. (1999). Genetic instability and darwinian selection in tumours. *Trends in Cell Biology, 9*(12), M57–M60. https://www.ncbi.nlm.nih.gov/pubmed/10611684.

Canugovi, C., Samaranayake, M., & Bhagwat, A. S. (2009). Transcriptional pausing and stalling causes multiple clustered mutations by human activation-induced deaminase. *The FASEB Journal, 23*(1), 34–44. https://doi.org/10.1096/fj.08-115352.

Caswell, D. R., Gui, P., Mayekar, M. K., Law, E. K., Pich, O., Bailey, C., et al. (2024). The role of APOBEC3B in lung tumor evolution and targeted cancer therapy resistance. *Nature Genetics, 56*(1), 60–73. https://doi.org/10.1038/s41588-023-01592-8.

Cen, S., Peng, Z. G., Li, X. Y., Li, Z. R., Ma, J., Wang, Y. M., et al. (2010). Small molecular compounds inhibit HIV-1 replication through specifically stabilizing APOBEC3G. *The Journal of Biological Chemistry, 285*(22), 16546–16552. https://doi.org/10.1074/jbc.M109.085308.

Chaudhuri, J., & Alt, F. W. (2004). Class-switch recombination: Interplay of transcription, DNA deamination and DNA repair. *Nature Reviews. Immunology, 4*(7), 541–552. https://doi.org/10.1038/nri1395.

Chelico, L., Pham, P., Calabrese, P., & Goodman, M. F. (2006). APOBEC3G DNA deaminase acts processively 3′ −> 5′ on single-stranded DNA. *Nature Structural & Molecular Biology, 13*(5), 392–399. https://doi.org/10.1038/nsmb1086.

Chelico, L., Pham, P., & Goodman, M. F. (2009). Stochastic properties of processive cytidine DNA deaminases AID and APOBEC3G. *Philosophical Transactions of the Royal Society of London. Series B, Biological Sciences, 364*(1517), 583–593. https://doi.org/10.1098/rstb.2008.0195.

Chelico, L., Sacho, E. J., Erie, D. A., & Goodman, M. F. (2008). A model for oligomeric regulation of APOBEC3G cytosine deaminase-dependent restriction of HIV. *The Journal of Biological Chemistry, 283*(20), 13780–13791. https://doi.org/10.1074/jbc.M801004200.

Conticello, S. G. (2008). The AID/APOBEC family of nucleic acid mutators. *Genome Biology, 9*(6), 229. https://doi.org/10.1186/gb-2008-9-6-229.

Conticello, S. G., Ganesh, K., Xue, K., Lu, M., Rada, C., & Neuberger, M. S. (2008). Interaction between antibody-diversification enzyme AID and spliceosome-associated factor CTNNBL1. *Molecular Cell, 31*(4), 474–484. https://doi.org/10.1016/j.molcel.2008.07.009.

Diamond, C. P., Im, J., Button, E. A., Huebert, D. N. G., King, J. J., Borzooee, F., et al. (2019). AID, APOBEC3A and APOBEC3B efficiently deaminate deoxycytidines neighboring DNA damage induced by oxidation or alkylation. *Biochimica et Biophysica Acta (BBA) – General Subjects, 1863*(11), 129415. https://doi.org/10.1016/j.bbagen.2019.129415.

Dickerson, S. K., Market, E., Besmer, E., & Papavasiliou, F. N. (2003). AID mediates hypermutation by deaminating single stranded DNA. *The Journal of Experimental Medicine, 197*(10), 1291–1296. https://doi.org/10.1084/jem.20030481.

DiMarco, A. V., Qin, X., McKinney, B. J., Garcia, N. M. G., Van Alsten, S. C., Mendes, E. A., et al. (2022). APOBEC mutagenesis inhibits breast cancer growth through induction of T cell-mediated antitumor immune responses. *Cancer Immunology Research, 10*(1), 70–86. https://doi.org/10.1158/2326-6066.CIR-21-0146.

Dorner, T., Foster, S. J., Farner, N. L., & Lipsky, P. E. (1998). Somatic hypermutation of human immunoglobulin heavy chain genes: Targeting of RGYW motifs on both DNA strands. *European Journal of Immunology, 28*(10), 3384–3396. https://doi.org/10.1002/(SICI)1521-4141(199810)28:10<3384::AID-IMMU3384>3.0.CO;2-T.

Driscoll, C. B., Schuelke, M. R., Kottke, T., Thompson, J. M., Wongthida, P., Tonne, J. M., et al. (2020). APOBEC3B-mediated corruption of the tumor cell immunopeptidome induces heteroclitic neoepitopes for cancer immunotherapy. *Nature Communications, 11*(1), 790. https://doi.org/10.1038/s41467-020-14568-7.

Durfee, C., Temiz, N. A., Levin-Klein, R., Argyris, P. P., Alsoe, L., Carracedo, S., et al. (2023). Human APOBEC3B promotes tumor development in vivo including signature mutations and metastases. *Cell Reports Medicine, 4*(10), 101211. https://doi.org/10.1016/j.xcrm.2023.101211.

Feng, Y., Seija, N., Di Noia, J. M., & Martin, A. (2020). AID in antibody diversification: There and back again. *Trends in Immunology, 41*(7), 586–600. https://doi.org/10.1016/j.it.2020.04.009.

Fourati, S., Malet, I., Binka, M., Boukobza, S., Wirden, M., Sayon, S., et al. (2010). Partially active HIV-1 Vif alleles facilitate viral escape from specific antiretrovirals. *AIDS (London, England), 24*(15), 2313–2321. https://doi.org/10.1097/QAD.0b013e32833e515a.

Ghorbani, A. (2022). Ancestral reconstruction reveals catalytic inactivation of activation-induced cytidine deaminase concomitant with cold water adaption in the Gadiformes bony fish.

Ghorbani, A., King, J. J., & Larijani, M. (2022). The optimal pH of AID is skewed from that of its catalytic pocket by DNA-binding residues and surface charge. *The Biochemical Journal, 479*(1), 39–55. https://doi.org/10.1042/BCJ20210529.

Ghorbani, A., Quinlan, E. M., & Larijani, M. (2021). Evolutionary comparative analyses of DNA-editing enzymes of the immune system: From 5-dimensional description of protein structures to immunological insights and applications to protein engineering. *Frontiers in Immunology, 12*, 642343. https://doi.org/10.3389/fimmu.2021.642343.

Grillo, M. J., Jones, K. F. M., Carpenter, M. A., Harris, R. S., & Harki, D. A. (2022). The current toolbox for APOBEC drug discovery. *Trends in Pharmacological Sciences, 43*(5), 362–377. https://doi.org/10.1016/j.tips.2022.02.007.

Harjes, S., Jameson, G. B., Filichev, V. V., Edwards, P. J. B., & Harjes, E. (2017). NMR-based method of small changes reveals how DNA mutator APOBEC3A interacts with its single-stranded DNA substrate. *Nucleic Acids Research, 45*(9), 5602–5613. https://doi.org/10.1093/nar/gkx196.

Harjes, S., Kurup, H. M., Rieffer, A. E., Bayaijargal, M., Filitcheva, J., Su, Y., et al. (2023). Structure-guided inhibition of the cancer DNA-mutating enzyme APOBEC3A. *bioRxiv*. https://doi.org/10.1101/2023.02.17.528918.

Harris, R. S., Bishop, K. N., Sheehy, A. M., Craig, H. M., Petersen-Mahrt, S. K., Watt, I. N., et al. (2003). DNA deamination mediates innate immunity to retroviral infection. *Cell, 113*(6), 803–809. https://doi.org/10.1016/s0092-8674(03)00423-9.

Harris, R. S., Petersen-Mahrt, S. K., & Neuberger, M. S. (2002). RNA editing enzyme APOBEC1 and some of its homologs can act as DNA mutators. *Molecular Cell, 10*(5), 1247–1253. https://doi.org/10.1016/s1097-2765(02)00742-6.

Hellman, L. M., & Fried, M. G. (2007). Electrophoretic mobility shift assay (EMSA) for detecting protein-nucleic acid interactions. *Nature Protocols, 2*(8), 1849–1861. https://doi.org/10.1038/nprot.2007.249.

Holland, S. J., Berghuis, L. M., King, J. J., Iyer, L. M., Sikora, K., Fifield, H., et al. (2018). Expansions, diversification, and interindividual copy number variations of AID/APOBEC family cytidine deaminase genes in lampreys. *Proceedings of the National Academy of Sciences of the United States of America, 115*(14), E3211–E3220. https://doi.org/10.1073/pnas.1720871115.

Hu, W., Begum, N. A., Mondal, S., Stanlie, A., & Honjo, T. (2015). Identification of DNA cleavage- and recombination-specific hnRNP cofactors for activation-induced cytidine deaminase. *Proceedings of the National Academy of Sciences of the United States of America, 112*(18), 5791–5796. https://doi.org/10.1073/pnas.1506167112.

Isozaki, H., Sakhtemani, R., Abbasi, A., Nikpour, N., Stanzione, M., Oh, S., et al. (2023). Therapy-induced APOBEC3A drives evolution of persistent cancer cells. *Nature, 620*(7973), 393–401. https://doi.org/10.1038/s41586-023-06303-1.

Ito, F., Fu, Y., Kao, S. A., Yang, H., & Chen, X. S. (2017). Family-wide comparative analysis of cytidine and methylcytidine deamination by eleven human APOBEC proteins. *Journal of Molecular Biology, 429*(12), 1787–1799. https://doi.org/10.1016/j.jmb.2017.04.021.

Iwatani, Y., Takeuchi, H., Strebel, K., & Levin, J. G. (2006). Biochemical activities of highly purified, catalytically active human APOBEC3G: Correlation with antiviral effect. *Journal of Virology, 80*(12), 5992–6002. https://doi.org/10.1128/JVI.02680-05.

King, J. J., Borzooee, F., Im, J., Asgharpour, M., Ghorbani, A., Diamond, C. P., et al. (2021). Structure-based design of first-generation small molecule inhibitors targeting the catalytic pockets of AID, APOBEC3A, and APOBEC3B. *ACS Pharmacology & Translational Science, 4*(4), 1390–1407. https://doi.org/10.1021/acsptsci.1c00091.

King, J. J., & Larijani, M. (2017). A novel regulator of activation-induced cytidine deaminase/APOBECs in immunity and cancer: SchrodingerBs CATalytic pocket. *Frontiers in Immunology, 8*, 351. https://doi.org/10.3389/fimmu.2017.00351.

King, J. J., Manuel, C. A., Barrett, C. V., Raber, S., Lucas, H., Sutter, P., et al. (2015). Catalytic pocket inaccessibility of activation-induced cytidine deaminase is a safeguard against excessive mutagenic activity. *Structure (London, England: 1993), 23*(4), 615–627. https://doi.org/10.1016/j.str.2015.01.016.

Klemm, L., Duy, C., Iacobucci, I., Kuchen, S., von Levetzow, G., Feldhahn, N., et al. (2009). The B cell mutator AID promotes B lymphoid blast crisis and drug resistance in chronic myeloid leukemia. *Cancer Cell, 16*(3), 232–245. https://doi.org/10.1016/j.ccr.2009.07.030.

Krishnan, A., Iyer, L. M., Holland, S. J., Boehm, T., & Aravind, L. (2018). Diversification of AID/APOBEC-like deaminases in metazoa: Multiplicity of clades and widespread roles in immunity. *Proceedings of the National Academy of Sciences of the United States of America, 115*(14), E3201–E3210. https://doi.org/10.1073/pnas.1720897115.

Kurup, H. M., Kvach, M. V., Harjes, S., Barzak, F. M., Jameson, G. B., Harjes, E., et al. (2022). Design, synthesis, and evaluation of a cross-linked oligonucleotide as the first nanomolar inhibitor of APOBEC3A. *Biochemistry, 61*(22), 2568–2578. https://doi.org/10.1021/acs.biochem.2c00449.

Kvach, M. V., Barzak, F. M., Harjes, S., Schares, H. A. M., Jameson, G. B., Ayoub, A. M., et al. (2019). Inhibiting APOBEC3 activity with single-stranded DNA containing 2′-deoxyzebularine analogues. *Biochemistry, 58*(5), 391–400. https://doi.org/10.1021/acs.biochem.8b00858.

Kvach, M. V., Barzak, F. M., Harjes, S., Schares, H. A. M., Kurup, H. M., Jones, K. F., et al. (2020). Differential inhibition of APOBEC3 DNA-mutator isozymes by fluoro- and non-fluoro-substituted 2′-deoxyzebularine embedded in single-stranded DNA. *Chembiochem: A European Journal of Chemical Biology, 21*(7), 1028–1035. https://doi.org/10.1002/cbic.201900505.

Lada, A. G., Dhar, A., Boissy, R. J., Hirano, M., Rubel, A. A., Rogozin, I. B., et al. (2012). AID/APOBEC cytosine deaminase induces genome-wide kataegis. discussion 47 *Biology Direct, 7*, 47. https://doi.org/10.1186/1745-6150-7-47.

Lada, A. G., Krick, C. F., Kozmin, S. G., Mayorov, V. I., Karpova, T. S., Rogozin, I. B., et al. (2011). Mutator effects and mutation signatures of editing deaminases produced in bacteria and yeast. *Biochemistry (Mosc, 76*(1), 131–146. https://doi.org/10.1134/s0006297911010135.

Langmead, B., & Salzberg, S. L. (2012). Fast gapped-read alignment with Bowtie 2. *Nature Methods, 9*(4), 357–359. https://doi.org/10.1038/nmeth.1923.

Larijani, M., Frieder, D., Basit, W., & Martin, A. (2005). The mutation spectrum of purified AID is similar to the mutability index in Ramos cells and in ung(-/-)msh2(-/-) mice. *Immunogenetics, 56*(11), 840–845. https://doi.org/10.1007/s00251-004-0748-0.

Larijani, M., Frieder, D., Sonbuchner, T. M., Bransteitter, R., Goodman, M. F., Bouhassira, E. E., et al. (2005). Methylation protects cytidines from AID-mediated deamination. *Molecular Immunology, 42*(5), 599–604. https://doi.org/10.1016/j.molimm.2004.09.007.

Larijani, M., & Martin, A. (2007). Single-stranded DNA structure and positional context of the target cytidine determine the enzymatic efficiency of AID. *Molecular and Cellular Biology, 27*(23), 8038–8048. https://doi.org/10.1128/MCB.01046-07.

Larijani, M., & Martin, A. (2012). The biochemistry of activation-induced deaminase and its physiological functions. *Seminars in Immunology, 24*(4), 255–263. https://doi.org/10.1016/j.smim.2012.05.003.

Larijani, M., Petrov, A. P., Kolenchenko, O., Berru, M., Krylov, S. N., & Martin, A. (2007). AID associates with single-stranded DNA with high affinity and a long complex half-life in a sequence-independent manner. *Molecular and Cellular Biology, 27*(1), 20–30. https://doi.org/10.1128/MCB.00824-06.

Law, E. K., Levin-Klein, R., Jarvis, M. C., Kim, H., Argyris, P. P., Carpenter, M. A., et al. (2020). APOBEC3A catalyzes mutation and drives carcinogenesis in vivo. *The Journal of Experimental Medicine, 217*(12), https://doi.org/10.1084/jem.20200261.

Law, E. K., Sieuwerts, A. M., LaPara, K., Leonard, B., Starrett, G. J., Molan, A. M., et al. (2016). The DNA cytosine deaminase APOBEC3B promotes tamoxifen resistance in ER-positive breast cancer. *Science Advances, 2*(10), e1601737. https://doi.org/10.1126/sciadv.1601737.

Lecossier, D., Bouchonnet, F., Clavel, F., & Hance, A. J. (2003). Hypermutation of HIV-1 DNA in the absence of the Vif protein. *Science (New York, N. Y.), 300*(5622), 1112. https://doi.org/10.1126/science.1083338.

Leonard, B., Starrett, G. J., Maurer, M. J., Oberg, A. L., Van Bockstal, M., Van Dorpe, J., et al. (2016). APOBEC3G expression correlates with T-cell infiltration and improved clinical outcomes in high-grade serous ovarian carcinoma. *Clinical Cancer Research: An Official Journal of the American Association for Cancer Research, 22*(18), 4746–4755. https://doi.org/10.1158/1078-0432.CCR-15-2910.

Li, H., & Durbin, R. (2009). Fast and accurate short read alignment with Burrows-Wheeler transform. *Bioinformatics (Oxford, England), 25*(14), 1754–1760. https://doi.org/10.1093/bioinformatics/btp324.

Li, M., Shandilya, S. M., Carpenter, M. A., Rathore, A., Brown, W. L., Perkins, A. L., et al. (2012). First-in-class small molecule inhibitors of the single-strand DNA cytosine deaminase APOBEC3G. *ACS Chemical Biology, 7*(3), 506–517. https://doi.org/10.1021/cb200440y.

Liddament, M. T., Brown, W. L., Schumacher, A. J., & Harris, R. S. (2004). APOBEC3F properties and hypermutation preferences indicate activity against HIV-1 in vivo. *Current Biology, 14*(15), 1385–1391. https://doi.org/10.1016/j.cub.2004.06.050.

Ma, L., Zhang, Z., Liu, Z., Pan, Q., Wang, J., Li, X., et al. (2018). Identification of small molecule compounds targeting the interaction of HIV-1 Vif and human APOBEC3G by virtual screening and biological evaluation. *Scientific Reports, 8*(1), 8067. https://doi.org/10.1038/s41598-018-26318-3.

Mangeat, B., Turelli, P., Caron, G., Friedli, M., Perrin, L., & Trono, D. (2003). Broad antiretroviral defence by human APOBEC3G through lethal editing of nascent reverse transcripts. *Nature, 424*(6944), 99–103. https://doi.org/10.1038/nature01709.

Martin, A. S., Salamango, D. J., Serebrenik, A. A., Shaban, N. M., Brown, W. L., & Harris, R. S. (2019). A panel of eGFP reporters for single base editing by APOBEC-Cas9 editosome complexes. *Scientific Reports, 9*(1), 497. https://doi.org/10.1038/s41598-018-36739-9.

Martin, M. (2011). Cutadapt removes adapter sequences from high-throughput sequencing reads next generation sequencing. *small RNA; microRNA; Adapter Removal, 17*(1), 3. https://doi.org/10.14806/ej.17.1.200.

Mechtcheriakova, D., Svoboda, M., Meshcheryakova, A., & Jensen-Jarolim, E. (2012). Activation-induced cytidine deaminase (AID) linking immunity, chronic inflammation, and cancer. *Cancer Immunology, Immunotherapy: CII, 61*(9), 1591–1598. https://doi.org/10.1007/s00262-012-1255-z.

Meng, F. L., Du, Z., Federation, A., Hu, J., Wang, Q., Kieffer-Kwon, K. R., et al. (2014). Convergent transcription at intragenic super-enhancers targets AID-initiated genomic instability. *Cell, 159*(7), 1538–1548. https://doi.org/10.1016/j.cell.2014.11.014.

Monajemi, M., Woodworth, C. F., Zipperlen, K., Gallant, M., Grant, M. D., & Larijani, M. (2014). Positioning of APOBEC3G/F mutational hotspots in the human immunodeficiency virus genome favors reduced recognition by CD8+ T cells. *PLoS One, 9*(4), e93428. https://doi.org/10.1371/journal.pone.0093428.

Nambu, Y., Sugai, M., Gonda, H., Lee, C. G., Katakai, T., Agata, Y., et al. (2003). Transcription-coupled events associating with immunoglobulin switch region chromatin. *Science (New York, N. Y.), 302*(5653), 2137–2140. https://doi.org/10.1126/science.1092481.

Olson, M. E., Li, M., Harris, R. S., & Harki, D. A. (2013). Small-molecule APOBEC3G DNA cytosine deaminase inhibitors based on a 4-amino-1,2,4-triazole-3-thiol scaffold. *ChemMedChem, 8*(1), 112–117. https://doi.org/10.1002/cmdc.201200411.

Pasqualucci, L., Bhagat, G., Jankovic, M., Compagno, M., Smith, P., Muramatsu, M., et al. (2008). AID is required for germinal center-derived lymphomagenesis. *Nature Genetics, 40*(1), 108–112. https://doi.org/10.1038/ng.2007.35.

Pavri, R., Gazumyan, A., Jankovic, M., Di Virgilio, M., Klein, I., Ansarah-Sobrinho, C., et al. (2010). Activation-induced cytidine deaminase targets DNA at sites of RNA polymerase II stalling by interaction with Spt5. *Cell, 143*(1), 122–133. https://doi.org/10.1016/j.cell.2010.09.017.

Petersen-Mahrt, S. K., & Neuberger, M. S. (2003). In vitro deamination of cytosine to uracil in single-stranded DNA by apolipoprotein B editing complex catalytic subunit 1 (APOBEC1). *The Journal of Biological Chemistry, 278*(22), 19583–19586. https://doi.org/10.1074/jbc.C300114200.

Pham, P., Landolph, A., Mendez, C., Li, N., & Goodman, M. F. (2013). A biochemical analysis linking APOBEC3A to disparate HIV-1 restriction and skin cancer. *The Journal of Biological Chemistry, 288*(41), 29294–29304. https://doi.org/10.1074/jbc.M113.504175.

Pham, P., Malik, S., Mak, C., Calabrese, P. C., Roeder, R. G., & Goodman, M. F. (2019). AID-RNA polymerase II transcription-dependent deamination of IgV DNA. *Nucleic Acids Research, 47*(20), 10815–10829. https://doi.org/10.1093/nar/gkz821.

Polevoda, B., McDougall, W. M., Bennett, R. P., Salter, J. D., & Smith, H. C. (2016). Structural and functional assessment of APOBEC3G macromolecular complexes. *Methods (San Diego, Calif.), 107*, 10–22. https://doi.org/10.1016/j.ymeth.2016.03.006.

Polevoda, B., McDougall, W. M., Tun, B. N., Cheung, M., Salter, J. D., Friedman, A. E., et al. (2015). RNA binding to APOBEC3G induces the disassembly of functional deaminase complexes by displacing single-stranded DNA substrates. *Nucleic Acids Research, 43*(19), 9434–9445. https://doi.org/10.1093/nar/gkv970.

Qian, J., Wang, Q., Dose, M., Pruett, N., Kieffer-Kwon, K. R., Resch, W., et al. (2014). B cell super-enhancers and regulatory clusters recruit AID tumorigenic activity. *Cell, 159*(7), 1524–1537. https://doi.org/10.1016/j.cell.2014.11.013.

Qiao, Q., Wang, L., Meng, F. L., Hwang, J. K., Alt, F. W., & Wu, H. (2017). AID recognizes structured DNA for class switch recombination. e364 *Molecular Cell, 67*(3), 361–373. https://doi.org/10.1016/j.molcel.2017.06.034.

Quinlan, E. M., King, J. J., Amemiya, C. T., Hsu, E., & Larijani, M. (2017). Biochemical regulatory features of activation-induced cytidine deaminase remain conserved from lampreys to humans. *Molecular and Cellular Biology, 37*(20), https://doi.org/10.1128/MCB.00077-17.

Ravi, R. K., Walton, K., & Khosroheidari, M. (2018). MiSeq: A next generation sequencing platform for genomic analysis. *Methods in Molecular Biology, 1706*, 223–232. https://doi.org/10.1007/978-1-4939-7471-9_12.

Rebhandl, S., Huemer, M., Greil, R., & Geisberger, R. (2015). AID/APOBEC deaminases and cancer. *Oncoscience, 2*(4), 320–333. https://doi.org/10.18632/oncoscience.155.

Roberts, S. A., Lawrence, M. S., Klimczak, L. J., Grimm, S. A., Fargo, D., Stojanov, P., et al. (2013). An APOBEC cytidine deaminase mutagenesis pattern is widespread in human cancers. *Nature Genetics, 45*(9), 970–976. https://doi.org/10.1038/ng.2702.

Rogozin, I. B., Iyer, L. M., Liang, L., Glazko, G. V., Liston, V. G., Pavlov, Y. I., et al. (2007). Evolution and diversification of lamprey antigen receptors: Evidence for involvement of an AID-APOBEC family cytosine deaminase. *Nature Immunology, 8*(6), 647–656. https://doi.org/10.1038/ni1463.

Salter, J. D., & Smith, H. C. (2018). Modeling the embrace of a mutator: APOBEC selection of nucleic acid ligands. *Trends in Biochemical Sciences, 43*(8), 606–622. https://doi.org/10.1016/j.tibs.2018.04.013.

Sanchez, A., Ortega, P., Sakhtemani, R., Manjunath, L., Oh, S., Bournique, E., et al. (2024). Mesoscale DNA features impact APOBEC3A and APOBEC3B deaminase activity and shape tumor mutational landscapes. *Nature Communications, 15*(1), 2370. https://doi.org/10.1038/s41467-024-45909-5.

Sasaki, T., Kudalkar, S. N., Bertoletti, N., & Anderson, K. S. (2018). DRONE: Direct tracking of DNA cytidine deamination and other DNA modifying activities. *Analytical Chemistry, 90*(20), 11735–11740. https://doi.org/10.1021/acs.analchem.8b01405.

Seplyarskiy, V. B., Soldatov, R. A., Popadin, K. Y., Antonarakis, S. E., Bazykin, G. A., & Nikolaev, S. I. (2016). APOBEC-induced mutations in human cancers are strongly enriched on the lagging DNA strand during replication. *Genome Research, 26*(2), 174–182. https://doi.org/10.1101/gr.197046.115.

Serrano, J. C., von Trentini, D., Berrios, K. N., Barka, A., Dmochowski, I. J., & Kohli, R. M. (2022). Structure-guided design of a potent and specific inhibitor against the genomic mutator APOBEC3A. *ACS Chemical Biology, 17*(12), 3379–3388. https://doi.org/10.1021/acschembio.2c00796.

Sharma, S., Patnaik, S. K., Taggart, R. T., Kannisto, E. D., Enriquez, S. M., Gollnick, P., et al. (2015). APOBEC3A cytidine deaminase induces RNA editing in monocytes and macrophages. *Nature Communications, 6*, 6881. https://doi.org/10.1038/ncomms7881.

Sheehy, A. M., Gaddis, N. C., Choi, J. D., & Malim, M. H. (2002). Isolation of a human gene that inhibits HIV-1 infection and is suppressed by the viral Vif protein. *Nature, 418*(6898), 646–650. https://doi.org/10.1038/nature00939.

Shen, H. M., Ratnam, S., & Storb, U. (2005). Targeting of the activation-induced cytosine deaminase is strongly influenced by the sequence and structure of the targeted DNA. *Molecular and Cellular Biology, 25*(24), 10815–10821. https://doi.org/10.1128/MCB.25.24.10815-10821.2005.

Shen, H. M., & Storb, U. (2004). Activation-induced cytidine deaminase (AID) can target both DNA strands when the DNA is supercoiled. *Proceedings of the National Academy of Sciences of the United States of America, 101*(35), 12997–13002. https://doi.org/10.1073/pnas.0404974101.

Shi, K., Carpenter, M. A., Banerjee, S., Shaban, N. M., Kurahashi, K., Salamango, D. J., et al. (2017). Structural basis for targeted DNA cytosine deamination and mutagenesis by APOBEC3A and APOBEC3B. *Nature Structural & Molecular Biology, 24*(2), 131–139. https://doi.org/10.1038/nsmb.3344.

Sohail, A., Klapacz, J., Samaranayake, M., Ullah, A., & Bhagwat, A. S. (2003). Human activation-induced cytidine deaminase causes transcription-dependent, strand-biased C-to-U deaminations. *Nucleic Acids Research, 31*(12), 2990–2994. https://doi.org/10.1093/nar/gkg464.

Swanton, C., McGranahan, N., Starrett, G. J., & Harris, R. S. (2015). APOBEC enzymes: Mutagenic fuel for cancer evolution and heterogeneity. *Cancer Discovery, 5*(7), 704–712. https://doi.org/10.1158/2159-8290.CD-15-0344.

Taylor, B. J., Nik-Zainal, S., Wu, Y. L., Stebbings, L. A., Raine, K., Campbell, P. J., et al. (2013). DNA deaminases induce break-associated mutation showers with implication of APOBEC3B and 3A in breast cancer kataegis. *Elife, 2*, e00534. https://doi.org/10.7554/eLife.00534.

Uriu, K., Kosugi, Y., Ito, J., & Sato, K. (2021). The Battle between retroviruses and APOBEC3 genes: Its past and present. *Viruses, 13*(1), https://doi.org/10.3390/v13010124.

Wang, B., Zhang, X., Wang, Y., Chen, K., Wang, F., Weng, X., et al. (2021). One-pot fluorescent assay for sensitive detection of APOBEC3A activity. *RSC Chemical Biology, 2*(4), 1201–1205. https://doi.org/10.1039/d1cb00076d.

Xu, W. K., Byun, H., & Dudley, J. P. (2020). The role of APOBECs in viral replication. *Microorganisms, 8*(12), https://doi.org/10.3390/microorganisms8121899.

Yebra, G., & Holguin, A. (2011). Mutation Vif-22H, which allows HIV-1 to use the APOBEC3G hypermutation to develop resistance, could appear more quickly in certain non-B variants. *The Journal of Antimicrobial Chemotherapy, 66*(4), 941–942. https://doi.org/10.1093/jac/dkr012.

Yu, Q., Chen, D., Konig, R., Mariani, R., Unutmaz, D., & Landau, N. R. (2004). APOBEC3B and APOBEC3C are potent inhibitors of simian immunodeficiency virus replication. *The Journal of Biological Chemistry, 279*(51), 53379–53386. https://doi.org/10.1074/jbc.M408802200.

Zhang, H., Yang, B., Pomerantz, R. J., Zhang, C., Arunachalam, S. C., & Gao, L. (2003). The cytidine deaminase CEM15 induces hypermutation in newly synthesized HIV-1 DNA. *Nature, 424*(6944), 94–98. https://doi.org/10.1038/nature01707.

Zheng, S., Vuong, B. Q., Vaidyanathan, B., Lin, J. Y., Huang, F. T., & Chaudhuri, J. (2015). Non-coding RNA generated following lariat debranching mediates targeting of AID to DNA. *Cell, 161*(4), 762–773. https://doi.org/10.1016/j.cell.2015.03.020.

CHAPTER EIGHT

An in vitro cytidine deaminase assay to monitor APOBEC activity on DNA

Ambrocio Sanchez[a,b,c] **and Rémi Buisson**[a,b,c,d,*]

[a]Department of Biological Chemistry, School of Medicine, University of California Irvine, Irvine, California, United States

[b]Chao Family Comprehensive Cancer Center, University of California Irvine, Irvine, California, United States

[c]Center for Virus Research, University of California Irvine, Irvine, California, United States

[d]Department of Pharmaceutical Sciences, School of Pharmacy & Pharmaceutical Sciences, University of California Irvine, Irvine, California, United States

*Corresponding author. e-mail address: rbuisson@uci.edu

Contents

Abstract

APOBEC enzymes promote the deamination of cytosine (C) to uracil (U) in DNA to defend cells against viruses but also serve as a predominant source of mutations in cancer genomes. This protocol describes an assay to monitor APOBEC deaminase activity in vitro on a synthetic DNA oligonucleotide. The method described here focuses specifically on APOBEC3B to illustrate the different steps of the assay. However, the protocol can be applied to monitor the DNA deaminase activity of any other member of the APOBEC family, such as APOBEC3A. This assay involves

Methods in Enzymology, Volume 713
ISSN 0076-6879, https://doi.org/10.1016/bs.mie.2024.11.037
Copyright © 2025 Elsevier Inc. All rights are reserved, including those for text and data mining, AI training, and similar technologies.

preparing APOBEC3B-expressing cell extract or purifying APOBEC3B by immunoprecipitation, followed by incubation with a single-stranded DNA containing a TpC motif. The deaminated cytosine is then removed by recombinant Uracil DNA Glycosylase present in the reaction to form an abasic site. The abasic site creates a weakness in the DNA's backbone, causing the DNA to be cleaved under high temperatures and alkaline conditions. Denaturing gel electrophoresis is used to separate cleaved DNA from full-length DNA, enabling the quantification of the percentage of deamination induced by APOBEC3B. This protocol can be used to determine the presence of APOBEC and the regulation of APOBEC activity in specific cell lines, to study substrate preference targeted by different members of the APOBEC family and different APOBEC mutants, or to determine the efficiency and specificity of inhibitor compounds against APOBEC enzymes.

1. Introduction

The apolipoprotein B mRNA-editing enzyme catalytic polypeptide-like (APOBEC) proteins are essential components of the innate immune system that induce mutations through the deamination of cytosine (C) into uracil (U) and act as defense mechanisms against DNA or RNA viruses and retroelements by causing mutations in their genomes (Harris & Dudley, 2015; Harris & Liddament, 2004; Harris et al., 2003; Manjunath et al., 2023; Pecori et al., 2022; Petljak, Green, et al., 2022). In humans, the APOBEC family is encoded by eleven genes, including APOBEC1 (A1), APOBEC2 (A2), APOBEC3A-H (A3B, A3C, A3D, A3F, A3G, and A3H), APOBEC4 (A4), and AID (Conticello et al., 2005). Many viruses such as Epstein–Barr virus (EBV), hepatitis B virus (HBV), human immunodeficiency virus type 1 (HIV-1), rubella virus, and severe acute respiratory syndrome coronavirus 2 (SARS-CoV-2), have been found to accumulate APOBEC-driven hypermutations in their genomes, highlighting the prevalence of APOBEC enzymes in targeting viral genomes to block their replication while also contributing to viral evolution (Alteri et al., 2015; Anwar et al., 2013; Sato et al., 2014).

Cancer-focused genomic studies have identified APOBEC3-associated mutations in over 70 % of cancer types, implicating them as one of the predominant causes of genomic mutations in cancer (Burns, Temiz, et al., 2013; Leonard et al., 2013, 2015; Petljak, Dananberg, et al., 2022; Roberts et al., 2013; Swanton et al., 2015). APOBEC3-induced mutations are especially prevalent in breast, lung, cervical, bladder, and head & neck cancer genomes. Two of the eleven members of the APOBEC family, APOBEC3A and APOBEC3B, are responsible for the majority of the

APOBEC mutational signatures identified in tumor cells (Buisson et al., 2019; DeWeerd et al., 2022; Isozaki et al., 2023; Jarvis et al., 2022; Langenbucher et al., 2021; Petljak et al., 2019; Petljak, Dananberg, et al., 2022). APOBEC3A and APOBEC3B are present in the nucleus of the cells, where they have direct access to genomic DNA and cause mutations in cell genomes alongside their normal role of protecting cells against viral infections (Auerbach et al., 2022; Lackey et al., 2013; Landry et al., 2011). This ability to alter genomic information has made APOBEC3A and APOBEC3B two of the major contributors to tumor heterogeneity (Durfee et al., 2023; Law et al., 2020; Roper et al., 2019; Venkatesan et al., 2021). Beyond inducing mutations, overexpression of APOBEC3A and APOBEC3B in cancer cells leads to increased replication stress and the formation of DNA double-strand breaks (Buisson et al., 2017; Burns, Lackey, et al., 2013; Green et al., 2017; Landry et al., 2011; Leonard et al., 2013; Martínez-Ruiz et al., 2023; Oh et al., 2021; Venkatesan et al., 2021). Additionally, emerging evidence has linked APOBEC3A and APOBEC3B activity to cancer drug resistance (Caswell et al., 2024; Gupta et al., 2024; Isozaki et al., 2023; Law et al., 2016).

APOBEC3A and APOBEC3B both target thymidine followed by a cytidine (TpC motif) on RNA and single-stranded DNA (ssDNA) to induce the deamination of cytosine to uracil (C > U) (Alonso De La Vega et al., 2023; Jalili et al., 2020; Sharma et al., 2015; Shi, Carpenter, et al., 2017; Silvas et al., 2018; Stenglein et al., 2010). On DNA, APOBEC3A favors cytidine deamination on a YTC motif, while APOBEC3B prefers an RTC sequence motif (where Y is a pyrimidine and R is a purine)(Chan et al., 2015; Sanchez et al., 2024). Recently, studies revealed that both APOBEC3A and APOBEC3B preferentially target DNA motifs that adopt specific stem-loop structure configurations, causing hotspot mutations in tumor genomes (Brown et al., 2021; Buisson et al., 2019; Butt et al., 2024; Langenbucher et al., 2021; Sanchez et al., 2024). However, APOBEC3A and APOBEC3B have significant differences in the types of DNA stem-loops that they target (Butt et al., 2024; Sanchez et al., 2024). Mutations induced by APOBEC3A occur predominantly in DNA that features TpC sites in hairpin structures with 3- or 4-nucleotide (nt) loops where the cytosine is located at the 3′ end of the loop (Buisson et al., 2019; Langenbucher et al., 2021; Sanchez et al., 2024). In contrast, APOBEC3B preferentially induces mutations in hairpins with larger loops of 4 to 6 nt, and with the TpC site also located at the 3′ end of the loop. Moreover, APOBEC3A and APOBEC3B exhibit preferences for specific sequences surrounding the targeted cytosine (Butt et al., 2024; Sanchez et al., 2024). APOBEC3A and APOBEC3B predilection for certain types of DNA structures and sequences directly impacts the mutational signatures found in the genomes of cancer patients, which can be monitored to distinguish patient tumors that accumulate mutations driven by APOBEC3A or APOBEC3B (Sanchez et al., 2024).

In addition to their function of protecting cells from viral infections and promoting cancer mutagenesis, APOBEC enzymes are used to generate base editing tools (Gehrke et al., 2018; Jin et al., 2020; St. Martin et al., 2018; Wang et al., 2018). Base editors consist of the recruitment of an APOBEC enzyme or other types of DNA deaminases (e.g., TadA) to a specific location in the genome using the catalytically inactive CRISPR/dCas9 systems to modify a specific base (Gaudelli et al., 2017; Komor et al., 2016). Base editing technologies have recently transformed the potential to correct genetic diseases that were hard to treat with conventional methods by generating specific and precise point mutations in the genomic DNA of these patients (Rees & Liu, 2018). Therefore, it is important to better understand how different APOBEC enzymes recognize specific types of DNA sequences and structures to improve base editor targeting efficiency.

The in vitro APOBEC deaminase assay described in this protocol can be used to study the deaminase activity of the different APOBEC family members and better understand their function against viral infection or promoting mutations in tumors, and their application in base editor technologies.

2. Materials and reagents

2.1 Cell culture

1. CO_2 cell culture incubator
2. Dulbecco's Modified Eagle's Medium (DMEM)
3. Fetal Bovine Serum (FBS)
4. HEK-293T cell line
5. Penicillin/streptomycin
6. Phosphate Buffered Saline (PBS)
7. Tissue culture hood
8. Tissue culture plate (10 cm)
9. Trypsin-EDTA (0.05 %)
10. U2OS cell line

2.2 Reagents

1. Acrylamide/Bis 19:1 (40 %, Fisher Scientific, #BP1406-1)
2. Ammonium persulfate (30 %, Sigma Aldrich, #A682-500)
3. Anti-FLAG M2 affinity gel (Millipore-Sigma, #A2220)
4. Benzonase nuclease (≥250 units/μL, Millipore-Sigma, #E1014-25KU)
5. Bio-Rad Protein Assay Dye Reagent Concentrate (BioRad, #5000006)
6. Bromophenol blue (Fisher Scientific, #B392-5)
7. EDTA (0.5 M, pH 8, Fisher Scientific, #AM9260G)
8. Formamide (Fisher Scientific, #BP227-500)
9. Glacial acetic acid (Fisher Scientific, #A38-212)
10. Glycerol (Fisher Scientific, #G33-1)
11. H_2O (UltraPure Distilled water DNase, RNase free, Invitrogen #10977)
12. HEPES (1 M, pH 7.4, Sigma Aldrich, #7365-45-9)
13. Igepal CA-630 (MP Biomedicals, #198596)
14. Lipofectamine 2000 transfection reagent (Invitrogen, #11668019)
15. $MgCl_2$ (1 M, Fisher Scientific, #AM9530G)
16. Na_3VO_4 (1 M, Sigma Aldrich, #S6508)
17. NaCl (5 M, Fisher Scientific, #J60434. AK)
18. NaF (0.5 M, Sigma Aldrich, #67414-mL-F)
19. NaOH (10 N, Sigma Aldrich #SX0607N)
20. TEMED (Fisher Scientific, #BP150-100)
21. Protease inhibitor cocktail (100X, Sigma Aldrich, #P8340)
22. RNase A (20 mg/mL, Invitrogen, #12091-021)
23. Tris-HCl (1 M, pH 7.5, Invitrogen, #15567027)
24. Triton X-100 (Sigma Aldrich, #X100-1L)
25. Uracil DNA Glycosylase (5000 units/mL New England BioLabs, #M0280S)
26. Urea (Fisher Scientific, #U15-3)
27. $ZnCl_2$ (0.5 M, Sigma Aldrich, #793523)
28. 3X FLAG Peptide (5 mg/mL, Sigma Aldrich, #SAE0194)

2.3 Equipment and materials

1. Centrifuge
2. Gel imaging system (with 6-FAM fluorescence detection setup)
3. Conical tubes (15 mL)
4. Bottle-top vacuum filter system (0.45 μm)
5. Heated circulating bath
6. High voltage power supply

7. Magnetic hotplate stirrer
8. Microfuge tubes (1.5 mL and 200 μL)
9. Needle 25-gauge, 5/8-inch
10. Rotator-shaker
11. Sonicator
12. Syringe (1.0 mL)
13. Thermocycler
14. Vertical gel electrophoresis apparatus (large format: 16 × 20 cm gel)

3. Methods

3.1 Overview

The following protocol describes, step by step, how to monitor the DNA deaminase activity of APOBEC enzymes on a DNA oligonucleotide labeled with a fluorophore. To illustrate the assay, the protocol focuses specifically on APOBEC3B deaminase activity. However, this protocol can also be applied to monitor the DNA deaminase activity of other members of the APOBEC family.

The first step of the protocol involves generating a cell extract that contains either endogenous APOBEC3B or overexpressed APOBEC3B. As an alternative to the cell extract method, APOBEC3B can be purified to separate APOBEC3B from other cellular factors. Next, the cell extract is mixed with a DNA oligonucleotide that has a single TpC motif in its sequence. APOBEC3B specifically deaminates the TpC motif on the DNA oligonucleotide, and the presence of recombinant Uracil DNA Glycosylase (UDG) in the reaction will remove the uracil from the DNA, producing an abasic site. The resulting abasic site creates a weakness in the DNA's backbone, causing the DNA to be cleaved under alkaline and high-temperature conditions (Fig. 1A). Finally, the DNA is loaded onto a denaturing polyacrylamide gel to separate the cleaved DNA oligonucleotide from the full-length DNA. The quantification of the percentage of cleaved DNA relative to the total amount of DNA determines the deaminase activity levels of APOBEC3B present in the sample (Fig. 1B).

3.2 Design of the DNA oligonucleotides

APOBEC3B targets the TpC motif on single-stranded DNA (ssDNA) or within the loop of a DNA hairpin structure, with the cytosine located at the 3′ position of the loop (Sanchez et al., 2024). Moreover, APOBEC3B

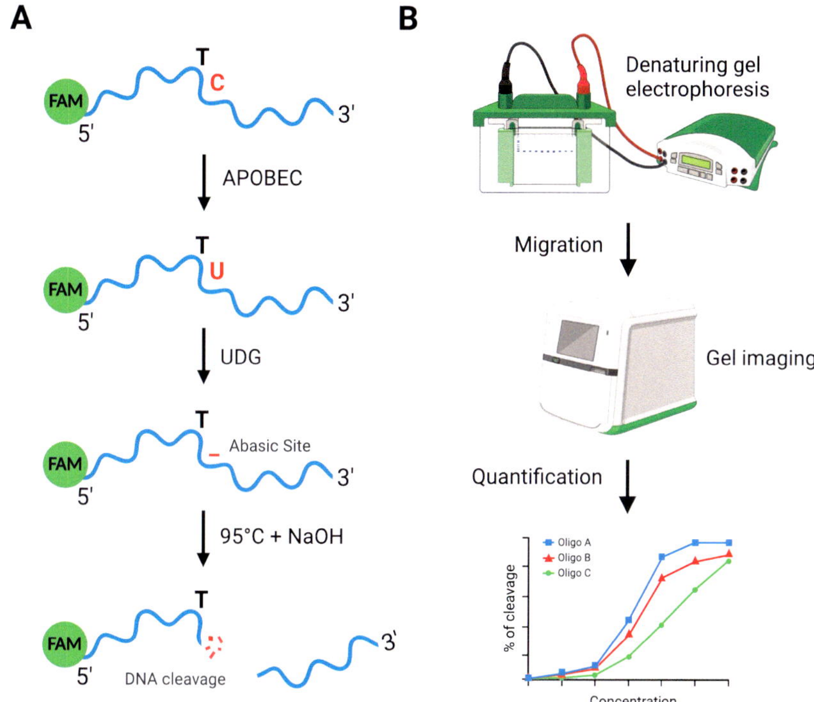

Fig. 1 Schematic of methodology to perform the cytidine deaminase assay. (A) Schematic of the in vitro assay to measure APOBEC cytidine deamination activity. (B) Schematic of the methodology to process and quantify samples after performing the in vitro APOBEC deaminase activity assay. Created with BioRender.com.

preferentially targets stem-loop structures with a loop of 5 nucleotides and 5'-CCG or 5'-CCA sequence preceding the TpC site (Butt et al., 2024; Sanchez et al., 2024). In this protocol, both a ssDNA and a DNA hairpin substrate are used to illustrate how to perform and quantify APOBEC activity with the in vitro cytidine DNA deaminase assay.

The synthetic DNA oligonucleotide sequences used in this protocol are:

ssDNA: 5′-(6-FAM)-GCAAGCTGG**TC**GGAAAATGA-3′.

Stem-loop DNA: 5′-(6-FAM)-GCAAGCCCCG**TC**GGCTTGCTGA-3′.

The recommended length of the ssDNA is between 20 to 25 nucleotides with a TpC motif located in the middle (highlighted in green in the sequences above) for optimal separation between the full-length and the cleaved DNA after the deamination assay (see ***section 3.6, step 8***). For the stem-loop DNA substrate, the sequence that forms the stem is underlined.

A 6-FAM (6-Carboxyfluorescein) fluorophore is used for the detection of the DNA after performing the DNA deaminase assay reactions (see ***section 3.6, step 11***). The FAM fluorophore is added to the 5' end of the DNA oligonucleotide rather than the 3' end to prevent the degradation of the oligonucleotide by 5' exonucleases present in the cell extract, which are found to be more potent than 3' exonucleases. Using this strategy, little to no DNA degradation is observed after 1-hour incubation at 37 °C of the DNA with cell extract.

3.3 Preparation of APOBEC3B expressing cell extracts or purified APOBEC3B

The decision between selecting cell extract or purified APOBEC3B depends on the specific context in which APOBEC enzyme is being studied. Not all cell lines express APOBEC3B, or the endogenous levels of APOBEC3B may be too low to be detected using the cell extract method. In such cases, ectopic expression of APOBEC3B is required. Additionally, the cell line used should not express other APOBEC members that could potentially compete and deaminate the TpC site or other cytosines within the DNA oligonucleotide described above. This is especially important when APOBEC3A is also present because it also targets the TpC motif on DNA (Kouno et al., 2017; Shi et al., 2016). Therefore, in such a situation, the purification method would be recommended to separate APOBEC3B from other APOBEC enzymes present in the cells. For this protocol, we selected U2OS cells that express high levels of endogenous APOBEC3B and HEK-293T cells that do not express either APOBEC3A or APOBEC3B.

3.3.1 Option 1: cell lysate preparation expressing endogenous APOBEC3B

1. U2OS cells are cultured in DMEM supplemented with 10 % fetal bovine serum (FBS) and 1 % penicillin/streptomycin.
2. Collect U2OS cells by trypsinization and add the cells into a 15 mL conical tube.
3. Pellet the cells by centrifuging at 300 g for 3 min at room temperature and aspirate the media.
4. Wash the cells with PBS 1X (4 °C) and centrifuge at 300 g for 3 min

5. Repeat the wash and centrifugation steps and carefully aspirate the supernatant without disturbing the pellet.
6. Resuspend the cells in 300 μL of lysis buffer (25 mM HEPES (pH 7.9), 10 % glycerol, 150 mM NaCl, 1 mM EDTA, 1 mM $MgCl_2$, and 1 mM $ZnCl_2$, 0.5 % Triton X-100, and 1X of Protease Inhibitor Cocktail) and incubate at 4 °C for 5 min.
7. Sonicate the sample twice at 60 % amplitude for 10 s at 4 °C.
8. Centrifuge the sample at 20,000 g for 5 min at 4 °C to remove insoluble material.
9. Transfer the supernatant to a 1.5 mL microfuge tube.
10. Add 6 μL of RNase A (20 mg/mL) and incubate for 20 min at 4 °C. (see **note 4.1**)
11. Centrifuge the sample at 20,000 g for 10 min at 4 °C to remove insoluble material and transfer the supernatant to a new 1.5 mL microfuge tube.
12. Perform the Bradford protein assay as described per the manufacturer's instructions to quantify cell lysate protein concentration.
13. Aliquot 20 μL samples into new 1.5 mL microfuge tubes and freeze on dry ice or liquid nitrogen.
14. Store samples at −80 °C until ready to perform the deaminase reaction assay. (see **note 4.2**)
15. Proceed to **section 3.4, step 1**.

3.3.2 Option 2: ectopic expression of APOBEC3B in HEK-293T cells

1. HEK-293T cells are maintained and grown in DMEM supplemented with 10 % FBS, and 1 % penicillin/streptomycin.
2. Seed HEK-293T cells in two 10 cm tissue culture plates at 30-40 % confluency and culture them for 16-20 h to achieve ~60-70 % confluency.
3. Transfect each plate with an APOBEC3B-expressing vector (e.g., pcDNA3.1-APOBEC3B-Flag) using Lipofectamine 2000 per manufacturer's instruction. (see **note 4.3**)
4. Replace the transfected cells' media with fresh media containing penicillin/streptomycin 8 h following transfection.
5. The next day (16–20 h post-transfection), collect the transfected HEK-293T cells expressing APOBEC3B into a 15 mL conical tube.
6. Pellet the cells by centrifuging at 300 g for 3 min at room temperature and aspirate the media.
7. Wash the cells with PBS 1X (4 °C) and centrifuge at 300 g for 3 min

8. Repeat the wash and centrifugation step and aspirate all PBS 1X without disturbing the pellet.
9. Proceed as described in **Section 3.3.1, steps 6 to 15**.

3.3.3 Option 3: APOBEC3B purification by immunoprecipitation (IP)

1. Transfect six 10 cm tissue culture plates of HEK-293T cells with APOBEC3B-expressing vector as described in **Section 3.3.2**.
2. Collect the HEK-293T cells into a 15 mL conical tube.
3. Pellet the cells by centrifuging at 300 g for 3 min at room temperature and aspirate the media.
4. Wash the cells with PBS 1X (4 °C) and centrifuge at 300 g for 3 min
5. Repeat the wash and centrifugation steps and carefully aspirate the supernatant without disturbing the pellet.
6. Resuspend the cells in 2 mL of IP lysis buffer (50 mM Tris-HCl [pH 7.5], 150 mM NaCl, 1 mM EDTA, and 0.5 % Igepal containing 1X of Protease Inhibitor Cocktail and protease phosphatase inhibitors [5 mM NaF, and 1 mM Na_3VO_4]) and incubate at 4 °C for 5 min.
7. Sonicate the sample twice at 60 % amplitude for 10 s at 4 °C.
8. Centrifuge the samples at 20,000 g for 5 min at 4 °C to remove insoluble material.
9. Transfer the supernatant to a 1.5 mL microfuge tube.
10. Add 40 μL of RNase A (20 mg/mL) and 2 μL of Benzonase (≥250 units/μL) and incubate for 30 min at 4 °C.
11. Centrifuge the samples at 20,000 g for 10 min at 4 °C to remove insoluble material and transfer the supernatant to a new 1.5 mL microfuge tube.
12. Add 150 μL of anti-flag, M2 affinity gel beads (equilibrated in the IP lysis buffer beforehand by washing twice the beads with PBS) to the cell lysates and then incubate for 2.5 h at 4 °C on a rotator shaker.
13. Centrifuge the beads for 1 min at 400 g at 4 °C
14. Wash the beads with 1 mL of IP wash buffer (50 mM Tris-HCl (pH 7.5), 350 mM NaCl, 1 mM EDTA, and 0.5 % Igepal, 5 mM NaF, and 1 mM Na_3VO_4) and centrifuge 1 min at 400 g at 4 °C.
15. Repeat the wash step three times.
16. After the final centrifugation step, remove all the IP washing buffer (use a 1.0 mL syringe with a 25-gauge, 5/8 in. needle to remove residual buffer).
17. Add 53 μL of IP elution buffer (25 mM HEPES (pH 7.9), 10 % glycerol, 150 mM NaCl, 1 mM EDTA, 1 mM $MgCl_2$, and 1 mM $ZnCl_2$),

1 μL of Protease Inhibitor Cocktail 100X, and 6 μL of 3X flag peptide (5 mg/mL) to the protein-bound beads and incubate for 3 h at 4 °C on a rotator shaker.

18. Centrifuge the samples at 400 g for 2 min at 4 °C.
19. Collect the supernatant and aliquot the samples (10 μL) in 1.5 mL microfuge tubes.
20. Evaluate the pulldown efficiency by western blot using an anti-Flag antibody.
21. Store the samples at −80 °C until ready to perform the deaminase reaction assay. (see **note 4.2**).

3.4 *In vitro* deaminase reaction assay

1. In 1.5 mL microfuge tubes, add 42 μL of deaminase reaction buffer (50 mM Tris-HCl (pH 7.5), 10 mM EDTA (pH 8), 30 mM NaCl, 0.25 mM $ZnCl_2$) containing 1.5 units of uracil DNA glycosylase and 20 pmol of DNA oligonucleotide (see **note 4.4**) and 8 μL of cell extract (5 to 30 μg) or purified APOBEC3B. Adjust the amount of cell extract or purified APOBEC3B depending on the cell lysate protein concentration, IP efficiency, or the percentage of cleavage product obtained at the end of the method. If less than 8 μL of cell extract or purified APOBEC3B is used in the reaction, complement the reaction up to 8 μL with lysis buffer or IP elution buffer. (see **note 4.5**)
2. Incubate the samples for 1 h at 37 °C using a thermocycler.
3. Add 0.5 μL of NaOH (10 N), mix, and incubate the samples for 40 min at 95 °C using a thermocycler.
4. Add 50 μL of formamide (50 % final) containing bromophenol blue.
5. Store the samples at −20 °C or −80 °C until ready for the next step.

3.5 Denaturing urea polyacrylamide gel electrophoresis

6. To prepare a TAE-Urea (8 M) 20 % polyacrylamide solution, add 24 g of urea powder to 25 mL of a 40 % bis-Acrylamide 19:1 solution.
7. Heat the solution using a magnetic hotplate stirrer until the urea is completely dissolved.
8. Add 1 mL of 50X TAE buffer (2 M Tris-base, 1 M Glacial Acetic Acid, and 50 mM EDTA) and complete the solution to 50 mL with H_2O.
9. Filter the urea polyacrylamide solution using a bottle-top vacuum filter system (0.45 μm).
10. Cool the TAE-Urea (8 M) 20 % polyacrylamide solution to 25 °C or store the solution at room temperature until use.

11. To cast the gel, mix 50 mL of TAE-Urea (8 M) 20 % polyacrylamide solution with 500 µL of 30 % APS, and 50 µL TEMED.
12. Add the solution to the gel casting system (e.g., BIO-RAD Protean II xi #1651815 or equivalent) per the manufacturer's instruction. After pouring the solution, insert the comb into the gel. (see **note 4.6**)
13. Allow gel polymerization for 30 min at room temperature.

3.6 Sample separation and visualization

14. After removing the comb and washing the wells several times, install the polymerized gel into a vertical electrophoresis system (e.g., BIO-RAD Protean II xi Cell apparatus system #1651815 or equivalent) per the manufacturer's instructions.
15. Add 1X TAE buffer to the buffer chambers of the electrophoresis apparatus system.
16. Connect the central core of the electrophoresis apparatus system to a circulating bath (e.g., ThermoFisher Scientific SAHARA S13 / SC150 #153-1138 or equivalent) set up at 60 °C and pre-heat the system for 10 min
17. Pre-run the gel at 250 V for 20 min using a high-voltage power supply (e.g., Bio-Rad PowerPac Universal Power Supply #1645070 or equivalent).
18. 10 min before the end of the pre-run, heat the samples (from **section 3.4, step 5**) for 5 min at 95 °C using a thermocycler and then incubate the samples at 4 °C before loading.
19. After the pre-run, extensively wash the wells of the polyacrylamide gel with 1X-TAE buffer.
20. Load 8 to 12 µL of the samples into the wells and add loading buffer (50 % formamide) to the empty wells next to the samples (this improves the quality of the samples' migration).
21. Run the samples at 250 V at 60 °C for 1 h. (see **note 4.7**)
22. Remove the polyacrylamide gel from the glasses and incubate the gel in H_2O for 5 min at room temperature.
23. To detect DNA cleavage mediated by APOBEC3B deaminase activity, place the polyacrylamide gel into a gel imaging system that can detect FAM fluorescence (e.g., Bio-Rad Chemidoc MP Imaging System #12003154 or equivalent). (see **note 4.8**)
24. Set the gel imaging system to Fluorescein (excitation 495 nm, Emission 517 nm) or Alexa-488 setting (excitation 495 nm, emission 519 nm) to detect FAM-labeled DNA (full-length and/or cleaved) (Fig. 2A). (see **note 4.9**)

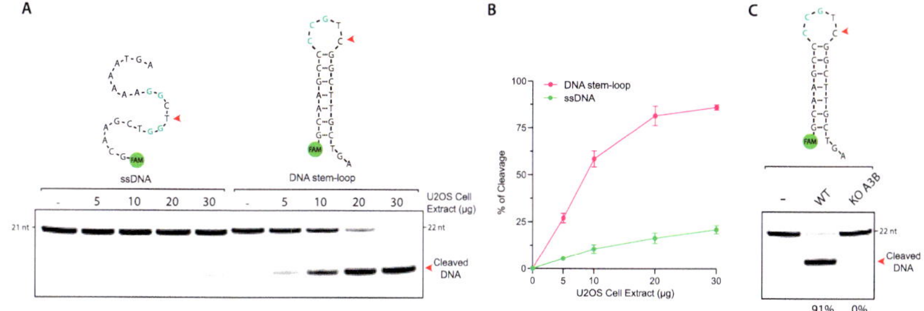

Fig. 2 In vitro cytidine deaminase assay to monitor APOBEC3B activity. (A) APOBEC3B deamination activity was monitored on the ssDNA or DNA stem-loop substrates using the indicated amount of U2OS whole-cell extract expressing endogenous APOBEC3B. (B) Quantification of the APOBEC3B deamination activity shown in (A). Data are presented as mean values ± S.D. (C) APOBEC3B deamination activity assay using 30 μg of U2OS whole cell extract of wild type or APOBEC3B knockout cells on the DNA stem-loop oligonucleotide shown. The percentage of cleavage is indicated.

25. Quantify the percentage of DNA deamination by measuring the intensity of the DNA band corresponding to the cleaved DNA divided by the total amount of substrate in the reaction (sample without APOBEC3B) (Fig. 2B and C). (see **note 4.10**)

4. Notes

4.1 In ***section 3.3.1, step 10***, RNase A is added to the reaction because RNA inhibits APOBEC3B deaminase activity (Cortez et al., 2019).

4.2 In ***section 3.3.1, step 14*** and ***section 3.3.3, step 21,*** cell extracts or purified APOBEC can be stored at −80 °C for at least a month without compromising the results.

4.3 In ***section 3.3.2, step 3***, other types of transfection methods can be used, including calcium phosphate transfection for HEK-293T cells, as an alternative to the Lipofectamine 2000 method. The transfection protocol has to be optimized for different cell lines.

4.4 In ***section 3.4, step 1***, if studying other APOBEC family members, the DNA oligonucleotide sequence used in the in vitro DNA deamination assay needs to be adapted accordingly to the enzyme sequence preference (e.g., CCC motif for APOBEC3G (Yu et al., 2004)).

4.5 In ***section 3.4, step 1***, the amount of cell extract used in the assay needs to be adapted depending on the levels of expression of APOBEC or which APOBEC family member is expressed.

4.6 In ***section 3.5, step*** 7, larger gel casting systems are recommended to ensure optimal migration of the DNA oligonucleotide in the denaturing polyacrylamide gel. In this protocol, a 16 cm × 20 cm denaturing polyacrylamide gel with a thickness of 1 mm was used.

4.7 In ***section 3.6, step 8***, the migration time should be adapted based on the size of the DNA oligonucleotide used in the reaction.

4.8 In ***section 3.6, step 10***, in function on the gel imager system's capability, other types of fluorophores, such as TAMRA, Cy3, or Cy5, can be used as an alternative to FAM-labeled DNA oligonucleotide.

4.9 In ***section 3.6, step 11***, unlabeled DNA substrates can be detected by incubating the gel in a SYBR Green (Invitrogen, #S7563) or SYBR Gold (Invitrogen, #S11494) solution (1/10,000 in TAE 1X) for 10 min at room temperature. However, the quality of the staining and the sensitivity of detection will be strongly decreased compared to fluorophore-labeled DNA oligonucleotides.

4.10 In ***section 3.6, step 12***, it is recommended to perform the DNA deaminase assay with control cell extract that does not express any APOBEC3B or other APOBEC family members, depending on the type of DNA oligonucleotide used.

5. Conclusions

This protocol provides a detailed, step-by-step method of an in vitro cytidine deaminase assay to monitor APOBEC activity on DNA. Successful detection of APOBEC deaminase activity should result in a clear separation of a faster migrating band corresponding to the cleaved DNA compared to the full-length DNA oligonucleotide. The percentage of DNA deamination on the substrate may highly vary depending on the cell line used, the levels of APOBEC expression in cells, the amount of pull-down APOBEC protein, or which APOBEC protein is studied. For example, APOBEC3A is significantly more active than APOBEC3B (Cortez et al., 2019; Shi, Demir, et al., 2017; Shi et al., 2015, 2016). Therefore, a decreased amount of cell extract expressing APOBEC3A or purified proteins is required to perform the cytidine deaminase assay. Moreover, the types of DNA substrates used in the in vitro assay can also considerably affect the levels of DNA deamination. Indeed, the DNA sequence surrounding the cytidine and the DNA conformation may enhance or prevent APOBEC activity (Sanchez et al., 2024). The sequence of DNA oligonucleotides used in the reaction has to be selected and adapted depending on which APOBEC member is studied.

The in vitro cytidine deaminase assay described in this protocol can be used to detect APOBEC activity and its regulation in different cell lines, investigate substrate preferences of various APOBEC family members, and evaluate the effectiveness and specificity of inhibitors targeting APOBEC enzymes. This assay, which does not require high-end equipment, is accessible to any molecular biology laboratory and can be used to better understand APOBEC functions against viral infection and promoting tumor mutations or to optimize the use of APOBEC in base editor technologies.

Acknowledgments

We thank Dr. Elodie Bournique and Dr. Pedro Ortega for their comments on the manuscript. A.S. was supported by a National Institutes of Health Research Supplements to Promote Diversity in Health-Related Research (NCI R37-CA252081-S1). R.B. was supported by a MERIT Award from the National Institutes of Health (NCI R37-CA252081) and a Research Scholar Grant from the American Cancer Society (RSG-24–1249960-01-DMC).

References

Alonso De La Vega, A., Temiz, N. A., Tasakis, R., Somogyi, K., Salgueiro, L., Zimmer, E., ... Sotillo, R. (2023). Acute expression of human APOBEC3B in mice results in RNA editing and lethality. *Genome Biology, 24*(1), 267. https://doi.org/10.1186/S13059-023-03115-4.

Alteri, C., Surdo, M., Bellocchi, M. C., Saccomandi, P., Continenza, F., Armenia, D., ... Svicher, V. (2015). Incomplete APOBEC3G/F neutralization by HIV-1 Vif mutants facilitates the genetic evolution from CCR5 to CXCR4 usage. *Antimicrobial Agents and Chemotherapy, 59*(8), 4870–4881. https://doi.org/10.1128/AAC.00137-15.

Anwar, F., Davenport, M. P., & Ebrahimi, D. (2013). Footprint of APOBEC3 on the genome of human retroelements. *Journal of Virology, 87*(14), 8195–8204. https://doi.org/10.1128/JVI.00298-13.

Auerbach, A. A., Becker, J. T., Moraes, S. N., Moghadasi, S. A., Duda, J. M., Salamango, D. J., & Harris, R. S. (2022). Ancestral APOBEC3B nuclear localization is maintained in humans and apes and altered in most other old world primate species. *MSphere*. https://doi.org/10.1128/MSPHERE.00451-22.

Brown, A. L., Collins, C. D., Thompson, S., Coxon, M., Mertz, T. M., & Roberts, S. A. (2021). Single-stranded DNA binding proteins influence APOBEC3A substrate preference. *Scientific Reports, 11*(1), https://doi.org/10.1038/S41598-021-00435-Y.

Buisson, R., Langenbucher, A., Bowen, D., Kwan, E. E., Benes, C. H., Zou, L., & Lawrence, M. S. (2019). Passenger hotspot mutations in cancer driven by APOBEC3A and mesoscale genomic features. *Science (New York, N. Y.), 364*(6447), https://doi.org/10.1126/science.aaw2872.

Buisson, R., Lawrence, M. S., Benes, C. H., & Zou, L. (2017). APOBEC3A and APOBEC3B activities render cancer cells susceptible to ATR inhibition. *Cancer Research,* 77(17), https://doi.org/10.1158/0008-5472.CAN-16-3389.

Burns, M. B., Lackey, L., Carpenter, M. A., Rathore, A., Land, A. M., Leonard, B., ... Harris, R. S. (2013). APOBEC3B is an enzymatic source of mutation in breast cancer. *Nature, 494*(7437), 366–370. https://doi.org/10.1038/nature11881.

Burns, M. B., Temiz, N. A., & Harris, R. S. (2013). Evidence for APOBEC3B mutagenesis in multiple human cancers. *Nature Genetics, 45*(9), 977–983. https://doi.org/10.1038/ng.2701.

Butt, Y., Sakhtemani, R., Mohamad-Ramshan, R., Lawrence, M. S., & Bhagwat, A. S. (2024). Distinguishing preferences of human APOBEC3A and APOBEC3B for cytosines in hairpin loops, and reflection of these preferences in APOBEC-signature cancer genome mutations. *Nature Communications, 15*(1), https://doi.org/10.1038/S41467-024-46231-W.

Caswell, D. R., Gui, P., Mayekar, M. K., Law, E. K., Pich, O., Bailey, C., ... Swanton, C. (2024). The role of APOBEC3B in lung tumor evolution and targeted cancer therapy resistance. *Nature Genetics, 56*(1), 60–73. https://doi.org/10.1038/S41588-023-01592-8.

Chan, K., Roberts, S. A., Klimczak, L. J., Sterling, J. F., Saini, N., Malc, E. P., ... Gordenin, D. A. (2015). An APOBEC3A hypermutation signature is distinguishable from the signature of background mutagenesis by APOBEC3B in human cancers. *Nature Genetics, 47*(9), 1067–1072. https://doi.org/10.1038/ng.3378.

Conticello, S. G., Thomas, C. J. F., Petersen-Mahrt, S. K., & Neuberger, M. S. (2005). Evolution of the AID/APOBEC family of polynucleotide (deoxy)cytidine deaminases. *Molecular Biology and Evolution, 22*(2), 367–377. https://doi.org/10.1093/MOLBEV/MSI026.

Cortez, L. M., Brown, A. L., Dennis, M. A., Collins, C. D., Brown, A. J., Mitchell, D., ... Roberts, S. A. (2019). APOBEC3A is a prominent cytidine deaminase in breast cancer. *PLoS Genetics, 15*(12), e1008545. https://doi.org/10.1371/journal.pgen.1008545.

DeWeerd, R. A., Németh, E., Póti, Á., Petryk, N., Chen, C. L., Hyrien, O., ... Green, A. M. (2022). Prospectively defined patterns of APOBEC3A mutagenesis are prevalent in human cancers. *Cell Reports, 38*(12), https://doi.org/10.1016/j.celrep.2022.110555.

Durfee, C., Temiz, N. A., Levin-Klein, R., Argyris, P. P., Alsøe, L., Carracedo, S., ... Harris, R. S. (2023). Human APOBEC3B promotes tumor development in vivo including signature mutations and metastases. *Cell Reports. Medicine, 4*(10), https://doi.org/10.1016/J.XCRM.2023.101211.

Gaudelli, N. M., Komor, A. C., Rees, H. A., Packer, M. S., Badran, A. H., Bryson, D. I., & Liu, D. R. (2017). Programmable base editing of A•T to G•C in genomic DNA without DNA cleavage. *Nature, 551*(7681), 464–471. https://doi.org/10.1038/NATURE24644.

Gehrke, J. M., Cervantes, O., Clement, M. K., Wu, Y., Zeng, J., Bauer, D. E., ... Joung, J. K. (2018). An APOBEC3A-Cas9 base editor with minimized bystander and off-target activities. *Nature Biotechnology, 36*(10), 977. https://doi.org/10.1038/NBT.4199.

Green, A. M., Budagyan, K., Hayer, K. E., Reed, M. A., Savani, M. R., Wertheim, G. B., & Weitzman, M. D. (2017). Cytosine deaminase APOBEC3A sensitizes leukemia cells to inhibition of the DNA replication checkpoint. *Cancer Research,* 77(17), 4579–4588. https://doi.org/10.1158/0008-5472.CAN-16-3394.

Gupta, A., Gazzo, A., Selenica, P., Safonov, A., Pareja, F., da Silva, E. M., ... Chandarlapaty, S. (2024). APOBEC3 mutagenesis drives therapy resistance in breast cancer. *BioRxiv: The Preprint Server for Biology.* https://doi.org/10.1101/2024.04.29.591453.

Harris, R. S., Bishop, K. N., Sheehy, A. M., Craig, H. M., Petersen-Mahrt, S. K., Watt, I. N., ... Malim, M. H. (2003). DNA deamination mediates innate immunity to retroviral infection. *Cell, 113*(6), 803–809. https://doi.org/10.1016/S0092-8674(03)00423-9.

Harris, R. S., & Dudley, J. P. (2015). *APOBECs and virus restriction. Virology, Vols. 479–480,* Academic Press Inc, 131–145. https://doi.org/10.1016/j.virol.2015.03.012.

Harris, R. S., & Liddament, M. T. (2004). Retroviral restriction by APOBEC proteins. *Nature Reviews Immunology, 4*(11), 868–877. https://doi.org/10.1038/nri1489.

Isozaki, H., Sakhtemani, R., Abbasi, A., Nikpour, N., Stanzione, M., Oh, S., ... Hata, A. N. (2023). Therapy-induced APOBEC3A drives evolution of persistent cancer cells. *Nature*. https://doi.org/10.1038/S41586-023-06303-1.

Jalili, P., Bowen, D., Langenbucher, A., Park, S., Aguirre, K., Corcoran, R. B., ... Buisson, R. (2020). Quantification of ongoing APOBEC3A activity in tumor cells by monitoring RNA editing at hotspots. *Nature Communications, 11*(1), 2971. https://doi.org/10.1038/s41467-020-16802-8.

Jarvis, M. C., Carpenter, M. A., Temiz, N. A., Brown, M. R., Richards, K. A., Argyris, P. P., ... Harris, R. S. (2022). Mutational impact of APOBEC3B and APOBEC3A in a human cell line. *BioRxiv, 04*(26), 489523. https://doi.org/10.1101/2022.04.26.489523.

Jin, S., Fei, H., Zhu, Z., Luo, Y., Liu, J., Gao, S., ... Gao, C. (2020). Rationally designed APOBEC3B cytosine base editors with improved specificity. *Molecular Cell, 79*(5), 728–740.e6. https://doi.org/10.1016/j.molcel.2020.07.005.

Komor, A. C., Kim, Y. B., Packer, M. S., Zuris, J. A., & Liu, D. R. (2016). Programmable editing of a target base in genomic DNA without double-stranded DNA cleavage. *Nature, 533*(7603), 420–424. https://doi.org/10.1038/NATURE17946.

Kouno, T., Silvas, T. V., Hilbert, B. J., Shandilya, S. M. D., Bohn, M. F., Kelch, B. A., ... Schiffer, C. A. (2017). Crystal structure of APOBEC3A bound to single-stranded DNA reveals structural basis for cytidine deamination and specificity. *Nature Communications, 8,* 15024. https://doi.org/10.1038/ncomms15024.

Lackey, L., Law, E. K., Brown, W. L., & Harris, R. S. (2013). Subcellular localization of the APOBEC3 proteins during mitosis and implications for genomic DNA deamination. *Cell Cycle (Georgetown, Tex.), 12*(5), 762–772. https://doi.org/10.4161/cc.23713.

Landry, S., Narvaiza, I., Linfesty, D. C., & Weitzman, M. D. (2011). APOBEC3A can activate the DNA damage response and cause cell-cycle arrest. *EMBO Reports, 12*(5), 444–450. https://doi.org/10.1038/embor.2011.46.

Langenbucher, A., Bowen, D., Sakhtemani, R., Bournique, E., Wise, J. F., Zou, L., ... Lawrence, M. S. (2021). An extended APOBEC3A mutation signature in cancer. *Nature Communications, 12*(1), https://doi.org/10.1038/s41467-021-21891-0.

Law, E. K., Levin-Klein, R., Jarvis, M. C., Kim, H., Argyris, P. P., Carpenter, M. A., ... Harris, R. S. (2020). APOBEC3A catalyzes mutation and drives carcinogenesis in vivo. *The Journal of Experimental Medicine, 217*(12), https://doi.org/10.1084/jem.20200261.

Law, E. K., Sieuwerts, A. M., LaPara, K., Leonard, B., Starrett, G. J., Molan, A. M., ... Harris, R. S. (2016). The DNA cytosine deaminase APOBEC3B promotes tamoxifen resistance in ER-positive breast cancer. *Science Advances, 2*(10), e1601737. https://doi.org/10.1126/sciadv.1601737.

Leonard, B., Hart, S. N., Burns, M. B., Carpenter, M. A., Temiz, N. A., Rathore, A., ... Bentley, D., ... Harris, R. S. (2013). APOBEC3B upregulation and genomic mutation patterns in serous ovarian carcinoma. *Cancer Research, 73*(24), 7222–7231. https://doi.org/10.1158/0008-5472.CAN-13-1753.

Leonard, B., McCann, J. L., Starrett, G. J., Kosyakovsky, L., Luengas, E. M., Molan, A. M., ... Harris, R. S. (2015). The PKC/NF-κB signaling pathway induces APOBEC3B expression in multiple human cancers. *Cancer Research, 75*(21), 4538–4547. https://doi.org/10.1158/0008-5472.CAN-15-2171-T.

Manjunath, L., Oh, S., Ortega, P., Bouin, A., Bournique, E., Sanchez, A., ... Buisson, R. (2023). APOBEC3B drives PKR-mediated translation shutdown and protects stress granules in response to viral infection. *Nature Communications, 14*(1), https://doi.org/10.1038/S41467-023-36445-9.

Martínez-Ruiz, C., Black, J. R. M., Puttick, C., Hill, M. S., Demeulemeester, J., Larose Cadieux, E., ... McGranahan, N. (2023). Genomic-transcriptomic evolution in lung cancer and metastasis. *Nature, 616*(7957), 543–552. https://doi.org/10.1038/S41586-023-05706-4.

Oh, S., Bournique, E., Bowen, D., Jalili, P., Sanchez, A., Ward, I., ... Buisson, R. (2021). Genotoxic stress and viral infection induce transient expression of APOBEC3A and pro-inflammatory genes through two distinct pathways. *Nature Communications, 12*(1), https://doi.org/10.1038/S41467-021-25203-4.

Pecori, R., Di Giorgio, S., Paulo Lorenzo, J., & Nina Papavasiliou, F. (2022). Functions and consequences of AID/APOBEC-mediated DNA and RNA deamination. *Nature Reviews. Genetics.* https://doi.org/10.1038/S41576-022-00459-8.

Petljak, M., Alexandrov, L. B., Brammeld, J. S., Price, S., Wedge, D. C., Grossmann, S., ... Stratton, M. R. (2019). Characterizing mutational signatures in human cancer cell lines reveals episodic APOBEC mutagenesis. *Cell, 176*(6), 1282–1294.e20. https://doi.org/10.1016/j.cell.2019.02.012.

Petljak, M., Dananberg, A., Chu, K., Bergstrom, E. N., Striepen, J., von Morgen, P., ... Maciejowski, J. (2022). Mechanisms of APOBEC3 mutagenesis in human cancer cells. *Nature, 607*(7920), 799–807. https://doi.org/10.1038/S41586-022-04972-Y.

Petljak, M., Green, A. M., Maciejowski, J., & Weitzman, M. D. (2022). Addressing the benefits of inhibiting APOBEC3-dependent mutagenesis in cancer. *Nature Genetics, 54*(11), 1599–1608. https://doi.org/10.1038/S41588-022-01196-8.

Rees, H. A., & Liu, D. R. (2018). Base editing: Precision chemistry on the genome and transcriptome of living cells. *Nature Reviews. Genetics, 19*(12), 770. https://doi.org/10.1038/S41576-018-0059-1.

Roberts, S. A., Lawrence, M. S., Klimczak, L. J., Grimm, S. A., Fargo, D., Stojanov, P., ... Gordenin, D. A. (2013). An APOBEC cytidine deaminase mutagenesis pattern is widespread in human cancers. *Nature Genetics, 45*(9), 970–976. https://doi.org/10.1038/ng.2702.

Roper, N., Gao, S., Maity, T. K., Banday, A. R., Zhang, X., Venugopalan, A., ... Khan, J., ... Guha, U. (2019). APOBEC mutagenesis and copy-number alterations are drivers of proteogenomic tumor evolution and heterogeneity in metastatic thoracic tumors. *Cell Reports, 26*(10), 2651–2666.e6. https://doi.org/10.1016/J.CELREP.2019.02.028.

Sanchez, A., Ortega, P., Sakhtemani, R., Manjunath, L., Oh, S., Bournique, E., ... Buisson, R. (2024). Mesoscale DNA features impact APOBEC3A and APOBEC3B deaminase activity and shape tumor mutational landscapes. *Nature Communications, 15*(1), https://doi.org/10.1038/S41467-024-45909-5.

Sato, K., Takeuchi, J. S., Misawa, N., Izumi, T., Kobayashi, T., Kimura, Y., ... Koyanagi, Y. (2014). APOBEC3D and APOBEC3F potently promote HIV-1 diversification and evolution in humanized mouse model. *PLoS Pathogens, 10*(10), https://doi.org/10.1371/JOURNAL.PPAT.1004453.

Sharma, S., Patnaik, S. K., Taggart, R. T., Kannisto, E. D., Enriquez, S. M., Gollnick, P., & Baysal, B. E. (2015). APOBEC3A cytidine deaminase induces RNA editing in monocytes and macrophages. *Nature Communications, 6*, 6881. https://doi.org/10.1038/ncomms7881.

Shi, K., Carpenter, M. A., Banerjee, S., Shaban, N. M., Kurahashi, K., Salamango, D. J., ... Aihara, H. (2016). Structural basis for targeted DNA cytosine deamination and mutagenesis by APOBEC3A and APOBEC3B. *Nature Structural & Molecular Biology, 24*(2), 131–139. https://doi.org/10.1038/nsmb.3344.

Shi, K., Carpenter, M. A., Banerjee, S., Shaban, N. M., Kurahashi, K., Salamango, D. J., ... Aihara, H. (2017). Structural basis for targeted DNA cytosine deamination and mutagenesis by APOBEC3A and APOBEC3B. *Nature Structural & Molecular Biology, 24*(2), 131–139. https://doi.org/10.1038/nsmb.3344.

Shi, K., Carpenter, M. A., Kurahashi, K., Harris, R. S., & Aihara, H. (2015). Crystal structure of the DNA deaminase APOBEC3B catalytic domain. *The Journal of Biological Chemistry, 290*(47), 28120–28130. https://doi.org/10.1074/JBC.M115.679951.

Shi, K., Demir, Ö., Carpenter, M. A., Wagner, J., Kurahashi, K., Harris, R. S., ... Aihara, H. (2017). Conformational switch regulates the DNA cytosine deaminase activity of human APOBEC3B. *Scientific Reports,* 7(1), https://doi.org/10.1038/S41598-017-17694-3.

Silvas, T. V., Hou, S., Myint, W., Nalivaika, E., Somasundaran, M., Kelch, B. A., ... Schiffer, C. A. (2018). Substrate sequence selectivity of APOBEC3A implicates intra-DNA interactions. *Scientific Reports, 8*(1), 7511. https://doi.org/10.1038/s41598-018-25881-z.

St. Martin, A., Salamango, D., Serebrenik, A., Shaban, N., Brown, W. L., Donati, F., ... Harris, R. S. (2018). A fluorescent reporter for quantification and enrichment of DNA editing by APOBEC–Cas9 or cleavage by Cas9 in living cells. *Nucleic Acids Research, 46*(14), e84. https://doi.org/10.1093/NAR/GKY332.

Stenglein, M. D., Burns, M. B., Li, M., Lengyel, J., & Harris, R. S. (2010). APOBEC3 proteins mediate the clearance of foreign DNA from human cells. *Nature Structural & Molecular Biology, 17*(2), 222–229. https://doi.org/10.1038/nsmb.1744.

Swanton, C., McGranahan, N., Starrett, G. J., & Harris, R. S. (2015). APOBEC enzymes: Mutagenic fuel for cancer evolution and heterogeneity. *Cancer Discovery, 5*(7), 704–712. https://doi.org/10.1158/2159-8290.CD-15-0344.

Venkatesan, S., Angelova, M., Puttick, C., Zhai, H., Caswell, D. R., Lu, W. T., ... Swanton, C. (2021). Induction of APOBEC3 exacerbates DNA replication stress and chromosomal instability in early breast and lung cancer evolution. *Cancer Discovery, 11*(10), 2456–2473. https://doi.org/10.1158/2159-8290.CD-20-0725.

Wang, X., Li, J., Wang, Y., Yang, B., Wei, J., Wu, J., ... Yang, L. (2018). Efficient base editing in methylated regions with a human APOBEC3A-Cas9 fusion. *Nature Biotechnology, 36*(10), 946–949. https://doi.org/10.1038/nbt.4198.

Yu, Q., König, R., Pillai, S., Chiles, K., Kearney, M., Palmer, S., ... Landau, N. R. (2004). Single-strand specificity of APOBEC3G accounts for minus-strand deamination of the HIV genome. *Nature Structural & Molecular Biology, 11*(5), 435–442. https://doi.org/10.1038/NSMB758.

## CHAPTER NINE

Profiling rare C-to-U editing events via direct RNA sequencing

Adriano Fonzino[a,*], Pietro Luca Mazzacuva[b,c], Graziano Pesole[a,b], and Ernesto Picardi[a,b,*]

[a]Department of Biosciences, Biotechnology and Environment, University of Bari Aldo Moro, Bari BA, Italy
[b]Institute of Biomembranes, Bioenergetics and Molecular Biotechnology, National Research Council, Bari, Italy
[c]Department of Engineering, University Campus Bio-Medico of Rome, RM, Italy
*Corresponding authors. e-mail address: adriano.fonzino@uniba.it; ernesto.picardi@uniba.it

Contents

Abstract

In mammals, RNA editing involves the hydrolytic deamination of adenosine (A) to inosine (I) or of cytosine (C) to uracil (U) by the ADAR and APOBEC families of enzymes, respectively. Direct RNA (dRNA) sequencing by Oxford Nanopore Technology (ONT) allows the detection of Us and, thus, facilitates the unveiling of edited Cs avoiding Reverse Transcription and PCR amplification steps. However, dRNA data are noisy, and very rare events such as C-to-U conversions cannot be easily distinguished from background noise or mutation errors. To overcome this issue, we developed a novel machine-learning strategy based on the Isolation Forest (iForest) algorithm to denoise the signal deriving from dRNA highly-informative ONT data. Here we present a step-by-step protocol illustrating the usage of the C-to-U-Classifier package and how to apply its pretrained iForest models for ameliorating the detection of C-to-U events in mammalian transcriptomes. As an example, we show here the whole pipeline in action on data deriving from wild-type

Methods in Enzymology, Volume 713
ISSN 0076-6879, https://doi.org/10.1016/bs.mie.2024.11.040
Copyright © 2025 Elsevier Inc. All rights are reserved, including those for text and data mining, AI training, and similar technologies.

(WT) and APOBEC1 knock-out (KO) macrophagic cell lines. Additionally, the polishing power of our algorithm is proved through a synthetic in-vitro transcribed (IVT) sample in which C-to-U events are not present.

1. Introduction

Epitranscriptomics focuses on the study of chemical modifications occurring in cellular RNA molecules. So far, more than 170 different RNA modifications have been identified and characterized (Boccaletto et al., 2022). Non-transient RNA modifications are also known as RNA editing events and the most common in humans are represented by the conversion of adenosine (A) to inosine (I), catalyzed by the adenosine deaminases acting on double-stranded RNA (ADARs). The hydrolytic deamination of cytidine (C) to uridine (U) in single-stranded RNAs is instead carried out by members of the AID (activation-induced cytidine deaminase)/APOBEC (apolipoprotein B mRNA editing enzyme, catalytic polypeptide) proteins family (Lerner et al., 2018; Pecori et al., 2022). The best-known example in humans involves the apolipoprotein B (apoB) transcript in which APOBEC1 operates in a tissue-specific manner leading to a truncated apoB-48 protein in the small intestine and a full unedited protein in the liver (Teng et al., 1993; Navaratnam et al., 1993). While C-to-U editing is rare in humans, a comparative analysis of murine RNAseq data involving wild-type and APOBEC1-deficient mutants, revealed numerous C-to-U changes, mostly in AU-rich motifs in 3′ untranslated regions (3′UTRs) (Rosenberg et al., 2011; Blanc et al., 2014).

Advancements in third-generation sequencing platforms, such as Oxford Nanopore Technologies (ONT), have enabled the direct sequencing of transcripts in their native form. ONT sequencing protocols do not need reverse transcription (RT) and PCR amplification steps leaving the ribonucleotide modifications profile unaffected (Workman et al., 2019; Garalde et al., 2018). Raw current signal generated by ONT can be harnessed by a plethora of computational algorithms to unveil canonical sequences but also pinpoint different types of RNA modifications, such as N6-methyladenosine (m6A), pseudouridine (Ψ), or inosine (Begik et al., 2022). In this context, the sequencing of native RNA molecules provides the unique opportunity to profile C-to-U RNA editing at a transcriptomic scale overcoming short-read related issues. However, the error rate of native ONT reads (at least those produced with the 002 pore version) is still relatively high (Delahaye et al., 2021), especially when compared with the expected frequency of C-to-U editing events. To tackle this (Fonzino et al., 2024), we developed a machine learning approach based on anomaly detection by the usage of the iForest algorithm (Liu et al., 2008; Liu et al., 2012). We demonstrated that many low-frequency C-to-U changes detectable in dRNA data could be due to sequencing or alignment artifacts hampering the identification of high-confidence editing events. The iForest model embedded in our C-to-U-Classifier package was trained on a bonafide set of reads and is easily able to classify, at the per-read level, all the C-to-U changes as either "errors" or putative editing, without the need for computationally demanding analysis of ionic current signals. Here, using synthetic polyA+ RNAs (as a negative control) and real ONT data from murine RAW 264.7 cells, we illustrate how to use our C-to-U-Classifier package for ameliorating the detection of C-to-U editing events.

2. Materials and equipment

The following protocol is based on a set of Python scripts released with the C2U-Classifier (Fonzino et al., 2024) package, freely accessible at the GitHub repository: https://github.com/F0nz0/C_to_U_classifier. The C2U-Classifier is a machine learning-based package for Nanopore denoising and is useful for improving the C-to-U editing signal in dRNA sequencing experiments. It mainly relies on a semi-supervised machine learning technique (Liu et al., 2008; Liu et al., 2012) to classify read spans as either sequencing "errors" or "real" C-to-U-contexts. The model was trained on sequencing data obtained by the ONT SQK-RNA002 kit for direct-RNA sequencing. Consequently, the protocol described in this chapter can only be used to analyze reads produced with this sequencing kit. Although several aligners are available for mapping ONT reads, we use here minimap2 (Li, 2021) with default presets for noisy dRNA reads to perform both spliced (on murine data) and unspliced (on IVTs) alignments. Since ONT is a fast-evolving sequencing platform, our protocol and all the scripts of the C2U-Classifier package are based on specific versions of minimap2 and Guppy basecaller.

2.1 Prerequisites

2.1.1 Hardware

A 64-bit computer running Linux or other Unix-based operating system with at least 64 GB of RAM (for genome mapping using minimap2 and SNPs filtering) and at least 500 GB of free disk space to download reference sequences and fast5 files is needed. Guppy basecalling requires a GPU NVIDIA A100 with 40 GB RAM.

2.1.2 Software

The following programs are required to run the entire procedure correctly:

Anaconda ecosystem to create working environments and install relevant software

tar, **wget**, **gunzip** programs to download and decompress the data

C2U-Classifier Python package (Fonzino et al., 2024) is freely available and downloadable at https://github.com/F0nz0/C_to_U_classifier for the denoising of CU-contexts reads. It requires a Python3 interpreter (at least version 3.8) and all the dependencies and external modules listed in the requirements.txt file (see **Section 3.1**). C2U-Classifier is based on the pysam module (version 0.18.0 or higher) for retrieving basecalling features and scikit-learn (at least version 1.0.1) to run pre-trained machine learning models

Samtools version 1.6 or superior (Li et al., 2009) for handling SAM/BAM files and indexed textual tables or sequence files

Minimap2 version 2.24 or superior to align dRNA reads onto a reference genome and/or transcriptome (Li, 2021)

An **ANNOVAR** version released after the 2020-06-07 (Wang et al., 2010)

2.1.3 Reference sequences and annotations

The murine genome primary assembly mm10 (in fasta format) and the related gene annotations (in gtf format) can be obtained from GENCODE (https://www.gencodegenes.org) (Frankish et al., 2019). Known murine SNPs (version 142), compliant with the mm10 murine genome version, can be downloaded from UCSC using the link: https://hgdownload.soe.ucsc.edu/goldenPath/mm10/database/snp142.txt.gz. The in-vitro transcribed (IVT, gBlocks) reference sequences can be retrieved from Supplementary Table 5 in Nguyen et al. (2022). For the sake of simplicity, these synthetic sequences have been uploaded as a fasta file into a shared link to facilitate the downloading procedures (see **Section 3.2**).

2.1.4 dRNA Data

Raw data of 3 ONT direct-RNA runs in fast5 format can be downloaded from the SRA database. Two of them, **(1)** a wild-type RAW 264.7 macrophagic mouse cell line physiologically expressing APOBEC1 (WT) and **(2)** a derived APOBEC1 knockout (KO) cell line, are compared to differentially detect C-to-U editing sites (BioProject accession PRJNA949094 with SRA accessions WT: SRR23975979, KO: SRR23975978) (Fonzino et al., 2024).

The third run is an IVT fast5 of a dRNA run where C-to-U events are not present. It can be downloaded from SRA with the accession SRR18402565 (Nguyen et al., 2022).

3. Methods: step-by-step details

3.1 Environment setup

1. Create a new folder named "Book_chapter_CtoU_2024" and move inside it.

```
mkdir Book_chapter_CtoU_2024
cd Book_chapter_CtoU_2024
```

2. Download from the GitHub repository the `C to U classifier` source code into the working directory and decompress pre-trained models. It will be used to run the iForest algorithm and denoise dRNA data.

```
git clone https://github.com/F0nz0/C_to_U_classifier
cd C_to_U_classifier/
tar -xf iForest_pretrained_model.tar.gz
rm iForest_pretrained_model.tar.gz
```

3. Create a new virtual environment using anaconda and the Python venv package to install all the required dependencies.

```
#create a new conda virtual environment
conda create --name CtoU_book_chapter python= 3.8

#activate the conda env
conda activate CtoU_book_chapter

#install samtools
conda install -c bioconda samtools==1.6

#install minimap2
conda install -c bioconda minimap2==2.24

#install f5c (optimized re-implementation of Nanopolish)
conda install -c bioconda f5c==1.1

#create virtual environment inside thew anaconda CtoU_book_chapter
env
python3 -m venv venv

#4. Activate the venv:
source venv/bin/activate

#5. Upgrade pip version:
python3 -m pip install --upgrade pip

#6. Install wheel package via pip:
pip install wheel
```

```
#7. Install required Python packages using the requirements.txt file:
python -m pip install -r requirements.txt

#8. Test installation and package
python pipe_basecalling.py
```

If everything worked well, a message about the mandatory arguments to run the C2U-Classifier basecalling pipeline should be shown:

```
usage: pipe_basecalling.py [-h] -B BAM_FILEPATH -R REFERENCE_FILEP-
ATH [-T THRESHOLD] [-threads THREADS] [-aligner ALIGNER] [-O ORGANISM]
pipe_basecalling.py: error: the following arguments are required:-
B/--bam_filepath, -R/--reference_filepath
```

3.2 Downloading and preprocessing of reference sequences and dRNA raw data

1. Create a new directory in which reference files will be stored. Then, download the mouse mm10 reference genome from GENCODE and the related dbSNP142 collection from the UCSC "golden path". The murine genomic annotations can be retrieved from UCSC as well. Reference sequences for the synthetic construct produced by Nguyen et al. (2022) can be easily retrieved by a shared link:

```
#move back to the parent directory
cd..
#make directory for references and download references and move within
mkdir refs
cd refs

#download mm10 murine reference genome and decompress
wget https://ftp.ebi.ac.uk/pub/databases/gencode/Gencode_mouse/release_M10/GRCm38.
primary_assembly.genome.fa.gz
gunzip GRCm38.primary_assembly.genome.fa.gz

#download dbSNP142 and decompress
wget https://hgdownload.soe.ucsc.edu/goldenPath/mm10/database/snp142.txt.gz
gunzip snp142.txt.gz

#download annotations and decompress
wget https://ftp.ebi.ac.uk/pub/databases/gencode/Gencode_mouse/release_M10/gencode.
vM10.annotation.gtf.gz
gunzip gencode.vM10.annotation.gtf.gz

#download reference sequence for IVTs from shared link
wget --no-check-certificate 'https://docs.google.com/uc?export=download&id=1-
O7WbVyidenmc-EXouhsZX0RUa5uJ7lK&' -O gBlock_ref.fa
```

2. The C2U-Classifier package looks for reference chromosome names starting with *chr* inside the alignment files. If needed, the reference fasta file downloaded from GENECODE should be preprocessed to maintain reference chromosomes only (i.e. >chr1, >chr2, >chr3 and so on). Reference files are then indexed by Samtools.

```
# open Python3 interpreter
python
>>> import pysam
>>> out_fasta = open("GRCm38.primary_assembly.genome.filtered.fa",
"w")
>>> ref = pysam.FastaFile("GRCm38.primary_assembly.genome.fa")
>>> for seq in ref.references:
...   if seq.startswith("chr"):
...        out_fasta.write(f">{seq}\n")
...        out_fasta.write(ref.fetch(seq)+"\n")
>>> out_fasta.close()
>>> exit()

# remove not useful reference files
rm GRCm38.primary_assembly.genome.fa*

# indexing filtered reference files
samtools faidx gBlock_ref.fa
samtools faidx GRCm38.primary_assembly.genome.filtered.fa
```

3. Next, download raw data of direct-RNA sequencing runs (WT, KO, and IVT) from the NCBI-SRA database in the fast5 format. Fast5 files will be organized in different directories, one per run/sample.

```
# download data in fast5 formats and decompress
# move to parent directory
cd..
# create a directory for each sample
# (WT and KO: murine data; gBlocks: synthetic IVT)
mkdir WT
mkdir KO
mkdir gBlocks

### WT sample: download and decompression of fast5 archives ###
cd WT
wget https://sra-pub-src-2.s3.amazonaws.com/SRR23975979/APOBEC_WT_FAST5.tar.gz.1
tar -xvzf APOBEC_WT_FAST5.tar.gz.1
```

```
### KO sample: download and decompression of fast5 archives ###
cd ../KO
wget https://sra-pub-src-2.s3.amazonaws.com/SRR23975978/APOBEC_KO_FAST5.tar.gz.1
tar -xvzf APOBEC_KO_FAST5.tar.gz.1

### IVT/gBlocks sample: download and decompression of fast5 archives ###
cd ../gBlocks
wget https://sra-pub-src-2.s3.amazonaws.com/SRR18402565/20191124_gBlock2_G.fast5.tar.
gz.1
tar -xvzf 20191124_gBlock2_G.fast5.tar.gz.1

# move to the previous folder and remove all downloaded compressed fast5
archives.
cd ..
rm */*5.tar.gz.*
```

3.3 Basecalling and alignments

1. The C2U-Classifier provides pre-trained iForest models fed with basecalling features retrieved from ground-truth data (Fonzino et al., 2024). These models are then exploited to perform CU-context reads denoising, classifying them either into putative "real" or "false" C-to-U changes. Because of that, it is pivotal to use the same basecalling and alignment steps and software used to produce the training dataset. More in detail, the basecalling of raw signals has to be carried out utilizing Guppy version 5.0.11, which should be retrieved and installed autonomously (please visit https://community.nanoporetech.com/downloads for further details). After the basecalling step via Guppy and the genome alignment with minimap2, the output SAM files can be filtered, converted to binary alignment files (BAM), and indexed using samtools.

```
# define guppy path
guppy_basecaller=/path/to/your/guppy/installation/executable

############# WT murine sample #############
# define variables
SAMPLE=WT
FAST5_DIR=WT/lustrehome/epicardi/home/NANOPORE/EDITING/APOBEC_WT/
FAST5input
OUTPUT_DIR=$SAMPLE
REFERENCE_FILEPATH=refs/GRCm38.primary_assembly.genome.
filtered.fa
FASTQ_FILEPATH=$SAMPLE/pass/*fastq
SAM_FILEPATH=$SAMPLE/$SAMPLE.sam
BAM_FILEPATH=$SAMPLE/$SAMPLE.bam
```

```
# basecalling fast5 files
$guppy_basecaller -c rna_r9.4.1_70bps_hac.cfg -i $FAST5_DIR -s
$OUTPUT_DIR -r -x "auto"
# alignment to reference genome with minimap2 in splice-aware mode
minimap2 -t 12 -ax splice -uf -k14 --secondary=no $REFERENCE_FILEPATH
$FASTQ_FILEPATH > $SAM_FILEPATH

# filtering of unmapped reads sorting and conversion to binary align-
ment file
samtools view -b -F 2308 $SAM_FILEPATH | samtools sort -O
BAM > $BAM_FILEPATH

# indexing the BAM file
samtools index $BAM_FILEPATH

# remove unused SAM file
rm $SAM_FILEPATH

############# KO murine sample #############
# define variables
SAMPLE=KO
FAST5_DIR=KO/lustrehome/epicardi/home/NANOPORE/EDITING/APOBEC_KO/
FAST5input
OUTPUT_DIR=$SAMPLE
REFERENCE_FILEPATH=refs/GRCm38.primary_assembly.genome.
filtered.fa
FASTQ_FILEPATH=$SAMPLE/pass/*fastq
SAM_FILEPATH=$SAMPLE/$SAMPLE.sam
BAM_FILEPATH=$SAMPLE/$SAMPLE.bam

# basecalling fast5 files
$guppy_basecaller -c rna_r9.4.1_70bps_hac.cfg -i $FAST5_DIR -s
$OUTPUT_DIR -r -x "auto"
# alignment to reference genome with minimap2 in splice mode.
minimap2 -t 12 -ax splice -uf -k14 --secondary=no $REFERENCE_FILEPATH
$FASTQ_FILEPATH > $SAM_FILEPATH

# filtering of unmapped reads sorting and conversion to binary align-
ment file
samtools view -b -F 2308 $SAM_FILEPATH | samtools sort -O
BAM > $BAM_FILEPATH

# indexing the BAM file
samtools index $BAM_FILEPATH

# remove unused SAM file
rm $SAM_FILEPATH

############# gBlocks IVT sample #############
# define variables
SAMPLE=gBlocks
FAST5_DIR=gBlocks/20191124_gBlock2_G/191124_Gmod9_G_noACN_withRT/
20191124_0741_MN23396_FAK64959_fe7b9371/fast5
```

```
OUTPUT_DIR=$SAMPLE
REFERENCE_FILEPATH=refs/gBlock_ref.fa
FASTQ_FILEPATH=$SAMPLE/pass/*fastq
SAM_FILEPATH=$SAMPLE/$SAMPLE.sam
BAM_FILEPATH=$SAMPLE/$SAMPLE.bam

# basecalling fast5 files
$guppy_basecaller -c rna_r9.4.1_70bps_hac.cfg -i $FAST5_DIR -s
$OUTPUT_DIR -r -x "auto"
# alignment to reference sequence with minimap2 (no polarity expected)
minimap2 -t 12 -ax map-ont -k14 --secondary=no --for-only $REFERE-
NCE_FILEPATH $FASTQ_FILEPATH > $SAM_FILEPATH

# filtering of unmapped reads sorting and conversion to binary align-
ment file
samtools view -b -F 2308 $SAM_FILEPATH | samtools sort -O
BAM> $BAM_FILEPATH

# index the BAM file
samtools index $BAM_FILEPATH

# remove unused SAM file
rm $SAM_FILEPATH
```

For the Guppy command line, **-c** is the configuration file to use in high accuracy mode for direct RNA essays (r9.4.1. pore with an expected translocation speed for RNA of 70 bps), **-i** is the path to the directory containing the fast5 files to be basecalled, **-s** the save-path where the output fastq, log, and sequencing summary files will be saved, **-r** to work recursively within the input folder looking for all *.fast5 files, **-x** for selecting the device to use, either a CPU or a GPU (set to "auto" to use GPU when available). By default, Guppy creates two output folders in the save-path (-s), a "passed" and a "failed" directory for reads passing Guppy quality controls or not, respectively. For the minimap2 command line, to map basecalled passed reads to the reference genome, different presets for murine data (WT and KO) and IVT-gBlocks are used. Spliced-aware alignments against the reference genome are needed for murine data. More in detail, we map only passed reads against the reference sequences setting with **-t** the number of parallel threads (12), **-ax** the alignment modality to use ("splice" for murine data, and "map-ont" for IVT), **-uf** set only for murine data, indicating to find GT-AG junctions in the transcript, **-k** the k-mer size (decreased from the default value of 15 to 14 allowing minimap2 to align also noisy dRNA reads), **–secondary=no** to avoid secondary alignments in the output SAM file and, for IVTs, the **–for-only** flag to eliminate unexpected reverse mapped reads.

3.4 C2U-classifier basecalling pipeline and denoising

1. The C2U-Classifier provides a denoising pipeline for the classification of "real" and "false" (errors) CU changes based on basecalling features via pre-trained iForest models. The basecalling pipeline can be run through a single Python script. The program iterates across alignments focusing on all well-covered genomic positions of the BAM files, looking for CU-context reads. After that, the program extracts the basecalling features at a per-read level for every interval, encoding the information about base qualities, deletions, and insertions in a meaningful fashion and then consolidating these on disk. The basecalling features were used to classify every CU-context read interval (−3, +3 nt) either as an "error" or a "real" CU change putatively due to genomic variations and/or C-to-U changes, eliminating a huge amount of noise due to sequencing and alignment errors.

```
# launch C2U-Classifier basecalling pipeline for WT sample
echo Launching WT...
python C_to_U_classifier/pipe_basecalling.py \
    -B WT/WT.bam \
    -R refs/GRCm38.primary_assembly.genome.filtered.fa \
    -threads 24 \
    -O murine #<-- used for the computation of the APOBEC1 signature
    with a Likelihood-ratio test

# launch C2U-Classifier basecalling pipeline for KO sample
echo Launching KO...
python C_to_U_classifier/pipe_basecalling.py \
    -B KO/KO.bam \
    -R refs/GRCm38.primary_assembly.genome.filtered.fa \
    -threads 24 \
    -O murine #<-- used for the computation of the APOBEC1 signature
    with a Likelihood-ratio test

# # launch C2U-Classifier basecalling pipeline for gBlocks IVT sample
echo Launching gBlocks IVT...
python C_to_U_classifier/pipe_basecalling.py
    -B gBlocks/gBlocks.bam \
    -R refs/gBlock_ref.fa \
    -threads 3
```

The C2U-Classifier script requires the following arguments: **-B** path for the input BAM file produced as previously described, **-R** path for the reference sequence used to align reads, **-threads** number of parallel threads to use (every thread will iterate over a given genomic region/chromosome),

-O an optional argument which can be set alternatively to "murine" or "human" for the computation of an APOBEC1 signature probability via a Likelihood-ratio Test (provided with the C2U-Classifier package and computed using Illumina derived bonafide editing sites) (Fonzino et al., 2024). At the end of computations, the pipeline yields different output folders and different output tables with iForest predictions. The two main output directories are produced at the same level as the input BAM file and have the same root base name. For instance, for the input of gBlocks-IVT BAM file at `./gBlocks/gBlocks.bam`, the two main output directories generated by the software will be saved at:

(1) `./gBlocks/gBlocks.basecalling_features/`
(2) `./gBlocks/gBlocks.model_iForest_pretrained_results/`

The former contains three subfolders with all the basecalling features tables for every read-interval with contexts CC, CT (meaning CU, for reads mapping on both strands), or TT, respectively. The latter contains two tsv files for iForest per-read (`df_CT_predicted.tsv`) and genome-space aggregated predictions (`df_CT_predicted_aggregated.tsv`) based on CT(CU)-contexts basecalling features. The folder tree for the gBlocks sample should look like this:

```
./gBlocks/
├─ gBlocks.basecalling_features/
│  ├─ gBlocks.CCcontext_reads_features_forward/
│  │  ├─ CCcontext_reads_features_forward_chr1.tsv
│  │  ├─ CCcontext_reads_features_forward_chr2.tsv
│  │  ├─ CCcontext_reads_features_forward_chr3.tsv
│  ├─ gBlocks.CTcontext_reads_features_forward_rev/
│  │  ├─ CTcontext_reads_features_forward_rev_chr1.tsv
│  │  ├─ CTcontext_reads_features_forward_rev_chr2.tsv
│  │  ├─ CTcontext_reads_features_forward_rev_chr3.tsv
│  ├─ gBlocks.TTcontext_reads_features_forward/
│  │  ├─ TTcontext_reads_features_forward_chr1.tsv
│  │  ├─ TTcontext_reads_features_forward_chr2.tsv
│  │  ├─ TTcontext_reads_features_forward_chr3.tsv
├─ gBlocks.model_iForest_pretrained_results/
│  ├─ df_CT_predicted_aggregated.tsv
│  ├─ df_CT_predicted.tsv
├─ gBlocks.bam
├─ gBlocks.bam.bai
```

A deeper explanation of all the output (tsv) tables is provided in the next paragraph.

3.5 Analysis of the output tables

The following Python code needs to be executed from the main working directory after the activation of the conda and venv environments (see Section 3.1):

1. Import all the required Python modules and libraries:

```
# import required python modules and libraries
import os, sys
import pandas as pd
import matplotlib.pyplot as plt
import seaborn as sn
sys.path.append("C_to_U_classifier")
from C_to_U_classifier.predict_CT_iForest_cc import reduce_
basecalling_features
```

2. Load the tsv output tables for the three samples obtained with the C2U-Classifier pipeline as Pandas DataFrames objects. The tables contain the iForest predictions on both per-read and genome-space levels along with basecalling features extracted from the alignment files. For WT and KO samples, the exploratory data analysis performed in this section was restricted only to reads mapping onto a well-known C-to-U editing site on the 3′-UTR of the *B2m* gene at the position chr2:122152740(+) (Fonzino et al., 2024):

```
# load dataframes
ivt_basecall_feat = pd.read_table("gBlocks/gBlocks.
basecalling_features/gBlocks.CTcontext_reads_features_forward_rev/
CTcontext_reads_features_forward_rev_chr2.tsv",
                                        names=["region", "position",
"read_id", "strand"]+[f"pos{i}" for i in range(-3,4)]+ ["ins"])
ivt_read = pd.read_table("gBlocks/gBlocks.model_iForest_pretrained_
results/df_CT_predicted.tsv")
ivt_aggr = pd.read_table("gBlocks/gBlocks.model_iForest_pretrained_
results/df_CT_predicted_aggregated.tsv")

# load KO (focusing on B2M region on site chr2:122152740)
ko_basecall_feat = pd.read_table("KO/KO.basecalling_features/KO.
CTcontext_reads_features_forward_rev/CTcontext_reads_features_
forward_rev_chr2.tsv",
                                        names=["region", "position",
"read_id", "strand"]+[f"pos{i}" for i in range(-3,4)]+ ["ins"])
ko_read  =  pd.read_table("KO/KO.model_iForest_pretrained_results/
df_CT_predicted.tsv")
ko_aggr  =  pd.read_table("KO/KO.model_iForest_pretrained_results/
df_CT_predicted_aggregated.tsv")
# focus on chr2:122152740
ko_basecall_feat = ko_basecall_feat.query("region == 'chr2'").query
```

```
("position == 122152740")
ko_read = ko_read.query("region == 'chr2'").query("position ==
122152740")

# load WT (focusing on B2M region on site chr2:122152740)
wt_basecall_feat = pd.read_table("WT/WT.basecalling_features/
WT.CTcontext_reads_features_forward_rev/CTcontext_reads_features_
forward_rev_chr2.tsv",
                                        names=["region", "position",
"read_id", "strand"]+[f"pos{i}" for i in range(-3,4)]+["ins"])
wt_read = pd.read_table("WT/WT.model_iForest_pretrained_results/
df_CT_predicted.tsv")
wt_aggr = pd.read_table("WT/WT.model_iForest_pretrained_results/
df_CT_predicted_aggregated.tsv")
# focus on chr2:122152740
wt_basecall_feat = wt_basecall_feat.query("region == 'chr2'").query
("position == 122152740")
wt_read = wt_read.query("region == 'chr2'").query("position ==
122152740")
```

3. Visualize a few rows of the 3 loaded tables for the gBlocks/IVT sample, for the sake of example:

```
# see some example outputs for IVTs/gBlocks
display(ivt_basecall_feat.head(3)) ### -->basecalling features
display(ivt_read.head(3))          ### -->per-read predictions
display(ivt_aggr.head(3))          ### -->genome-space aggregated
                                          predictions
```

In the output tsv Table 1 (which can be found at the generic file-path: `*.basecalling_features/*.CTcontext_reads_features_forward_rev/CTcontext_reads_features_forward_rev_chr*.tsv`) are stored basecalling features extracted and encoded from the BAM file of all the reads with a CU-context to the reference sequence. For every read (row) a −3, +3 interval surrounding the central C query site is explored and the base qualities were extracted: the mismatches were encoded as a negative quality, deletion as a 0 while the insertions count is stored in the last column as a positive integer.

Table 1 IVT basecalling features.

Region	Position	Read_id	Strand	pos−3	pos−2	pos−1	pos0	pos+1	pos+2	pos+3	ins
chr2	13	0eb…5c6	+	16.0	10.0	4.0	3.0	6.0	−0.0	17.0	0.0
chr2	26	35f…5c4	+	17.0	−0.0	16.0	9.0	19.0	24.0	14.0	0.0
chr2	26	4c0…71a	+	16.0	−0.0	21.0	3.0	18.0	21.0	13.0	0.0

The basecalling features were then further preprocessed and classified by the iForest model assigning to each CU-context read interval a class with two levels (0: CU error; 1: high probability of a real CU context). Table 2 shows a few example rows where the last column (pred) in the `*.model_iForest_pretrained_results/df_CT_predictions.tsv` contains the prediction.

Table 3 shows the results of the C2U-Classifier aggregation step for the computation of CU substitution frequencies before and after the iForest denoising. This table (which can be retrieved at `*.model_iForest_pretrained_results/df_CT_predictions_aggregated.tsv`) is obtained by aggregating per-read predictions and calculating the number of CU-context reads predicted as "1"s in Table 2 by the C2U-Classifier. So, for each well-covered genomic position identified by the first three columns, the remaining columns indicate **T_native**: the number of reads carrying natively a U mapped on that position, **T_corrected**: the number of Us after the denoising, i.e. eliminating the reads predicted as 0 s (putative errors), **depth_stranded**: the number of aligned reads on the query site, **Tfreq_native** and **Tfreq_corrected** are the ratios between **T_native** and **T_corrected** on the **depth_stranded**, respectively. The **5mer** is a column indicating the reference kmer and **y_hat** is the prediction after the denoising of the current site, **0:** unaltered site, **1:** C-to-U editing site candidate.

Table 2 IVT per-read predictions.

Region	Position	Read_name	Strand	Pred
chr2	13	0eb2581d-b8b1-4ee7-a028-f361186ef5c6	+	0.0
chr2	26	35f2bbfb-1b36-4468-a095-700bb26eb5c4	+	0.0
chr2	26	4c03919a-0a9a-4c73-af48-eccfb50f271a	+	0.0

Table 3 IVT genome-space aggregated predictions.

Region	Position	Strand	T native	T corrected	Depth stranded	Tfreq native	Tfreq corrected	5mer	y hat
chr2	13	+	1	0	3035	0.000329	0.0	CACGA	0
chr2	26	+	6	0	3823	0.001569	0.0	AACAA	0
chr2	30	+	18	0	3816	0.004717	0.0	ATCAG	0

4. Reduce basecalling features retrieved by the C2U-Classifier (see Table 1) and down-sample overrepresented IVT reads for visualization purposes:

```
# reduce basecalling features and merge ivt,ko and wt per read predic-
tions+
ivt_basecall_feat_red = reduce_basecalling_features
(ivt_basecall_feat)
ivt_basecall_feat_red["sample"] = "IVT"
ivt_basecall_feat_red = pd.merge(ivt_basecall_feat_red, ivt_read,
how="inner", on=["region", "position", "read_name", "strand"])

ko_basecall_feat_red = reduce_basecalling_features
(ko_basecall_feat)
ko_basecall_feat_red["sample"] = "KO"
ko_basecall_feat_red = pd.merge(ko_basecall_feat_red, ko_read,
how="inner", on=["region", "position", "read_name", "strand"])

wt_basecall_feat_red = reduce_basecalling_features
(wt_basecall_feat)
wt_basecall_feat_red["sample"] = "WT"
wt_basecall_feat_red = pd.merge(wt_basecall_feat_red, wt_read,
how="inner", on=["region", "position", "read_name", "strand"])

basecall_fear_red = pd.concat([ivt_basecall_feat_red,
ko_basecall_feat_red, wt_basecall_feat_red], ignore_index=True)

# downsample ivt for visualization purposes and add categorical pre-
diction
# (1: true U - 0: false)
basecall_fear_red_sample = pd.concat([ivt_basecall_feat_red.sample
(300), ko_basecall_feat_red, wt_basecall_feat_red], ignore_index=True)
basecall_fear_red_sample["PRED"] = ["True" if i == 1 else "False" for i
in basecall_fear_red_sample["pred"].values]
display(basecall_fear_red_sample.head())
```

The basecalling features at the per-read level were "reduced" to **Tqual**: quality of the central U base, **MeanQual**: the average quality for the bases aligned for the current read-interval, **mismatches**, **dels**, **ins**: the number of mismatches, deletions and insertions detected, respectively. The per-read predictions from Table 2 merged with the reduced basecalled features are shown in the following Table 4:

5. Produce some plots for the reduced basecalling features and their corresponding per-read level predictions (Fig. 1):

```
# create a results folder
results_fp ="results"
```

Table 4 Reduced and summarized per-read basecalling features with iForest predictions.

Region	Position	Read_name	Strand	Tqual	Mean quality	Mismatches	Dels	Ins	Sample	Pred	PRED
chr2	897	548…f2f	+	6.0	18.0	0	0	0	IVT	0.0	False
chr2	38	992…fb0	+	5.0	15.7	0	0	0	IVT	0.0	False
chr2	895	9e5…8fc	+	6.0	8.7	1	0	0	IVT	0.0	False

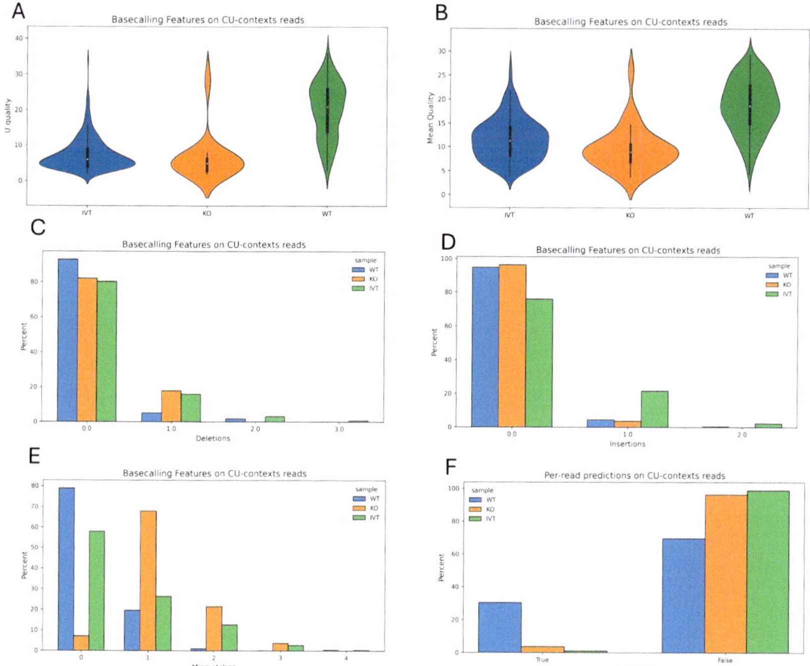

Fig. 1 Representation of the CU-contexts basecalling features retrieved by the C2U-Classifier and used to predict per-read level using a pre-trained iForest model. (A) central U base quality distribution in the three samples (IVT: gBlocks; WT/KO: APOBEC1 WT or KO murine macrophagic cell lines); (B) the Mean Quality for read intervals; distributions across the three runs of (C) deletions, (D) insertions and (E) mismatches counts. In (F) per-read predictions for IVT and murine CU-context reads. It is easy to note a higher percentage of reads predicted as errors (False) in both IVT and KO samples with respect to WT ones, in agreement with the genotype.

```
if not os.path.exists(results_fp):
    os.mkdir(results_fp)
for feat,label in zip(["Tqual", "MeanQual"], ["U quality", "Mean
Quality"]):
    plt.figure(figsize=(10,5))
    sn.violinplot(data=basecall_fear_red_sample,
                  y=feat, x="sample")
    plt.title(f"Basecalling Features on CU-contexts reads", fontsize=14)
    plt.ylabel(label, fontsize=12)
    plt.xlabel("")
    plt.savefig(os.path.join(results_fp, f"{feat}_basecalling_
    features.tiff"),
                  dpi=300, bbox_inches='tight', facecolor='white',
                  transparent=False)
    plt.show()

for feat,label in zip(["n_mismatches", "n_dels", "n_ins", "PRED"],
                      ["Mismatches", "Deletions", "Insertions",
                      "Prediction"]):
    plt.figure(figsize=(10,5))
    sn.histplot(data=basecall_fear_red_sample.sort_values
    ("sample", ascending=False),
                  x=feat, hue="sample", stat="percent",
                  multiple="dodge",
                  discrete=True, shrink=.7, common_norm=False)
    plt.xlabel(label, fontsize=12)
    plt.ylabel("Percent", fontsize=12)
    if not label == "Prediction":
        plt.title(f"Basecalling Features on CU-contexts reads",
        fontsize=14)
    else:
        plt.title(f"Per-read predictions on CU-contexts reads",
        fontsize=14)
    plt.savefig(os.path.join(results_fp, f"{feat}_basecalling_
    features.tiff"),
                  dpi=300, bbox_inches='tight', facecolor='white',
                  transparent=False)
    plt.show()
```

6. Dimensionality reduction via Principal Component Analysis and data visualization (Fig. 2):

```
# perform PCA
import numpy as np
from sklearn.preprocessing import StandardScaler
from sklearn.decomposition import PCA

# standardize dataset
sc = StandardScaler()
```

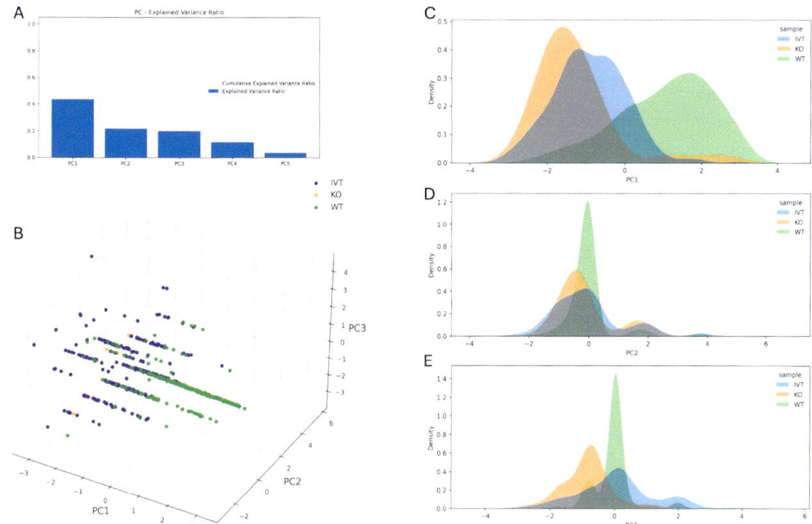

Fig. 2 Principal components (PCs) analysis of reduced basecalling features. (A) PCs Explained Variance Ratio as blue bars and the cumulative Explained Variance as a blue line. (B) a 3-dimensional scatterplot showing per-read intervals with CU-contexts of IVT, WT, and KO runs, segregating mainly along the (C) PC1 axis. IVT and KO read clustered together, while as the opposite, WT reads create a separate (but in part overlapping) cluster on the PC1. (D) PC2 (E) PC3 appeared less involved in this separation.

```
df_std = sc.fit_transform(basecall_fear_red_sample.iloc[:,4:9].
values)

# perform simple PCA
pca = PCA(n_components=df_std.shape[1])
df_std_pca = pd.DataFrame(pca.fit_transform(df_std), columns=
[f"PC{i}" for i in range(1,pca.n_components+1)])
df_std_pca["sample"] = basecall_fear_red_sample["sample"].tolist()
df_std_pca["PRED"] = basecall_fear_red_sample["PRED"].tolist()

# plot PCs explained variance
plt.figure(figsize=(10,5))
plt.bar([f"PC{i}" for i in range(1, pca.n_components+1)],
pca.explained_variance_ratio_, label="Explained Variance Ratio")
plt.plot(np.cumsum(pca.explained_variance_ratio_),
label="Cumulative Explained Variance Ratio", c="lightsteelblue")
plt.title("PC - Explained Variance Ratio")
plt.legend(loc="right")
plt.savefig(os.path.join(results_fp,
f"PCA_explained_Variance_Ratio_basecalling_features.tiff"),
                  dpi=300,  bbox_inches='tight',  facecolor='white',
                  transparent=False)
 plt.show()
```

```
# plot the first 3 Principal Components kernel density estimates
plt.figure(figsize=(10,4))
sn.kdeplot(data=df_std_pca, x="PC1", hue="sample",
common_norm=False, fill=True,
           alpha=.5, linewidth=0)
plt.savefig(os.path.join(results_fp, f"PCA_kde_PC1_basecalling_
features.tiff"),
                  dpi=300, bbox_inches='tight', facecolor='white',
                  transparent=False)
plt.show()

 plt.figure(figsize=(10,4))
sn.kdeplot(data=df_std_pca, x="PC2", hue="sample", common_norm=
False, fill=True,
           alpha=.5, linewidth=0)
plt.savefig(os.path.join(results_fp,
f"PCA_kde_PC2_basecalling_features.tiff"),
                  dpi=300, bbox_inches='tight', facecolor='white',
                  transparent=False)
 plt.show()

plt.figure(figsize=(10,4))
sn.kdeplot(data=df_std_pca, x="PC3", hue="sample",
common_norm=False, fill=True,
           alpha=.5, linewidth=0)
plt.savefig(os.path.join(results_fp,
f"PCA_kde_PC3_basecalling_features.tiff"),
                  dpi=300, bbox_inches='tight', facecolor='white',
                  transparent=False)
 plt.show()

# plot the first 3 Principal Components onto a 3d scatterplot
fig = plt.figure(figsize=(10,10))
ax = fig.add_subplot(projection='3d')
ax.scatter(df_std_pca.query("sample == 'IVT'")["PC1"],
           df_std_pca.query("sample == 'IVT'")["PC2"],
           df_std_pca.query("sample == 'IVT'")["PC3"], marker="o",
           color="blue", alpha=0.9,
           label="IVT")

ax.scatter(df_std_pca.query("sample == 'KO'")["PC1"],
           df_std_pca.query("sample == 'KO'")["PC2"],
           df_std_pca.query("sample == 'KO'")["PC3"], marker="o",
           color="orange", alpha=0.9,
           label="KO")

ax.scatter(df_std_pca.query("sample == 'WT'")["PC1"],
           df_std_pca.query("sample == 'WT'")["PC2"],
           df_std_pca.query("sample == 'WT'")["PC3"], marker="o",
           color="green", alpha=0.9,
           label="WT")
```

```
ax.set_xlabel('PC1', fontsize=14)
ax.set_ylabel('PC2', fontsize=14)
ax.set_zlabel('PC3', fontsize=14)

plt.legend(fontsize=12)

plt.savefig(os.path.join(results_fp, f"PCA_3d_basecalling_
features.tiff"),
                    dpi=300, bbox_inches='tight', facecolor='white',
                    transparent=False)
plt.show()
```

7. Evaluate genome-space aggregated data and predictions for IVT and murine reads on *B2m* transcripts. The CU substitution frequencies for every site on IVT were evaluated before and after the denoising (Fig. 3):

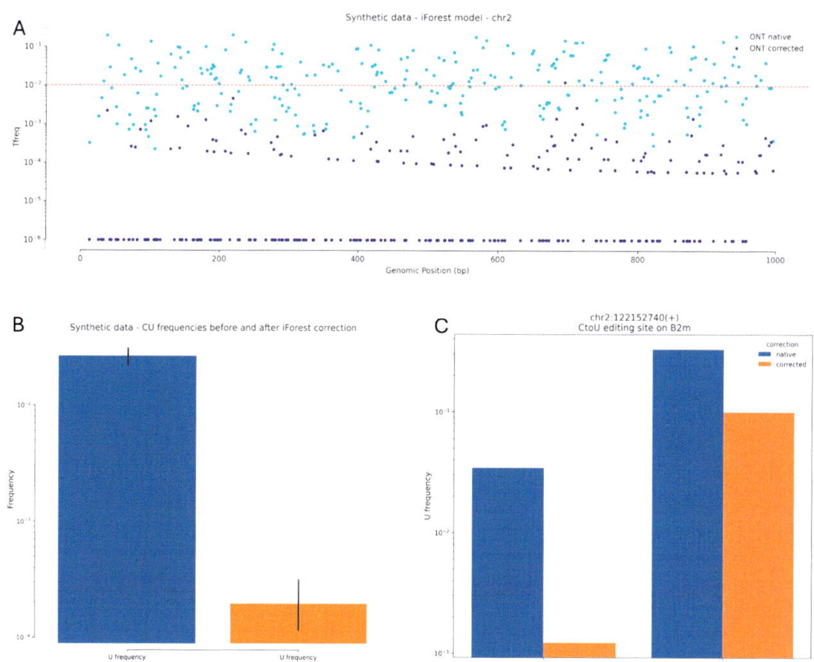

Fig. 3 Effect of the iForest denoising on the genome-space aggregated data. (A) CU substitutions frequencies along the entire gBlock construct. (B) the same data but depicted as a bar plot. (C) CU frequency on the *B2m* chr2:1221527340(+) editing site for KO and WT before and after the C2U-Classifier denoising via iForest model.

```
ivt_aggr_count = ivt_aggr[["Tfreq_native", "Tfreq_corrected"]]
ivt_aggr_count.columns =["U frequency\nnative", "U frequency\
ncorrected"]
display(ivt_aggr_count)
plt.figure(figsize=(10,9))
g = sn.barplot(data=ivt_aggr_count)
plt.yscale("log")
plt.title("Synthetic data - CU frequencies before and after iForest
correction", fontsize=14)
plt.ylabel("Frequency", fontsize=12)
sn.despine(offset=10, trim=True)
plt.savefig(os.path.join(results_fp,
f"IVT_CUfreqs_bef_aft_iforest_corr.tiff"),
                    dpi=300, bbox_inches='tight', facecolor='white',
                    transparent=False)
plt.show()

# load required function
from C_to_U_classifier.utils import plot_frequencies
plot_frequencies(reference_filepath="refs/gBlock_ref.fa",
df_aggregated_filepath="gBlocks/
gBlocks.model_iForest_pretrained_results/
df_CT_predicted_aggregated.tsv",
                    title="Synthetic data - iForest model",
                    save_folderpath=os.path.join(os.getcwd(),
                    "results"),
                    log=True,
                    offset=0.000001,
                    native_color="cyan",
                    corrected_color="blue",
                    native_size=15,
                    corrected_size=10,
                    strand="+",
                    freq_threshold=0.01)
# print the unique false positive after iForest correction
display(ivt_aggr.query("y_hat == 1"))
```

The IVT aggregated predictions were depicted along the whole synthetic constructs sequence in Fig. 3A, with the CU frequency (not expected in this sample) appearing to be strongly decreased after the iForest denoising and the filtering of per-read intervals classified as errors (Fig. 3B). After denoising, only a single site in the IVT dataset (Table 5) showed a residual CU corrected frequency above the 1 % (from a native frequency equal to about 15 %) from hundreds of starting candidate sites filtered out.

Table 5 An example of the **unique false positive** over hundreds of sites remained after the C2U-Classifier denoising on genome-space aggregated data.

Region	Position	Strand	T native	T corrected	Depth stranded	Tfreq native	Tfreq corrected	5mer	y hat
chr2	695	+	2124	173	14,242	0.149136	0.012147	CTCTT	1

8. Evaluating the denoising effect on the genome space for murine data on *B2m* transcripts:

```
# merge WT and KO preds aggregated on genome space
ko_aggr["sample"] ="KO"
wt_aggr["sample"] ="WT"
aggr = pd.concat([ko_aggr, wt_aggr], ignore_index=True)
# focus on well-known site chr2:122152740(+) B2m
aggr_site = aggr.query("region == 'chr2'").query("position ==
122152740")

# plot CU frequency before and after iforest correction
Tfreqs =[]
Tfreqs.append([aggr_site.query("sample == 'KO'")["Tfreq_native"].
values[0], "KO", "native"])
Tfreqs.append([aggr_site.query("sample == 'KO'")
["Tfreq_corrected"].values[0], "KO", "corrected"])
Tfreqs.append([aggr_site.query("sample == 'WT'")["Tfreq_native"].
values[0], "WT", "native"])
Tfreqs.append([aggr_site.query("sample == 'WT'")
["Tfreq_corrected"].values[0], "WT", "corrected"])
Tfreqs = pd.DataFrame(Tfreqs, columns=["CUfreq", "sample",
"correction"])

plt.figure(figsize=(10,9))
sn.barplot(data=Tfreqs, y="CUfreq", x="sample", hue="correction")
plt.yscale("log")
plt.title("chr2:122152740(+)\nCtoU editing site on B2m",
fontsize=14)
plt.xlabel("")
plt.ylabel("U frequency", fontsize=12)
plt.savefig(os.path.join(results_fp, f"KO_vs_WT_chr2_122152740_
B2m_CUfreqs_bef_aft_iforest_corr.tiff"),
                       dpi=300, bbox_inches='tight', facecolor='white',
                       transparent=False)
plt.show()
```

The denoising of KO murine reads recapitulates a similar effect noticeable also on IVT reads (Fig. 3C). On the *B2m* site at the position chr2:122152740(+), the CU frequency was pushed toward 0 for the KO sample while the WT sample appeared to be less involved by the denoising. This is an expected behavior since, at least on this site of interest, no C-to-U changes should be noticeable in the KO while strong editing evidence must be detectable on the same site for the WT sample.

3.6 Transcriptome-wide detection of C-to-U editing events after pre-trained iForest model denoising

1. This analysis was performed using Python3 code. All the required modules have to be imported, and some useful custom functions have to be defined to plot the results of the differential transcriptome-wide analysis on murine samples:

```
# Required python libraries
import pandas as pd
import numpy as np
import matplotlib.pyplot as plt
import os, pysam, sys, subprocess
from matplotlib.patches import ConnectionPatch
import logomaker as lm
from tqdm import tqdm
sys.path.append("./C_to_U_classifier")
from C_to_U_classifier.aggregate_results import
retrieve_depth_stranded

# Function to give chromosomes names in an VCF compatible format
def convert_chr(x):
      return x.replace("chr", "")

# Function to revert chromosomes names from VCF compatible format
def revert_chr(x):

      return "chr{}".format(x)

# Function to make pie charts
def make_pie_chart(series, title):
      path = os.path.abspath("")
      fig, (ax1, ax2) = plt.subplots(1, 2, figsize=(8, 5),
      gridspec_kw={'width_ratios': [3, 1.25]})
      fig.subplots_adjust(wspace=0)

      # Pie chart parameters
      secondary_ratios =[]
      for i in [i for i in series.index.tolist() if i not in ["exonic",
      "UTR3", "UTR5"]]:
          secondary_ratios.append(series.loc[i])
      overall_ratios =[sum(secondary_ratios)]
      for i in [ "exonic", "UTR5", "UTR3"]:
          overall_ratios.append(series.loc[i])
      labels =["Other Regions", "Exonic\n(Coding)", "UTR5", "UTR3"]
      explode =[0.1, 0.1, 0.1, 0.1]
```

```
# Rotate pie chart so that first wedge is split by the x-axis
angle =-180* overall_ratios[0]
wedges, *_ = ax1.pie(overall_ratios, autopct='%1.1f%%',
startangle=angle,
                     labels=labels, explode=explode,
                     wedgeprops = {"edgecolor": "black",
                     'linewidth': 1, 'antialiased': True})

# Bar chart parameters
secondary_labels =[i for i in series.index.tolist() if i not in
["exonic", "UTR3", "UTR5"]]
secondary_ratios =[i/sum(secondary_ratios) for i in
secondary_ratios]
bottom =1
width =.2
spacer =.0035

# Adding from the top matches the legend.
for j,(height,label) in enumerate(reversed([*zip(secondary_
ratios, secondary_labels)])):
    bottom -= height +(spacer*(len(secondary_labels)-1))
    bc = ax2.bar(0, height, width, bottom=bottom, color='C0',
    label=label,
                alpha=0.1 + 1/len(secondary_labels) * j,
                edgecolor="k")
    ax2.bar_label(bc,labels=[f"{height:.0%}"],label_type=
    'center', fontsize=10)
    for b in bc:
        h = b.get_height()
        # lower left vertex

        x0, y0 = b.xy
        # top left vertex
        x2, y2 = x0, y0+h
        if j == 0:
            xt, yt = x2, y2
        elif j == len(secondary_labels)-1:

            xb, yb = x0, y0

# Other bar chart parameters
ax2.set_title('Other Regions Stratification', fontsize=11)
ax2.legend(bbox_to_anchor=(1.0, 1.0))
ax2.axis('off')
ax2.set_xlim(- 2.5 * width, 2.5 * width)

# Use ConnectionPatch to draw lines between the two plots
theta1, theta2= wedges[0].theta1, wedges[0].theta2
center, r=wedges[0].center, wedges[0].r
bar_height = sum(secondary_ratios)

# Draw top connecting line
x = r * np.cos(np.pi / 180 * theta2) + center[0]
y = r * np.sin(np.pi / 180 * theta2) + center[1]
con = ConnectionPatch(xyA = (xt, yt), coordsA = ax2.transData,
xyB = (x, y), coordsB = ax1.transData)
                        con.set_color([0, 0, 0])
```

```
        con.set_linewidth(1)
        ax2.add_artist(con)

        # Draw bottom connecting line
        x = r * np.cos(np.pi / 180 * theta1) + center[0]
        y = r * np.sin(np.pi / 180 * theta1) + center[1]
        con = ConnectionPatch(xyA=(xb, yb), coordsA=ax2.transData,
                              xyB=(x, y), coordsB=ax1.transData)
        con.set_color([0, 0, 0])
        ax2.add_artist(con)
        con.set_linewidth(1)

        # Final figure parameters and plotting
        fig.suptitle(f"{title} Denoising", fontsize=15)
        plt.tight_layout()
        plt.savefig(os.path.join(path, f"{title}_Denoising_
        Pie_Chart.tiff"), dpi=600, bbox_inches="tight")
        plt.show()

# Function to make sequence logs
def make_logo(dataframe, title):
        # Parameters assignment
        path = os.path.abspath("")
        lenght =5
        genome=pysam.FastaFile(path+"/refs/
        GRCm38.primary_assembly.genome.filtered.fa")
        complement ={"A":"T", "C":"G", "G":"C", "T":"A"}

        # Extraction of cosensuses signatures from reference genome
        sequences =[]
        with tqdm(total=dataframe.shape[0], desc="Consensuses
        Extraction") as pbar:
             for row in dataframe.itertuples():
                 if row[3] == "+":
                    kmer =[]
                    for base in genome.fetch(row[1], row[2]-(lenght+1),
                    row[2]+lenght):
                        kmer.append(base)
                    sequences.append(kmer)
                 else:
                    kmer =[]
                    for base in genome.fetch(row[1], row[2]-3, row[2]+2):
                        kmer.append(complement.get(base))
                    kmer.reverse()
                           sequences.append(kmer)
                 pbar.update(1)

        # Computation of nucleotides frequencies from the consensus sig-
        natures
        cols=[f"Pos. {i}" for i in range(-lenght,1)]+[f"Pos. +{i}" for i in
        range(1,lenght+1)]
        sequences = pd.DataFrame(sequences, columns=cols)
```

```
        base_frequencies =[]
        for col in sequences.columns.tolist():
            frequencies = sequences.loc[:, col].value_counts
            (normalize=True)
            try:
                A = frequencies.loc["A"]
            except:
                A = 0.0
            try:
                C = frequencies.loc["C"]
            except:
                C = 0.0
            try:
                G = frequencies.loc["G"]
            except:
                G = 0.0
            try:
                T = frequencies.loc["T"]
            except:
                T = 0.0
            base_frequencies.append([A, C, G, T])
        base_frequencies = pd.DataFrame(base_frequencies, columns = ["A",
        "C", "G", "T"])
        # Logo parameters and plotting
        %matplotlib inline
        logo = lm.Logo(base_frequencies, font_name = 'DejaVu Sans',
        figsize=(10, 4))
        logo.ax.set_ylabel("Frequency", labelpad=1.5, fontsize=12)
        logo.ax.set_xticklabels('%+d'%x if x!=0 else '%d'%x for x in [i for
        i in range(-(lenght+2), (lenght+2), 2)])
        logo.ax.set_xlabel('Nucleotide Position', labelpad=1.5,
        fontsize=12)
        logo.ax.set_title(f"{title} Denoising", fontsize=15)
plt.tight_layout() plt.savefig(os.path.join(path,f"{title}_Denoising_
Logo.tiff"),dpi=600,bbox_inches="tight")
```

2. Load from disk full genome-space aggregated tables for both KO and WT samples, selecting only sites with coverage higher than 50 on both samples and with a minimum CU native frequency in WT equal to at least 1 %:

```
# Path to aggregated data files and working folder
notebook_path = os.path.abspath("")
ko_path = os.path.join(notebook_path, "KO/KO.model_iForest_
pretrained_results")
wt_path = os.path.join(notebook_path, "WT/WT.model_iForest_
pretrained_results")
files_name ="df_CT_predicted_aggregated.tsv"

# Upload WT aggregated data file and data selections
```

```
wt = pd.read_table(wt_path+"/"+files_name)
wt = wt[(wt.loc[:, "depth_stranded"] >= 50) &
(wt.loc[:, "Tfreq_native"] >= 0.01)]
wt.reset_index(drop=True, inplace=True)
copy = wt.loc[:, ["region", "position", "strand"]]
copy.loc[:, "index"] =[i for i in range(copy.shape[0])]
# Upload KO aggregated data file and data selection
ko = pd.read_table(ko_path+"/"+files_name)
ko = ko.merge(copy, how="inner", on=["region", "position", "strand"])
ko = ko[ko.loc[:, "depth_stranded"] >=50]
# Retrieve KO data information from its BAM file
row_to_drop = ko.loc[:, "index"].tolist()
ko.drop("index", axis=1, inplace=True)

copy = wt.copy()
copy.drop(row_to_drop, axis=0, inplace=True)
copy.reset_index(drop=True, inplace=True)

ko_bam = pysam.AlignmentFile(os.path.join(notebook_path, "KO/KO.bam"),
"rb")
new_data =[]
for i in tqdm(range(copy.shape[0]), desc=f"Remaining WT sites scan on KO
BAM"):
    r=copy.iloc[i, 0]
    p=copy.iloc[i, 1]
    s=copy.iloc[i, 2]
    dept = retrieve_depth_stranded(ko_bam, r, p, s)
    new_data.append([ r, p, s, 0, 0, dept, 0.0, 0.0, "-", 0, 0])
```

For sites well covered in WT but not in KO aggregated predictions, pysam is used to extract the stranded depth of coverage exploring the KO sample BAM file.

3. Merge filtered WT and KO data into a unique Pandas DataFrame and filter out all known dbSNP142 sites to eliminate possible genomic sources of CU variations:

```
# Adding retrieved discarded KO data to ko dataframe
new_data = pd.DataFrame(new_data, columns=wt.columns.tolist())
new_data = new_data[new_data.loc[:, "depth_stranded"] >=50]
new_data.reset_index(drop=True, inplace=True)
ko = pd.concat([ko, new_data], axis=0)

# WT and K0 dataframes merging and dbSNp142 filtering
final_table = wt.merge(ko, how ="inner", on=["region", "position",
"strand"], suffixes=('_WT', '_KO'))
dbsnps_path = os.path.join(notebook_path, "refs")
```

```
copy = final_table.loc[:, ["region", "position", "strand"]]
copy.loc[:, "index"] =[i for i in range(copy.shape[0])]
dbsnps  =  pd.read_table(os.path.join(dbsnps_path,  "snp142.txt"),
header=None, usecols=[1, 3, 6, 10])
dbsnps.columns =["region", "position", "strand", "gen."]
dbsnps = dbsnps[dbsnps.loc[:, "gen."] == "genomic"]
copy  =  copy.merge(dbsnps,  how="inner",  on=["region",  "position",
"strand"])
index_drop = copy.loc[:, "index"].tolist()
final_table.drop(index_drop, axis=0, inplace=True)

final_table.to_csv(os.path.join(notebook_path,
"df_CT_WT_KO_filtered_predicted_aggregated_no_snps.tsv"),
sep="\t", index=None)
```

4. Produce a VCF file from the DataFrame of C-to-U editing candidates and save it to disk. This file is then annotated using ANNOVAR and merged again with the C2U-Classifier information:

```
# VCF fle preparation from the merged dataframe
copy = final_table.loc[:, ["region", "position"]]
copy.loc[:, "region"] = copy.loc[:, "region"].apply(convert_chr)
copy.columns =["region", "start"]
copy.loc[:, "stop"] = copy.loc[:, "start"]
copy.loc[:, "ref"] ="C"
copy.loc[:, "var"] ="T"
copy.to_csv(os.path.join
(notebook_path,"df_CT_WT_KO_filtered_predicted_
aggregated_no_snps.vcf"), sep="\t", index=None, header=False)

# VCF file annotation with Annovar (change path to your annovar direc-
tory)
annovar_path ="/path/to/annovar"
cmd =[annovar_path+"/table_annovar.pl",
      notebook_path+"/df_CT_WT_KO_filtered_predicted_aggregated_
      no_snps.vcf",
      annovar_path+"/mousedb/", "-buildver", "mm10", "-out",
      notebook_path+"/df_CT_WT_KO_annovar_output", "-protocol",
      "refGene",
      "-operation", "g", "-nastring", ".", "--remove", "-polish"]

p = subprocess.Popen(cmd, stdout=subprocess.PIPE, stderr=subprocess.PIPE)

# Optional printing of Annovar output
result, err = p.communicate()
print(str(err.decode()))
# Retrieving and addition to the merged dataframe of annotated VCF information
```

```
annotated=pd.read_table(f"{notebook_path}/
df_CT_WT_KO_annovar_output.mm10_multianno.txt")
cols_remain =[i for i in annotated.columns.tolist() if i not in ["Chr",
"Start", "End"]]


annotated.drop("Start", axis=1, inplace=True)
annotated.columns =["region", "position"] + cols_remain
annotated.loc[:, "region"] = annotated.loc[:, "region"].apply(revert_chr)


not_annotated=pd.read_table(f"{notebook_path}/
df_CT_WT_KO_filtered_predicted_aggregated_no_snps.tsv")


final_table = not_annotated.merge(annotated, how="inner", on=["region",
"position"])
final_table.to_csv(f"{notebook_path}/
df_CT_WT_KO_filtered_predicted_aggregated_no_snps_annotated.tsv",
sep="\t", index=None)
```

5. Use the annotated table of C-to-U editing candidates. All the well-covered sites with CU-context reads in WT samples (at least 0.05 CU frequency) but not in KO are selected. The predicted C-to-U editing sites detected before and after the iForest denoising are compared and further evaluated:

```
# Data selection on the merged dataframe.
before = final_table[final_table.loc[:, "Tfreq_native_KO"]==0.0]
after = final_table[(final_table.loc[:, "y_hat_KO"]==0) & (final_
table.loc[:, "y_hat_WT"]==1)]

# Produce data for Table 6.
before_raw = before.query("Tfreq_native_WT >= 0.05").loc[:,
"Func.refGene"].value_counts()
after_raw = after.query("Tfreq_corrected_WT >= 0.05").loc[:,
"Func.refGene"].value_counts()

# Produce data for pie charts in Fig. 4.
before_zoom = before.query("Tfreq_native_WT >= 0.05").loc[:,
"Func.refGene"].value_counts(normalize=True)
after_zoom = after.query("Tfreq_corrected_WT >= 0.05").loc[:,
"Func.refGene"].value_counts(normalize=True)

# Porduce pie charts
make_pie_chart(before_zoom, "Before")
make_pie_chart(after_zoom, "After")
```

All the genomic regions annotated via ANNOVAR for all sites with editing evidence in WT but not in the KO sample are retrieved and enlisted in Table 6. In Fig. 4C and D, a graphic representation of

Table 6 Distributions of genomic regions for detected C-to-U editing sites before and after the iForest correction.

	Before	After
Exonic	810	12
UTR3	716	65
UTR5	61	2
Intergenic	55	4
Intronic	17	1
ncRNA_exonic	12	0
ncRNA_intronic	9	1
Downstream	2	1

distributions of these genomic regions is produced as pie-charts before and after the iForest correction, respectively.

6. Employing the pysam module, all the sequence + 5/−5 intervals surrounding edited sites are retrieved from the reference genome, and these are used to produce sequence logos to visualize the typical APOBEC1 signature noticeable after the iForest denoising of CU reads (Fig. 4A and B):

```
#Produce data for sequence logos in Fig. 4.
before_coordinates = before.query("Tfreq_native_WT >= 0.05").loc[:,
["region", "position", "strand"]]
after_coordinates = after.query("Tfreq_corrected_WT >= 0.05").loc
[:, ["region", "position", "strand"]]

# Produce sequence logos
make_logo(before_coordinates, "Before")
make_logo(after_coordinates, "After")
```

4. Notes, advantages and limitations

All the statements and discussions made in this protocol are unrelated to the newer RNA004 dRNA sequencing kit, pore, and chemistry recently released by ONT. As previously discussed, the ONT is a rapidly evolving technology, so theC2U-Classifier usage is restricted to data produced by SQK-RNA002 (or 001) sequencing kit and compatible pore. Because of that, and since it is based on pre-trained models fed with basecalling features produced by specific versions of Guppy and minimap2, we recommend limiting its usage only to alignment files retrieved using tools and versions suggested in this step-

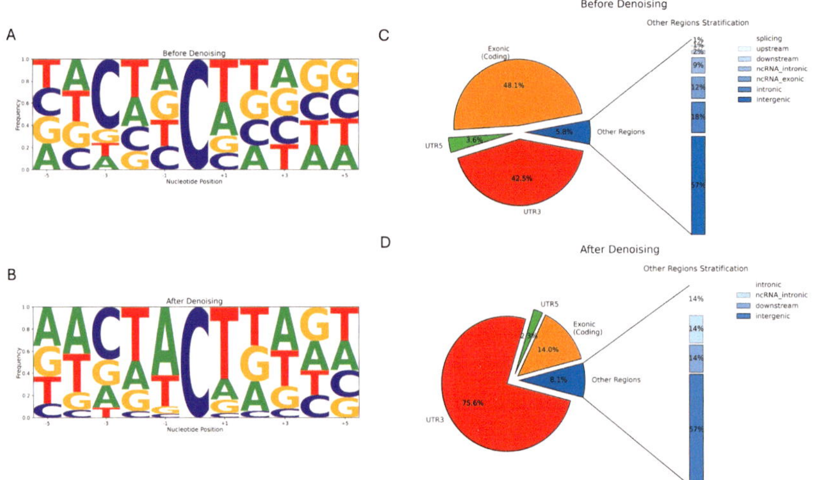

Fig. 4 Amelioration of the signal-to-noise ratio for the transcriptome-wide detection of C-to-U editing events via dRNA. (A, B) The application of the denoising pipeline led to a clear ACT signature compared to the random native one. The effect of the denoising pipeline was also clear after ANNOVAR annotation of all the detected editing sites before (C) and after (D) the iForest polishing. After the procedure, the fraction of sites falling 3′UTR regions was strongly enriched.

by-step guide. Despite that, a considerable amount of dRNA runs with older chemistry was released and deposited and can still be analyzed. Furthermore, the potentiality of ONT emphasized by the C2U-Classifier gave, for the first time, the possibility to improve the investigation of C-to-U events with a single-molecule resolution in parallel with the study of other concurrent epitranscriptomics modifications and their phasing, everything simply starting from the same raw electric signal stored into fast5 files and using other third-party software. To note, another limitation of our tool is the required filtering of the genomic source of variations, which can be achieved, in turn, in two ways: (1) as we carried out in this procedure, via the comparison with an APOBEC/KO cell line, or (2) alternatively, using data extracted from ONT DNAseq run of the same sample, eliminating all sites involved into actual C-U variations. Since this software acts as a denoising tool trying to eliminate most unexpected C-U substitutions, it cannot be used to detect other types of RNA editing events or base modifications. Further studies are needed to assess the suitability of our software with data from the latest pore because of the lower error rate of the new chemistry.

Funding

We kindly thank the Italian node of ELIXIR (Elixir-IIB) for the empowering project ELIXIRNextGenIT (Grant Code IR0000010) and the Bari ReCaS DataCenter for computing support. Authors are also grateful to the following National Research Centers: "High Performance Computing, Big Data and Quantum Computing" (Project no. CN_00000013) and "Gene Therapy and Drugs based on RNA Technology" (Project no. CN_00000041); and Extended Partnerships Age-It (Project no. PE_00000015) and MNESYS (Project no. PE_0000006 to G.P.). This work was also supported by projects: Life Science Hub Regione Puglia (LSH-Puglia, T4-AN-01 H93C22000560003) and INNOVA – Italian network of excellence for advanced diagnosis (PNC-EJ-2022–23683266 PNC-HLS-DA) – to G.P. and by ELIXIR-IT through the empowering project ELIXIRNextGenIT (Grant Code IR0000010 to G.P.).

References

Begik, O., Mattick, J. S., & Novoa, E. M. (2022). Exploring the epitranscriptome by native RNA sequencing. *RNA (New York, N. Y.), 28*(11), 1430–1439. https://doi.org/10.1261/rna.079404.122.

Blanc, V., Park, E., Schaefer, S., et al. (2014). Genome-wide identification and functional analysis of apobec-1-mediated C-to-U RNA editing in mouse small intestine and liver. *Genome Biology, 15*(6), R79. https://doi.org/10.1186/gb-2014-15-6-r79.

Boccaletto, P., Stefaniak, F., Ray, A., et al. (2022). MODOMICS: A database of RNA modification pathways. 2021 update. *Nucleic Acids Research, 50*(D1), D231–D235. https://doi.org/10.1093/nar/gkab1083.

Delahaye, C., Nicolas, J., & Andrés-León, E. (2021). Sequencing DNA with nanopores: Troubles and biases. *PLoS One, 16*(10), e0257521. https://doi.org/10.1371/journal.pone.0257521.

Fonzino, A., Manzari, C., Spadavecchia, P., Munagala, U., Torrini, S., Conticello, S., ... Picardi, E. (2024). Unraveling C-to-U RNA editing events from direct RNA sequencing. *RNA Biology, 21*(1), 1–14. https://doi.org/10.1080/15476286.2023.2290843.

Frankish, A., Diekhans, M., et al. (2019). GENCODE reference annotation for the human and mouse genomes. *Nucleic Acids Research, 47*(D1), D766–D773. https://doi.org/10.1093/nar/gky955.

Garalde, D. R., Snell, E. A., Jachimowicz, D., et al. (2018). Highly parallel direct RNA sequencing on an array of nanopores. *Nature Methods, 15*(3), 201–206. https://doi.org/10.1038/nmeth.4577.

Lerner, T., Papavasiliou, F. N., & Pecori, R. (2018). RNA editors, cofactors, and mRNA targets: An overview of the C-to-U RNA editing machinery and its implication in human disease. *Genes, 10*(1), 10. https://doi.org/10.3390/genes10010013.

Li, H., Handsaker, B., Wysoker, A., et al. (2009). The sequence alignment/Map format and SAMtools. *Bioinformatics (Oxford, England), 25*(16), 2078–2079. https://doi.org/10.1093/bioinformatics/btp352.

Li, H. (2021). New strategies to improve minimap2 alignment accuracy. *Bioinformatics (Oxford, England), 37*, 4572–4574. https://doi.org/10.1093/bioinformatics/btab705.

Liu, F. T., Ting, K. M., Zhou, Z.H. (2008). Isolation forest. In *2008 Eighth IEEE International Conference on Data Mining* (pp. 413–422). 2008 Dec 15–19; Pisa, Italy. https://doi.org/10.1109/ICDM.2008.17.

Liu, F. T., Ting, K. M., & Zhou, Z. H. (2012). Isolation-based anomaly detection. 39 *ACM Transactions on Knowledge Discovery, 6*(3), 1–3. https://doi.org/10.1145/2133360.2133363.

Navaratnam, N., Morrison, J. R., Bhattacharya, S., et al. (1993). The p27 catalytic subunit of the apolipoprotein B mRNA editing enzyme is a cytidine deaminase. *The Journal of Biological Chemistry, 268*(28), 20709–20712. https://doi.org/10.1016/S0021-9258(19)36836-X.

Nguyen, T. A., Heng, J. W. J., Kaewsapsak, P., et al. (2022). Direct identification of A-to-I editing sites with nanopore native RNA sequencing. *Nature Methods, 19*(7), 833–844. https://doi.org/10.1038/s41592-022-01513-3.

Pecori, R., Di Giorgio, S., Paulo Lorenzo, J., et al. (2022). Functions and consequences of AID/APOBEC-mediated DNA and RNA deamination. *Nature Reviews. Genetics, 23*(8), 505–518. https://doi.org/10.1038/s41576-022-00459-8.

Rosenberg, B. R., Hamilton, C. E., Mwangi, M. M., et al. (2011). Transcriptome-wide sequencing reveals numerous APOBEC1 mRNA-editing targets in transcript 3′ UTRs. *Nature Structural & Molecular Biology, 18*(2), 230–236. https://doi.org/10.1038/nsmb.1975.

Teng, B., Burant, C. F., & Davidson, N. O. (1993). Molecular cloning of an apolipoprotein B messenger RNA editing protein. *Science (New York, N. Y.), 260*(5115), 1816–1819. https://doi.org/10.1126/science.8511591.

Wang, K., Li, M., & Hakonarson, H. (2010). ANNOVAR: Functional annotation of genetic variants from next-generation sequencing data. *Nucleic Acids Research, 38*, e164. https://doi.org/10.1093/nar/gkq603.

Workman, R. E., Tang, A. D., Tang, P. S., et al. (2019). Nanopore native RNA sequencing of a human poly(A) transcriptome. *Nature Methods, 16*(12), 1297–1305. https://doi.org/10.1038/s41592-019-0617-2.

CHAPTER TEN

Global quantification of off-target activity by base editors

Michelle Eidelman[a,b], Eli Eisenberg[c,*], and Erez Y. Levanon[a,b,*]

[a]Mina and Everard Goodman Faculty of Life Sciences, Bar-Ilan University, Ramat Gan, Israel
[b]The Institute of Nanotechnology and Advanced Materials, Bar-Ilan University, Ramat Gan, Israel
[c]Raymond and Beverly Sackler School of Physics and Astronomy, Tel Aviv University, Tel Aviv, Israel
*Corresponding authors. e-mail address: elieis@post.tau.ac.il; erez.levanon@biu.ac.il

Contents

Abstract

Base editors are engineered deaminases combined with CRISPR components. These engineered deaminases are designed to target specific sites within DNA or RNA to make a precise change in the molecule. In therapeutics, they hold promise for correcting mutations associated with genetic diseases. However, a key challenge is minimizing unintended edits at off-target sites, which could lead to harmful mutations. Researchers are actively addressing this concern through a variety of optimization efforts that aim to improve the precision of base editors and minimize off-target activity. Here, we examine the various types of off-target activity, and the methods used to evaluate them. Current methods for finding off-target activity focus on identifying similar sequences in the genome or in the transcriptome, assuming the guide RNA misdirects the editor. The main method presented here, that was originally developed to quantify editing levels mediated by the ADAR enzyme, takes a different approach, investigating the inherent activity of base editors themselves, which might lead to off-target edits beyond sequence similarity. The editing index tool quantifies global off-target editing, eliminates the need to detect individual off-target sites, and allows for assessment of the global load of mutations.

1. Introduction

Single nucleotide variants (SNVs), the most prevalent form of genetic variation, are the most common cause of human genetic diseases. Base editing,

Methods in Enzymology, Volume 713
ISSN 0076-6879, https://doi.org/10.1016/bs.mie.2024.11.036
Copyright © 2025 Elsevier Inc. All rights are reserved, including those for text and data mining, AI training, and similar technologies.

a revolutionary technology, holds promise for correcting these disease-causing SNVs. This technology leverages the power of engineered deaminases combined with the precise targeting capabilities of the CRISPR system.

Deaminases are naturally occurring enzymes that can alter the sequence of either DNA or RNA by hydrolysis of either adenine or cytosine. Their engineering results in modified enzymes that target specific sites within DNA or RNA and make a precise change in the sequence. The CRISPR system comprises two key elements, the first one is the Cas protein (usually dead Cas or Cas nickase), and the second one is the guide RNA (gRNA) molecule. The gRNA molecule is complementary to the target sequence for the initial recognition, leading Cas protein to the exact location for editing. The Cas proteins bind and interact with the DNA or RNA molecule, ensuring precise targeting (Barrangou & Doudna, 2016).

There are several types of Cas proteins, each with unique functionalities and structural features. Cas9 and Cas12, for example, both target DNA but differ in their cleavage mechanisms and in their PAM sequence recognition (Li et al., 2023). Cas13, on the other hand, specifically targets and cleaves RNA molecules. Cas proteins rely on a short recognition sequence near the target site known as the PAM (protospacer adjacent motif). Different Cas proteins have different PAM sequences. The existence of PAM sequences is a limitation for base editors, thus researchers have engineered Cas proteins variants with less restrictive PAM compatibilities (Hu et al., 2018).

Base editors can be categorized as DNA base editors and RNA base editors (Porto, Komor, Slaymaker, & Yeo, 2020). The mechanism of DNA base editing involves the gRNA binding to the unedited DNA strand, creating an "R loop" at the target site (Anzalone et al., n.d.). The deaminase enzyme then modifies a specific base on the target strand. Additionally, the Cas protein introduces a nick in the unedited strand, initiating the repair mechanism. This process uses the modified target strand as a template to introduce the desired base change on the non-target strand. The RNA base editor mechanism involves the gRNA binding to the target site in the RNA molecule. This hybridization induces a structural conformation that enables the deaminase enzyme to precisely modify the target RNA base.

In addition to CRISPR-based systems, RNA base editing technologies offer an alternative Cas-free option. This approach involves A-to-I RNA base editing using an engineered deaminase and antisense guide RNA. The guide RNA is designed to base pair with the target mRNA, which subsequently results in the deaminase editing of the target adenosine (Merkle et al., 2019; Porto et al., 2020). An example of this system is λN, developed by removing the endogenous targeting domains from human ADAR2 and replacing them with an antisense RNA oligonucleotide. This modification resulted in a recombinant enzyme capable of being directed to edit any specific site within the RNA sequence (Montiel-Gonzalez, Vallecillo-Viejo, Yudowski, & Rosenthal, 2013; Montiel-Gonzalez, Diaz Quiroz, & Rosenthal, 2019). Another example is the SNAP-ADAR tool which involves fusing the deaminase domain to a SNAP-tag. This fusion enables covalent attachment to a guide RNA of approximately 20 nucleotides in length. The guide RNA requires chemical modifications, including the incorporation of a self-labeling moiety, to support the self-labeling reaction (Kiran Kumar et al., 2024; Stafforst & Schneider, 2012). However, the application of this method remains limited due to considerable off-target effects (Buchumenski et al., 2021) and the challenges associated with producing and introducing a protein-RNA complex into the cell.

The most known DNA base editors are cytosine base editors (CBEs) that are capable of conversion of the target cytosine (C) to uracil (U), and adenine base editors (ABEs) that are capable of conversion of adenosine (A) to inosine (I) (Jeong, Song, & Bae, 2020; Kantor, McClements, & Maclaren, 2020; Rees & Liu, 2018). During DNA replication or repair, the uracils and inosines are converted to thymines and guanines, respectively (Fig. 1).

These engineered deaminases are part of a new generation of tools that can be used to edit genes more precisely than ever before. Their precise capability to correct genetic mutations or rewrite specific genome sequences holds the promise of revolutionizing therapeutic strategies and biotechnological applications. Base editors have several advantages over classical gene editing techniques such as zinc fingers, mega nucleases, TALENs (Transcription activator-like effector nucleases), and CRISPR (Calos, 2017; Gaj, Gersbach, & Barbas, 2013). Base editors are more precise and modify a single nucleotide. They don't generate double strand DNA breaks (DSB) and do not rely on insertion or deletion of larger pieces of DNA. Additionally, the design and construction of base editors are generally easier and faster than designing zinc fingers, TALENs, or mega nucleases. Base editors often involve modifying existing Cas proteins, whereas the classical techniques require complex protein engineering or identification of rare naturally occurring enzymes. For example, zinc finger nucleases require the engineering of a new protein for each specific editing, making the process labor-intensive. In contrast, base editors only require modification of the guide RNA (gRNA) molecule to achieve target specificity, significantly simplifying the procedure (Anzalone et al., n.d.; Newby & Liu, 2021; Slesarenko, Lavrov, & Smirnikhina, 2022; Strauß & Lahaye, 2013).

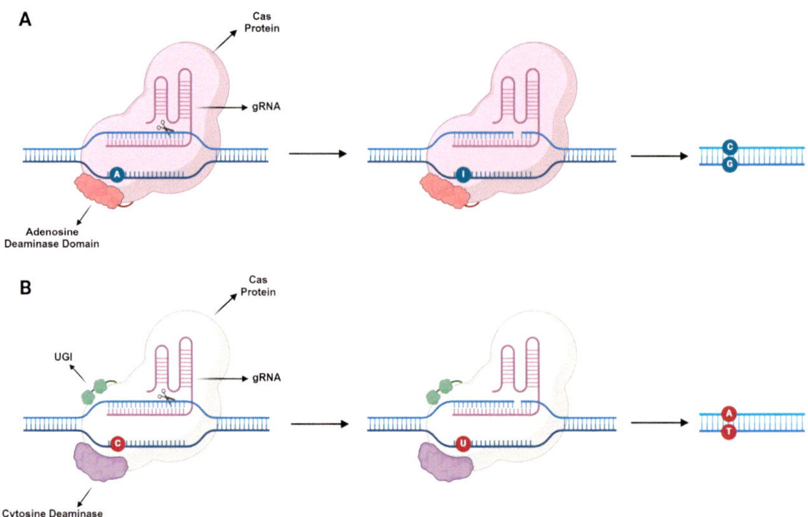

Fig. 1 A schematic representation of DNA base editors. **(A)** The gRNA binds to the non-edited strand at the target site, creating an "R loop". The adenosine deaminase domain then edits the target adenosine (A) into inosine (I). This process is followed by the Cas protein nicking the non-edited strand, initiating the DNA repair mechanism, which results in A-to-G editing and a G-C base pair. **(B)** A similar process occurs with CBEs. The cytosine deaminase domain edits the target cytosine (C) into uracil (U). Uracil glycosylase inhibitors (UGIs) prevent the uracil glycosylase enzymes from removing the uracil base that was introduced by the CBE from the DNA. This inhibition is essential to ensure that the uracil is not removed before it is converted into the desired base, resulting in C-to-T editing and a T-A base pair (Porto et al., 2020; Wang et al., 2017).

2. Off-target activity of base editors

Base editors have emerged as a revolutionary technology within the realm of human gene therapy due to their remarkable capacity for precise and efficient correction of disease-causing mutations. However, several studies have unveiled a concerning aspect of base editors: their propensity for non-specific activity, particularly in the first generations of these editors. This non-specificity manifests as off-target edits in both DNA and RNA, potentially leading to unintended consequences at both the targeted and untargeted loci (Grünewald et al., 2019; Lee, Smith, Liu, Willi, & Hennighausen, 2020; Zuo et al., 2019). Minimizing this off-target activity is a major challenge limiting potential therapeutic applications of these methods.

There are several types of unintended editing activity. Bystander editing of neighboring cytosine or adenine bases within the editing window may be modified by the respective deaminase domain. A narrow editing window is preferred to minimize these effects. Furthermore, undesired editing at off-target sites can be categorized into two distinct classes: Cas-dependent and Cas-independent off-targets. Cas-dependent off-targets arise when the guide RNA (gRNA) – the molecule responsible for directing the Cas enzyme to the target location – binds to unintended genomic sequences, leading to Cas nicking and subsequent base editing at these off-target sites. These unintended genomic sequences typically share similarities with the target site, allowing the gRNA to bind to them. To overcome this phenomenon, design strategies for high-fidelity gRNA are developed to ensure near-perfect complementarity with the target sequence and minimize the potential for Cas-dependent off-target edits (Allen, Rosenberg, & Hendel, 2020).

The second category, Cas-independent off-targets, stems from the inherent properties of the fused deaminase enzymes incorporated within base editors. These deaminases, by their very nature, are known to induce low-level, widespread RNA and DNA modifications, further expanding the potential pool of unintended edits.

The first generations of DNA base editors have demonstrated a significant amount of unintended RNA editing (Grünewald et al., 2019). A more comprehensive understanding of deaminase specificities and the development of more targeted deaminases are crucial to mitigate Cas-independent off-target effects.

There are several approaches to decrease such off-target activities. One approach involves the development of novel gRNAs that incorporate modifications to increase their stability and reduce off-target effects (Sakovina, Vokhtantsev, Vorobyeva, Vorobyev, & Novopashina, 2022; Whittaker et al., 2024; Zhang et al., 2023). Additionally, off-target effects can also be mitigated by identifying novel Cas homologs that utilize rarer PAM sequences, thereby decreasing the likelihood to edit non-target genomic DNA (Guo, Ma, Gao, & Guo, 2023). Furthermore, the specificity of base editors can be enhanced through the engineering of deaminase enzymes by optimizing the deaminase domain such that activity is increased on target sites while minimizing undesired off-target modifications (Chen et al., 2022; Gaudelli et al., 2020; Li et al., 2021; Li et al., 2023; Zhang et al., 2022). Another approach to minimizing off-target activity involves incorporating a nuclear localization signal (NLS) into the enzymes (Vallecillo-Viejo, Liscovitch-Brauer, Montiel-Gonzalez,

Eisenberg, & Rosenthal, 2018). Enzymes with an NLS exhibit reduced non-specific off-target levels compared with matched enzymes containing a nuclear export signal (NES) (Buchumenski et al., 2021).

Moreover, off-target activity of base editors may be suppressed through optimization of the delivery method. DNA-based delivery strategies, such as plasmid transfection or viral vectors, can lead to prolonged overexpression of genome-editing tools, thereby increasing the likelihood of off-target editing events. Ribonucleoprotein (RNP) complexes and mRNA delivery methods limit the exposure to base editors, consequently minimizing off-target activity (Rees & Liu, 2018; Rees et al., 2017).

Collectively, these advancements in gRNA design, Cas protein engineering, deaminase optimization, and delivery methods represent significant progress toward improving the accuracy and efficacy of base editors for precise genome editing applications.

3. Evaluating off-target effects of base editors

Detection of off-target editing activity may benefit from utilizing methods developed to identify naturally occurring editing sites. Application of bioinformatics tools such as the Genome Analysis Toolkit (GATK) (Depristo et al., 2011; McKenna et al., 2010) for germline and somatic variant discovery and REDITools (Picardi & Pesole, 2013) for detecting RNA editing events, for example, may enhance the ability to detect off-target sites. These tools facilitate the generation of comprehensive lists of variants for each sample, encompassing potential off-target editing events. Through meticulous filtering and refinement processes, these tools improve the accuracy and reliability of detected variants, thereby ensuring high confidence results. By integrating these methodologies, one can achieve a robust framework for accurately identifying and characterizing known and novel off-target RNA editing events, thus ensuring a thorough assessment of the specificity and safety of RNA base editors.

However, these methods present challenges such as the need for high sequencing coverage to ensure sufficient read depth for confident variant calling. Especially for DNA-editing, this is a major concern as deep-coverage Whole-genome sequencing (WGS) is quite costly.

Furthermore, accurate detection of off-target effects requires comprehensive knowledge of genomic diversity across different populations. In addition, incorrect alignment of sequencing reads to the reference genome can lead to false positives in variant detection. Many of these difficulties are partially solved by comparing RNA-seq data from samples treated with base editors to untreated controls of the same genetic background. Mismatch sites identified only in the treated samples are more likely to represent off-target editing activity.

In addition, one may improve accuracy by focusing on specific genomic or transcriptomic positions of interest, such as regions that are similar to the targeted region and thus are more likely to undergo Cas-dependent editing. For example, Basic Local Alignment Search Tool (BLAST) (Camacho et al., 2009) can be employed to look for genomic regions with sequences similar to the gRNA target sequence, thus identifying regions prone to unintended editing due to sequence similarity.

Alternatively, one may consider focusing on sites that are known to undergo endogenous editing (e.g., sites recorded in REDIportal (Mansi et al., 2021; Picardi, D'Erchia, Giudice, Lo, & Pesole, 2017; Pietro D'Addabbo et al., 2024), a comprehensive database of known RNA editing sites across various species and tissues). These sites are likely susceptible to Cas-independent targeting by the deaminases fused to the base editor. Observing an increased editing activity at these sites may indicate unintended activity of the base editor.

However, these methods cannot capture off-target activity occurring throughout the genome or transcriptome, at positions that we currently cannot predict. Furthermore, when editing efficiency is evaluated per position, its accuracy is limited by sequencing coverage, often preventing the detection of low levels of editing, thereby potentially overlooking significant off-target events at these low-activity sites.

The RNA Editing Index (Roth, Levanon, & Eisenberg, 2019), described in detail in the subsequent section, is another tool that provides a global assessment of editing activity across a set of regions. Its algorithm circumvents the need for high per-site coverage, thereby enabling the quantification of low editing activity in wide regions that might not be individually resolved. The Editing Index tool can be applied to DNA-seq data as well (Buchumenski et al., 2021), and is currently the main approach for studying this type of off-target editing, probably due to the inherent Cas-independent activity of the deaminase.

For the detection of DNA off-target activity, specialized techniques such as CIRCLE-seq, and Digenome-seq are highly effective in identifying off-target cleavage sites. CIRCLE-seq (Tsai et al., 2017) (Circularization for In Vitro Reporting of Cleavage Effects by Sequencing) involves the circularization of genomic DNA, followed by in vitro digestion and sequencing of the cleaved products. This method is particularly sensitive and does not require living cells. Digenome-seq (Kim et al., 2015; Liang & Huang, 2019) involves the in vitro digestion of genomic DNA with the editing enzyme, followed by high-throughput sequencing to pinpoint cleavage sites that align with regions similar to the gRNA target.

CRISPResso (Pinello et al., 2016) serves as an exemplary tool for identifying and quantifying Cas-dependent off-targets in one or more specific regions. It facilitates precise quantification and visualization of CRISPR-Cas9 editing outcomes, providing a comprehensive evaluation of its impact on amplicon sequences. This tool is particularly effective in detecting off-target editing events at selected loci. The methods mentioned here do not necessarily encompass all available techniques for detecting off-target activity.

Some of the methods presented above necessitate the execution of experimental protocols, whereas others are computational tools. Utilizing these advanced sequencing techniques and tools, researchers can achieve a detailed and accurate assessment of off-target activity, thereby enhancing the understanding and optimization of genome editing technologies.

4. RNA editing index

The RNA Editing Index tool (Roth et al., 2019) was designed to quantify levels of editing mediated by the ADAR (adenosine deaminase acting on RNA) enzyme. It computes the (weighted) average editing levels across all adenosines in a global manner for a set of regions. The editing index is defined as the ratio of A-to-G mismatches to the total adenosine coverage, with the result multiplied by 100 to express the editing percentage in the regions of interest. To ensure accuracy, genomic sites that overlap with common single nucleotide polymorphisms (SNPs) are excluded from the analysis. Averaging over multiple positions, the index accounts for editing activity in regions with low coverage, thereby eliminating the need to quantify the editing level at each individual site and allowing the detection of even small changes in low-level editing activity. The tool is versatile and capable of calculating editing levels for all types of mismatches. Using the RNA Editing index tool allows for the detection of differences in editing levels across different conditions (Karmon et al., 2023; Knebel et al., 2024; Mann, Kopel, Eisenberg, & Levanon, 2023; Merdler-Rabinowicz et al., 2023; Tsivion-Visbord et al., 2020; Lo Giudice et al., 2020).

Endogenous deaminases, induce abundant, low-level, and non-specific RNA modifications (Bazak et al., 2014). These are difficult to identify and quantify. This editing index provides a means to compare their editing level accurately. Similarly, it may be employed to evaluate the global off-target activity of DNA and RNA base editors (Buchumenski et al., 2021). The specific mismatch of interest can be adjusted based on the type of base editor being analyzed, allowing for tailored assessments that align with the unique characteristics and activities of different base editors.

A previously published study by Buchumenski et al. described a computational pipeline that is based on the RNA Editing index tool and evaluates global off-target activity, including low-level variations. It was applied to various ABEs and CBEs DNA and RNA base editors to assess their off-target activity. Using this approach, a novel type of stochastic unintended off-target edits, undetectable by established methods, was identified. Unlike other methods for assessing off-target activity, this approach calculates the average editing levels rather than identifying specific sites. Consequently, it complements existing techniques for identifying specific off-target sites and uncovers a different type of off-target activity of base editors, arising from the inherent activity of the deaminase enzyme fused to the Cas protein.

Applying this approach to RNA-seq samples treated with DNA or RNA base editors, and focusing on coding sequences (CDS) where editing is more likely to be functionally relevant, it has been demonstrated that even the most effective base editors exhibit off-target activity far exceeding the natural error rate (Fig. 2). Among the 37 active enzymes analyzed, 35 displayed significant off-target RNA editing, indicating widespread off-target activity for A-to-I (Fig. 2A) and C-to-U (Fig. 2B) base editors. This elevated off-target activity can potentially cause harmful effects due to the increased number of unintended mutations. The deamination rates observed in the coding regions of the human genome, measured as the excess index over the control baseline level, show significant variability across different active base editors, ranging from 0.004 % (Study H, BE3 [hA3AY130F]–site 3) to 3 % (Study I, BE3) (Fig. 2). In terms of the total transcriptome mutation load, the observed range of deamination rates (0.004–3 %) corresponds to approximately 675 to 658,000 heterozygous genomic mutations in the coding sequences alone. The safety standards in terms of these RNA mutations are yet to be determined, but it seems plausible that at least the high end of the above range may lead to harmful consequences.

The same approach has been used on DNA-seq data to quantify the global off-target activity within the genome. The off-target detected in the

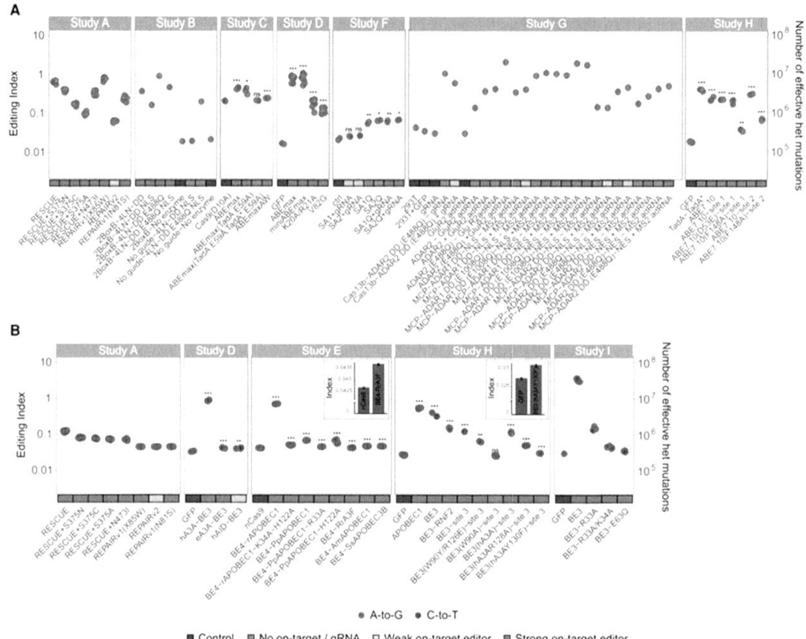

Fig. 2 This figure was reproduced from the paper by Buchumenski, et al. (Buchumenski et al., 2021), and presents data from several studies comparing samples treated with base editors to controls across multiple datasets. By applying the RNA editing index tool to coding sequence (CDS) regions, the figure demonstrates that editing activity is globally enhanced following the introduction of base editors. In some cases, such as study I in panel B, there is extensive editing on a large scale, with up to five percent of cytosines in the cell undergoing editing. **(A)** Analysis of adenine base editors (ABEs), A-to-G editing index. **(B)** Analysis of cytosine base editors (CBEs), C-to-T editing index.

genome was lower than that observed in RNA-seq data, however it remained higher than the control in several cases. It should be noted that the potential consequences of off-target activity in the genome are substantially more severe.

This approach provides a method for quantifying global transcriptomic and genomic off-target activity by DNA or RNA base editors.

4.1 Quantifying off-targets using the Editing index tool

The RNAEditingIndex tool is available at https://github.com/a2iEditing/RNAEditingIndexer. It requires installation of SAMtools (version 1.8 or higher), bedtools (version 2.26 or higher), bamUtils, Java (with SDK), and

Python 2.7. Installation of RNAEditingIndex is straightforward. A configuration script (configure.sh) is used to set up necessary environment variables and verify the presence of required dependencies.

The calculation uses a table of known SNPs (provided with the tool but may be adapted and updated at will). The input file is a sorted BAM file, which may be generated by the user using any alignment program. The analysis generates intermediate CMPileup files, which are numerical representations of pileup files. These files are stored in the designated output directory and subsequently deleted post-analysis, unless otherwise specified.

An example command line for a simple run:

RNAEditingIndex -d <BAM_files_directory> -f Aligned.sortedByCoord.out.bam -l < logs_directory > -o <cmpileup_output_directory> -os <summary_files_directory> -- genome hg38 -- regions <Path_to_bed_file_with_CDS_regions>.

The output file provides the weighted average mismatch rate over the coding region for A-to-G or C-to-T mismatches, depending on the base editor analyzed. These are possibly due to off-target activity. In addition, indices for mismatch types are calculated to establish the noise level. An excess of C-to-U mismatches (for C-to-U base editors) is an indication of off-target activity.

For example, in Buchumenski, et al. paper the RNAEditingIndex was used to study off-target activity in cells expressing BE4 C-to-U DNA base editor (Yu et al., 2020). RNA-seq data were downloaded from NCBI database (https://www.ncbi.nlm.nih.gov/) and aligned to the human reference genome hg38 using STAR (Dobin et al., 2013) v2.6.0 aligner with default parameters. The BAM files generated from the alignment were used as input for the Editing Index tool. The output reveals a significantly elevated level of C-to-U mismatches, and the excess over the control enables an estimate of the abundance of off-target events (Fig. 3).

Thus far, stochastic off-target activity was assessed globally, for the full genome or transcriptome, or for the coding region. It is possible to employ this methodology to investigate whether base editors demonstrate a preferential editing pattern towards specific regions or genes. Applying the RNA Editing index to specific regions, one may look for "hot-spots" that are more susceptible to off-target editing. By determining which genes are more likely to be edited unintentionally, researchers can better understand the risks involved in these editing events and develop strategies to minimize these off-target modifications, thereby improving the safety of gene editing therapies.

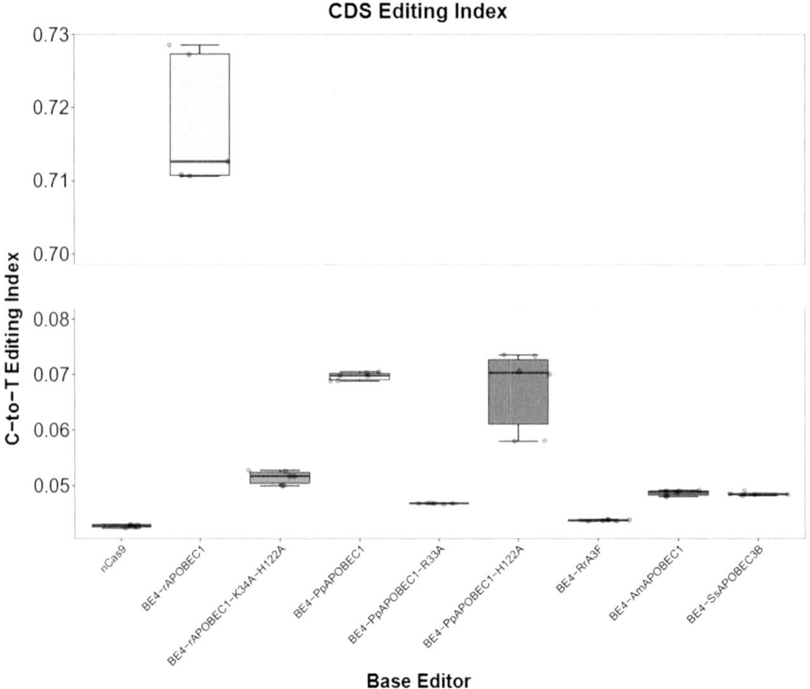

Fig. 3 Focus on Study E from Buchumenski et al. (Fig. 2B), where the RNA Editing Index output demonstrates the existence of off-target activity of the cytosine base editors due to the elevated level of C-to-T mismatches versus the control cohort.

In addition, finding a set of "hot-spots", which are more susceptible to off-target editing, may enhance the sensitivity of detection, allowing focus on these regions in order to quantify the level of off-target editing.

Furthermore, identifying gene editing preferences can provide valuable insights into the mechanisms of base editor activity and the factors influencing target selection. This knowledge may contribute to a broader understanding of editing technologies. Additionally, this information can be utilized to enhance the design of base editors, making them more efficient and specific. By understanding their editing preferences, scientists can engineer base editors with a reduced likelihood of causing unintended edits, thereby increasing their utility in both research and clinical settings.

Acknowledgements

EYL was supported by the ISF (grant number 2637/23).

References

Allen, D., Rosenberg, M., & Hendel, A. (2020). Using synthetically engineered guide RNAs to enhance CRISPR genome editing systems in mammalian cells. *Frontiers in Genome Editing, Vol. 2*, Frontiers Media S.A. https://doi.org/10.3389/fgeed.2020.617910.

Anzalone, A. V., Koblan, L. W., & Liu, D. R. (n.d.). *Genome editing with CRISPR–Cas nucleases, base editors, transposases and prime editors*. https://doi.org/10.1038/s41587-020-0561-9.

Barrangou, R., & Doudna, J. A. (2016). Applications of CRISPR technologies in research and beyond. *Nature Biotechnology, 34*(9), 933–941. https://doi.org/10.1038/nbt.3659.

Bazak, L., Haviv, A., Barak, M., Jacob-Hirsch, J., Deng, P., Zhang, R., ... Levanon, E. Y. (2014). A-to-I RNA editing occurs at over a hundred million genomic sites, located in a majority of human genes. *Genome Research, 24*(3), 365–376. https://doi.org/10.1101/gr.164749.113.

Buchumenski, I., Roth, S. H., Kopel, E., Katsman, E., Feiglin, A., Levanon, E. Y., & Eisenberg, E. (2021). Global quantification exposes abundant low-level off-target activity by base editors. *Genome Research, 31*(12), 2354–2361. https://doi.org/10.1101/GR.275770.121.

Calos, M. P. (2017). Genome editing techniques and their therapeutic applications. *Clinical Pharmacology & Therapeutics, 101*(1), 42–51. https://doi.org/10.1002/CPT.542.

Camacho, C., Coulouris, G., Avagyan, V., Ma, N., Papadopoulos, J., Bealer, K., & Madden, T. L. (2009). BLAST+: Architecture and applications. *BMC Bioinformatics, 10*(1), 1–9. https://doi.org/10.1186/1471-2105-10-421/FIGURES/4.

Chen, L., Zhang, S., Xue, N., Hong, M., Zhang, X., Zhang, D., ... Li, D. (2022). Engineering precise adenine base editor with infinitesimal rates of bystander mutations and off-target editing. *BioRxiv, 08*(12), 503700. https://doi.org/10.1101/2022.08.12.503700.

D'Addabbo, P., Cohen-Fultheim, R., Twersky, I., Fonzino, A., Silvestris, D. A., Prakash, A., ... Picardi, E. (2024). REDIportal: toward an integrated view of the A-to-I editing, *Nucleic Acids Research*, gkae1083, https://doi.org/10.1093/nar/gkae1083.

Depristo, M. A., Banks, E., Poplin, R., Garimella, K. V., Maguire, J. R., Hartl, C., ... Daly, M. J. (2011). A framework for variation discovery and genotyping using next-generation DNA sequencing data. *Nature Genetics, 43*(5), 491–498. https://doi.org/10.1038/ng.806.

Dobin, A., Davis, C. A., Schlesinger, F., Drenkow, J., Zaleski, C., Jha, S., ... Gingeras, T. R. (2013). STAR: Ultrafast universal RNA-seq aligner. *Bioinformatics (Oxford, England), 29*(1), 15–21. https://doi.org/10.1093/bioinformatics/bts635.

Gaj, T., Gersbach, C. A., & Barbas, C. F. (2013). ZFN, TALEN, and CRISPR/Cas-based methods for genome engineering. *Trends in Biotechnology, 31*(7), 397–405. https://doi.org/10.1016/J.TIBTECH.2013.04.004.

Gaudelli, N. M., Lam, D. K., Rees, H. A., Solá-Esteves, N. M., Barrera, L. A., Born, D. A., ... Ciaramella, G. (2020). Directed evolution of adenine base editors with increased activity and therapeutic application. *Nature Biotechnology, 38*(7), 892–900. https://doi.org/10.1038/s41587-020-0491-6.

Grünewald, J., Zhou, R., Garcia, S. P., Iyer, S., Lareau, C. A., Aryee, M. J., & Joung, J. K. (2019). Transcriptome-wide off-target RNA editing induced by CRISPR-guided DNA base editors. *Nature, 569*(7756), 433–437. https://doi.org/10.1038/s41586-019-1161-z.

Guo, C., Ma, X., Gao, F., & Guo, Y. (2023). Off-target effects in CRISPR/Cas9 gene editing. *Frontiers in Bioengineering and Biotechnology, 11*, 1143157. https://doi.org/10.3389/FBIOE.2023.1143157/BIBTEX.

Hu, J. H., Miller, S. M., Geurts, M. H., Tang, W., Chen, L., Sun, N., ... Liu, D. R. (2018). Evolved Cas9 variants with broad PAM compatibility and high DNA specificity. *Nature, 556*(7699), 57–63. https://doi.org/10.1038/nature26155.

Jeong, Y. K., Song, B., & Bae, S. (2020). Current status and challenges of DNA base editing tools. *Molecular Therapy, Vol. 28*, Cell Press, 1938–1952. https://doi.org/10.1016/j.ymthe.2020.07.021.

Kantor, A., McClements, M. E., & Maclaren, R. E. (2020). Crispr-cas9 dna base-editing and prime-editing. *International Journal of Molecular Sciences, Vol. 21*, MDPI AG1–22. https://doi.org/10.3390/ijms21176240.

Karmon, M., Kopel, E., Barzilai, A., Geva, P., Eisenberg, E., Levanon, E. Y., & Greenberger, S. (2023). Altered RNA editing in atopic dermatitis highlights the role of double-stranded RNA for immune surveillance. *Journal of Investigative Dermatology, 143*(6), 933–943.e8. https://doi.org/10.1016/J.JID.2022.11.010.

Kim, D., Bae, S., Park, J., Kim, E., Kim, S., Yu, H. R., ... Kim, J. S. (2015). Digenome-seq: Genome-wide profiling of CRISPR-Cas9 off-target effects in human cells. *Nature Methods, 12*(3), 237–243. https://doi.org/10.1038/nmeth.3284.

Kiran Kumar, K. D., Singh, S., Schmelzle, S. M., Vogel, P., Fruhner, C., Hanswillemenke, A., ... Stafforst, T. (2024). An improved SNAP-ADAR tool enables efficient RNA base editing to interfere with post-translational protein modification. *Nature Communications, 15*(1), https://doi.org/10.1038/s41467-024-50395-w.

Knebel, U. E., Peleg, S., Dai, C., Cohen-Fultheim, R., Jonsson, S., Poznyak, K., ... Dor, Y. (2024). Disrupted RNA editing in beta cells mimics early-stage type 1 diabetes. *Cell Metabolism, 36*(1), 48–61.e6. https://doi.org/10.1016/j.cmet.2023.11.011.

Lee, H. K., Smith, H. E., Liu, C., Willi, M., & Hennighausen, L. (2020). Cytosine base editor 4 but not adenine base editor generates off-target mutations in mouse embryos. *Communications Biology, 3*(1), 1–6. https://doi.org/10.1038/s42003-019-0745-3.

Li, G., Cheng, Y., Li, Y., Ma, H., Pu, Z., Li, S., ... Yao, Y. (2023). A novel base editor SpRY-ABE8eF148A mediates efficient A-to-G base editing with a reduced off-target effect. *Molecular Therapy – Nucleic Acids, 31*, 78–87. https://doi.org/10.1016/J.OMTN.2022.12.001.

Li, J., Yu, W., Huang, S., Wu, S., Li, L., Zhou, J., ... Qiao, Y. (2021). Structure-guided engineering of adenine base editor with minimized RNA off-targeting activity. *Nature Communications, 12*(1), 1–8. https://doi.org/10.1038/s41467-021-22519-z.

Li, T., Yang, Y., Qi, H., Cui, W., Zhang, L., Fu, X., ... Yu, T. (2023). CRISPR/Cas9 therapeutics: Progress and prospects. *Signal Transduction and Targeted Therapy, Vol. 8*. Springer Nature. https://doi.org/10.1038/s41392-023-01309-7.

Liang, P., & Huang, J. (2019). Off-target challenge for base editor-mediated genome editing. *Cell Biology and Toxicology, 35*(3), 185–187. https://doi.org/10.1007/S10565-019-09474-8/METRICS.

Lo Giudice, C., Silvestris, D. A., Roth, S. H., Eisenberg, E., Pesole, G., Gallo, A., & Picardi, E. (2020). Quantifying RNA editing in deep transcriptome datasets. *Front Genet, 6*(11), 194. https://doi.org/10.3389/fgene.2020.00194. PMID: 32211029; PMCID: PMC7069340.

Mann, T. D., Kopel, E., Eisenberg, E., & Levanon, E. Y. (2023). Increased A-to-I RNA editing in atherosclerosis and cardiomyopathies. *PLoS Computational Biology, 19*(4), https://doi.org/10.1371/journal.pcbi.1010923.

Mansi, L., Tangaro, M. A., Lo Giudice, C., Flati, T., Kopel, E., Schaffer, A. A., ... Picardi, E. (2021). REDIportal: Millions of novel A-to-I RNA editing events from thousands of RNAseq experiments. *Nucleic Acids Research, 49*(D1), D1012–D1019. https://doi.org/10.1093/nar/gkaa916.

McKenna, A., Hanna, M., Banks, E., Sivachenko, A., Cibulskis, K., Kernytsky, A., ... DePristo, M. A. (2010). The Genome Analysis Toolkit: A MapReduce framework for analyzing next-generation DNA sequencing data. *Genome Research, 20*(9), 1297–1303. https://doi.org/10.1101/GR.107524.110.

Merdler-Rabinowicz, R., Gorelik, D., Park, J., Meydan, C., Foox, J., Karmon, M., ... Levanon, E. Y. (2023). Elevated A-to-I RNA editing in COVID-19 infected individuals. *NAR Genomics and Bioinformatics, 5*(4), https://doi.org/10.1093/nargab/lqad092.

Merkle, T., Merz, S., Reautschnig, P., Blaha, A., Li, Q., Vogel, P., ... Stafforst, T. (2019). Precise RNA editing by recruiting endogenous ADARs with antisense oligonucleotides. *Nature Biotechnology, 37*(2), 133–138. https://doi.org/10.1038/s41587-019-0013-6.

Montiel-Gonzalez, M. F., Diaz Quiroz, J. F., & Rosenthal, J. J. C. (2019). *Current strategies for Site-Directed RNA Editing using ADARs. Methods, Vol. 156*, Academic Press Inc. 16–24. https://doi.org/10.1016/j.ymeth.2018.11.016.

Montiel-Gonzalez, M. F., Vallecillo-Viejo, I., Yudowski, G. A., & Rosenthal, J. J. C. (2013). Correction of mutations within the cystic fibrosis transmembrane conductance regulator by site-directed RNA editing. *Proceedings of the National Academy of Sciences of the United States of America, 110*(45), 18285–18290. https://doi.org/10.1073/pnas.1306243110.

Newby, G. A., & Liu, D. R. (2021). In vivo somatic cell base editing and prime editing. *Molecular therapy, Vol. 29*, Cell Press, 3107–3124. https://doi.org/10.1016/j.ymthe.2021.09.002.

Picardi, E., & Pesole, G. (2013). REDItools: High-throughput RNA editing detection made easy. *Bioinformatics (Oxford, England), 29*(14), 1813–1814. https://doi.org/10.1093/BIOINFORMATICS/BTT287.

Picardi, E., D'Erchia, A. M., Lo Giudice, C., & Pesole, G. (2017). REDIportal: A comprehensive database of A-to-I RNA editing events in humans. *Nucleic Acids Research, 45*(D1), D750–D757. https://doi.org/10.1093/NAR/GKW767.

Pinello, L., Canver, M. C., Hoban, M. D., Orkin, S. H., Kohn, D. B., Bauer, D. E., & Yuan, G. C. (2016). Analyzing CRISPR genome-editing experiments with CRISPResso. *Nature Biotechnology, 34*(7), 695–697. https://doi.org/10.1038/nbt.3583.

Porto, E. M., Komor, A. C., Slaymaker, I. M., & Yeo, G. W. (2020). Base editing: Advances and therapeutic opportunities. *Nature Reviews. Drug Discovery, 19*(12), 839–859. https://doi.org/10.1038/s41573-020-0084-6.

Rees, H. A., & Liu, D. R. (2018). Base editing: Precision chemistry on the genome and transcriptome of living cells. *Nature Reviews. Genetics, 19*(12), 770–788. https://doi.org/10.1038/s41576-018-0059-1.

Rees, H. A., Komor, A. C., Yeh, W. H., Caetano-Lopes, J., Warman, M., Edge, A. S. B., & Liu, D. R. (2017). Improving the DNA specificity and applicability of base editing through protein engineering and protein delivery. *Nature Communications, 8*(1), 1–10. https://doi.org/10.1038/ncomms15790.

Roth, S. H., Levanon, E. Y., & Eisenberg, E. (2019). Genome-wide quantification of ADAR adenosine-to-inosine RNA editing activity. *Nature Methods, 16*(11), 1131–1138. https://doi.org/10.1038/s41592-019-0610-9.

Sakovina, L., Vokhtantsev, I., Vorobyeva, M., Vorobyev, P., & Novopashina, D. (2022). Improving stability and specificity of CRISPR/Cas9 system by selective modification of guide RNAs with 2′-fluoro and locked nucleic acid nucleotides. *International Journal of Molecular Sciences, 23*(21), https://doi.org/10.3390/ijms232113460.

Slesarenko, Y. S. A. L., Lavrov, V., & Smirnikhina, S. A. (2022). *Off-target effects of base editors: What we know and how we can reduce it, 68*, 39–48. https://doi.org/10.1007/s00294-021-01211-1.

Stafforst, T., & Schneider, M. F. (2012). An RNA-deaminase conjugate selectively repairs point mutations. *Angewandte Chemie – International Edition, 51*(44), 11166–11169. https://doi.org/10.1002/anie.201206489.

Strauß, A., & Lahaye, T. (2013). Zinc fingers, TAL effectors, or Cas9-based DNA binding proteins: What's best for targeting desired genome loci? *Molecular Plant, 6*(5), 1384–1387. https://doi.org/10.1093/mp/sst075.

Tsai, S. Q., Nguyen, N. T., Malagon-Lopez, J., Topkar, V. V., Aryee, M. J., & Joung, J. K. (2017). CIRCLE-seq: A highly sensitive in vitro screen for genome-wide CRISPR–Cas9 nuclease off-targets. *Nature Methods, 14*(6), 607–614. https://doi.org/10.1038/nmeth.4278.

Tsivion-Visbord, H., Kopel, E., Feiglin, A., Sofer, T., Barzilay, R., Ben-Zur, T., ... Levanon, E. Y. (2020). Increased RNA editing in maternal immune activation model of neurodevelopmental disease. *Nature Communications, 11*(1), https://doi.org/10.1038/s41467-020-19048-6.

Vallecillo-Viejo, I. C., Liscovitch-Brauer, N., Montiel-Gonzalez, M. F., Eisenberg, E., & Rosenthal, J. J. C. (2018). Abundant off-target edits from site-directed RNA editing can be reduced by nuclear localization of the editing enzyme. *RNA Biology, 15*(1), 104–114. https://doi.org/10.1080/15476286.2017.1387711.

Wang, L., Xue, W., Yan, L., Li, X., Wei, J., Chen, M., ... Chen, J. (2017). Enhanced base editing by co-expression of free uracil DNA glycosylase inhibitor. *Cell Research, Vol. 27*, Nature Publishing Group 1289–1292. https://doi.org/10.1038/cr.2017.111.

Whittaker, M. N., Brooks, D. L., Quigley, A., Jindal, I., Said, H., Qu, P., ... Wang, X. (2024). Improved specificity and efficacy of base-editing therapies with hybrid guide RNAs. *BioRxiv*. https://doi.org/10.1101/2024.04.22.590531.

Yu, Y., Leete, T. C., Born, D. A., Young, L., Barrera, L. A., Lee, S. J., ... Gaudelli, N. M. (2020). Cytosine base editors with minimized unguided DNA and RNA off-target events and high on-target activity. *Nature Communications, 11*(1), https://doi.org/10.1038/s41467-020-15887-5.

Zhang, X., Wang, X., Lv, J., Huang, H., Wang, J., Zhuo, M., ... Rong, Z. (2023). Engineered circular guide RNAs boost CRISPR/Cas12a- and CRISPR/Cas13d-based DNA and RNA editing. *Genome Biology, 24*(1), 1–18. https://doi.org/10.1186/S13059-023-02992-Z/FIGURES/6.

Zhang, Z., Tao, W., Huang, S., Sun, W., Wang, Y., Jiang, W., ... Lin, C. P. (2022). Engineering an adenine base editor in human embryonic stem cells with minimal DNA and RNA off-target activities. *Molecular Therapy Nucleic Acids, 29*, 502–510. https://doi.org/10.1016/j.omtn.2022.07.026.

Zuo, E., Sun, Y., Wei, W., Yuan, T., Ying, W., Sun, H., ... Yang, H. (2019). Cytosine base editor generates substantial off-target single-nucleotide variants in mouse embryos. *Science (New York, N. Y.), 364*(6437), 289–292. https://doi.org/10.1126/SCIENCE.AAV9973/SUPPL_FILE/AAV9973_ZUO_SM.PDF.

CHAPTER ELEVEN

Restoration of cytidine to uridine genetic code using an MS2-APOBEC1 artificial enzymatic approach

Sonali Bhakta[a,b] **and Toshifumi Tsukahara**[a,c,*]

[a]Area of Bioscience and Biotechnology, School of Materials Science, Japan Advanced Institute of Science and Technology, 1–1 Asahidai, Nomicity, Ishikawa 923–1292, Japan

[b]Department of Anatomy and Histology, Bangladesh Agricultural University, Mymensingh 2202, Bangladesh

[c]GeCoRT Co., Ltd., Kanagawa, 220–0011, Japan

*Corresponding author: e-mail address: tukahara@jaist.ac.jp

Contents

Abstract

By employing site-directed RNA editing (SDRE) to restore point-mutated RNA molecules, it is possible to change gene-encoded information and synthesize proteins with different functionality from a single gene. Thymine (T) to cytosine (C) point mutations cause various genetic disorders, and when they occur in protein-coding regions, C-to-uridine (U) RNA changes can lead to non-synonymous alterations. By joining the deaminase domain of apolipoprotein B messenger RNA (mRNA) editing catalytic polypeptide 1 (APOBEC1) with a guide RNA (gRNA) complementary to a target mRNA,

Methods in Enzymology, Volume 713
ISSN 0076-6879, https://doi.org/10.1016/bs.mie.2024.11.034
Copyright © 2025 Elsevier Inc. All rights are reserved, including those for text and data mining, AI training, and similar technologies.

we created an artificial RNA editase. We used an mRNA encoding blue fluorescent protein (BFP), obtained from the green fluorescent protein (GFP) gene through the introduction of a T > C mutation, as our target RNA. In a proof of principle experiment, we reverted the T > C mutation at the RNA level using our APOBEC1 site-directed RNA editing system, recovering GFP signal. Sanger sequencing of cDNA from transfected cells and polymerase chain reaction-restriction length polymorphism analysis validated this result, indicating an editing of approximately 21 %. Our successful development of an artificial RNA editing system using the deaminase APOBEC1, in conjunction with the MS2 system, may lead to the development of treatments for genetic diseases based on the restoration of specific types of wild type sequences at the mRNA level.

1. Introduction

RNA editing is a biological process which alter RNA sequence irreversibly. Additionally, it can be harnessed as a tool for repairing DNA mutations at the RNA level. The term 'RNA editing' was first used to describe a process that occurs in trypanosomes in mitochondrion-encoded kinetoplastid messenger RNA (mRNA), involving the insertion and deletion of uridine monophosphate (UMP) into nascent transcripts after transcription (Wolf, Gerber, & Keller, 2002). Since the discovery of this type of post-transcriptional sequence modification, the number of processes associated with the term RNA editing has grown to include the insertion and deletion of nucleotides other than UMP, base deamination, and co-transcriptional insertion of non-template nucleotides. Further, RNA editing has been observed in mRNA, transfer RNA (tRNA), and ribosomal RNA molecules, including nuclear encoded RNAs and those derived from mitochondria and chloroplast organelles (Wolf et al., 2002). Examples of RNA editing have been identified in numerous Metazoan species, in addition to unicellular eukaryotes, such as trypanosomes and plants. In addition, RNA editing has been observed on a small scale in prokaryotes, including a detailed study of tRNA editing in *Escherichia coli* (Brennicke, Marchfelder, & Binder, 1999).

RNA editing can be categorized into two basic types, depending on their response mechanisms. Insertion/deletion RNA editing, involves the insertion or deletion of targeted nucleotides, with the aim of changing the codon sequence of a targeted mRNA (Wolf et al., 2002). Another form of RNA editing involves changing one coding nucleotide to a different nucleotide via base alteration or modification, without modifying the overall length of the RNA, ultimately changing the codon sequence; this approach is often applied to treat single nucleotide mutations. In this chapter, we focus on RNA editing of cytidine (C) to uridine (U). Thanks to rapid advances in the development of various tools, particularly next generation sequencing and bioinformatics methods, over the last two decades, many thousands of RNA editing events have been detected in both plants and mammals (Bahn et al., 2012; John, Weirick, Dimmeler, & Uchida, 2017).

In mammals, C-to-U alteration depends on apoB mRNA editing catalytic polypeptide-like (APOBEC) family proteins. In humans, eleven genes encoding members of the APOBEC family have been discovered to date, all of which have a zinc-dependent deaminase domain (Blanc & Davidson, 2010; Navaratnam et al., 1995). Among APOBEC protein subfamilies, only APOBEC1, 3 A, 3 B, and 3 G (Fig. 1) are proven to mediate C-to-U RNA editing (Moris, Murray, & Cardinaud, 2014; Patnaik & Kannisto, 2016; Sharma et al., 2015; Sharma, Patnaik, Taggart, & Baysal, 2016). The role of APOBEC enzymes in RNA editing is well-established, although these proteins also exhibit DNA editing capacity. APOBEC proteins are members of the cytidine deaminase family, which catalyze the conversion of C residues in RNA molecules to U residues. This mechanism is essential for producing variations in RNA sequences and can impact RNA translation efficiency, stability, and interactions with other molecules, among other aspects of RNA function (Bhakta & Tsukahara, 2022).

C-to-U RNA editing can also be used as a therapeutic approach that can convert specific C-to-U nucleotides in transcript sequences. C-to-U RNA editing can be applied to convert start or stop codons to those

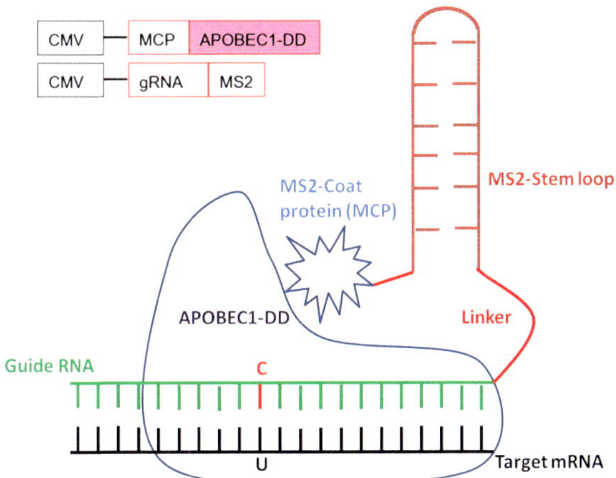

Fig. 1 Schematic diagram of the MS2-APOBEC1 system for restoration of the RNA genetic code from C-to-U.

encoding amino acids, depending on splice site preferences (Gott & Emeson, 2000). C-to-U type RNA editing was originally identified in vertebrates in mRNA encoding apolipoprotein B (apoB). Deamination through hydrolysis at the C4 site of C was later found to contribute to apoB editing (Chen et al., 1987; Powell et al., 1987). The presence of both cis-acting (tripartite regulatory sequences) and trans-acting elements around the altered C are required for this conversion process to occur, and conversion requires not only the catalytic C deaminase but also auxiliary proteins (Smith, 2016). Editing of C-to-U at the RNA level has also been detected in higher plants, particularly in their mitochondria and chloroplasts (Freyer, Kiefer-Meyer, & Kössel, 1997).

Previous studies have investigated artificial site-directed A-to-I RNA editing using ADAR family enzymes tethered to guide RNA (gRNA) molecules complementary to a target sequence (Katrekar et al., 2019; Maas, Rich, & Nishikura, 2003; Nishikura, 2010). For ADAR tethering, multiple options have been described: Stafforst and colleagues used the SNAP-tag (Stafforst & Schneider, 2012; Vogel & Stafforst, 2014; Vogel, Hanswollemenke, & Stafforst, 2017), Rosenthal and colleagues used the Lambda N protein and box B element RNA (Montiel-González, Vallecillo-Viejo, & Rosenthal, 2016; Montiel-Gonzalez, Quiroz & Rosenthal 2019), and our group has applied the MS2 system (Azad, Bhakta, & Tsukahara, 2017; Bhakta, Azad, & Tsukahara, 2018). There are some other tools which can be used for C to U editing such as RESCUE or RESCUE-S which are based on CRISPR/Cas proteins. Here we describe the use of an MS2-based system for APOBEC1 tethering and therefore, for the restoration of the genetic code in RNA containing a C-to-U point mutation using the restoration of GFP from BFP mRNA as an example (Fig. 1). Future research in this field should focus on improving editing efficiency using techniques including modification of the MS2 stem-loop and employing gRNA molecules of varying lengths. Further, evaluation of practical application of the MS2 system in animal models is strongly warranted.

2. Methods

The target blue fluorescent protein (*BFP*) gene was generated by introduction of a point mutation of thymine (T) residue 199 of the green fluorescent protein (*GFP*) gene in the pcDNA3.1 +eGFP plasmid (Addgene), using the methods described by Luyen et al. (Luyen, Nguyen, Azad, Suzuki,

& Tsukahara, 2012, Luyen et al., 2015). HEK293 cells were transfected with the BFP construct (Accession: SAMN14389947; ID: 14389947), and transfectants were selected by culture in medium containing 500 ng/mL G418. Selected positive colonies were confirmed by Sanger sequencing, using the following primers to amplify *BFP*:

Forward primer:
GCTTATCGAAATTAATACG.

Reverse primer:
CACGGGCAGCTTGCCGGTGGT

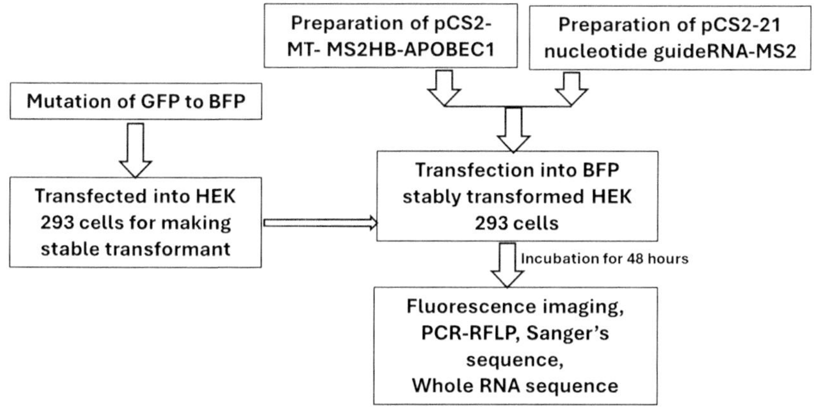

2.1 APOBEC 1 deaminase plasmid construction

Using the *Xho*I and *Xba*I restriction sites (Takara, Shiga, Japan), the sequence encoding the APOBEC1 deaminase domain was cloned downstream of MS2 in the pCS2 + MT vector under the control of the pol II CMV IE-94 promoter, to localize the enzyme to the desired *BFP* codon; the resulting plasmid was referred to as pCS2 + MT-MS2HB-APOBEC1. Primers containing the necessary restriction sites were used to amplify *APOBEC1* from HEK293 cells; the primer sequences were: *Xho*I catalytic *APOBEC1* Forward, tcca**ctcgag**atgccctgggagttt-gacgtctt; and *Xba*I catalytic *APOBEC1* Reverse, acgg**tctaga**ttaagggtgccgactcagaaactc. The trinucleotide, atg/tta, was used as a leader sequence to allow appropriate recognition by the restriction enzymes. Selected positive colonies were analyzed by Sanger sequencing to confirm the presence of the *APOBEC1* catalytic domain open reading frame. Sequences were verified using NCBI-BLAST and the ExPASY Bioinformatics resource portal.

2.2 Preparation of gRNA to direct the deaminase to the target sequence

A gRNA complementary to the target sequence was prepared by inserting a 21-nucleotide sequence with a mismatch adenine (A) at the target C position, into the pCS2 only vector plasmid upstream of MS2-RNA under control of the CMVIE-94 promoter. This was achieved by adding the guide sequence to a forward primer for PSL-MS2–6 × (pCS2 + guide-MS2-RNA), where MS2 6 × refers to six copies of the MS2 stem-loop sequence. Although 12 × and 24 × MS2 stem loops are available, the 6 × MS2 stem-loop was chosen for this study. The following sequence order was used for this purpose: atca**GAATTC***CACTGCACGCCGTTGGACAGG*GAATGGCC ATG, where the underlined region indicates the MS2–6 × forward primer, bold font identifies the restriction site, italic font indicates the 21-nucleotide guide sequence, and the atca tetra-nucleotide (lowercase) is leader sequence that allows correct recognition by the restriction enzyme. Positive colonies were confirmed by Sanger sequencing. The RNA Pol II promoter (CMV-IE94) drives gRNA expression and is expressed in human cells. The A:C mismatch is placed centrally in the gRNA.

2.3 Cell culture and transfection

Lipofectamine 2000 (Invitrogen, Carlsbad, CA, USA) was used to transfect 50–70% confluent HEK293 cells stably expressing BFP. Each well of a 12-well plate was seeded with 3×10^5 cells. Each well contained 1 mL Opti-MEM (Gibco) and 4 μl Lipofectamine 2000, and was transfected with 800 ng APOBEC1 deaminase and 700 ng gRNA plasmids. Samples were then cultivated in an incubator at 37 °C for 6 h, before replacement of Opti-MEM with D-MEM (Gibco), to promote cell growth, and culture at 37 °C for 48 h before observation.

2.4 Observation of fluorescence by confocal microscopy

Cells were observed under optimal conditions using an Olympus FV1000D confocal laser scanning microscope (Olympus, Shinjuku-ku, Tokyo). To obtain very clear images, conditions were designed to improve effective resolution, dye selection, and exposure time, and magnification was adjusted, allowing the exact location of fluorescence in cell samples to be pinpointed, particularly within individual cells (Fig. 2A).

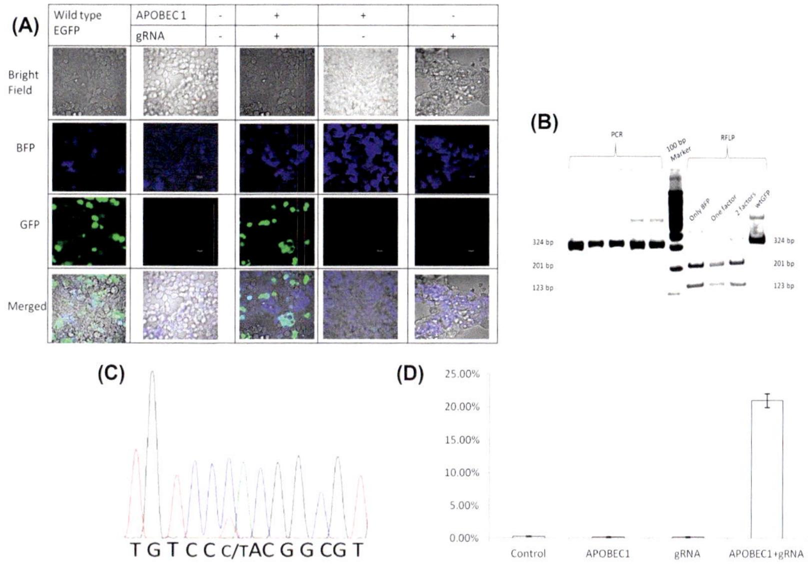

Fig. 2 (A) HEK 293 cells stably expressing BFP were transfected with wild type GFP, one factor (i.e., either APOBEC1 or guide RNA alone) or two factors (APOBEC1 + guide RNA). Green fluorescence expression was observed only when two factors were present, implying that both factors were necessary for C-to-U editing. Imaging was performed by laser scanning confocal microscopy. (B) PCR-restriction fragment length polymorphism analysis of cDNA extracted from transfected HEK 293 cells stably expressing BFP by restriction digestion with BtgI. The *BFP* sequence was cleaved into two fragments of 201 and 123 bp, whereas the restored *GFP* sequence was not cleaved and remained at 324 bp. (C) Confirmation of restoration of *BFP* to *GFP* (C-to-U) by Sanger sequencing. Using the forward or sense primer, dual peaks were observed (CCA to CTA), due to restoration of the genetic code from C to U (*BFP* to *GFP*), after application of the two editing factors, APOBEC1 deaminase and gRNA. (D) Edit-R analysis of sense and antisense chromatogram heights of peaks for the edited nucleotide and statistical analysis (mean ± SEM; n = 5). Analysis was conducted using Edit R Version 10 (https://moriaritylab.shinyapps.io/editr_v10/) based on Ab.1 files generated by Sanger sequencing.

2.5 Extraction of RNA from transfected cells and cDNA synthesis

Following microscopic analysis, cells were collected and total RNA samples extracted using TRIzol reagent (Invitrogen). Collected RNA was converted into cDNA using the Invitrogen SuperScript III First-Strand Synthesis System for reverse transcribed PCR.

2.6 Confirmation of sequence restoration by PCR-restriction fragment length polymorphism (RFLP)

To confirm successful restoration of the wild type sequence, PCR products amplified with GoTaq polymerase (Promega, Madison, WI, USA) on a GeneAmpPCR system 9700 (Applied Biosystems, Foster City, CA, USA) were digested with a restriction enzyme that distinguished between edited and unedited DNA sequences. Amplicons were then separated by electrophoresis in 6 % polyacrylamide gels and stained with SYBR Green dye (Invitrogen). Each PCR reaction (20 μl) included 100 ng cDNA.

Following PCR amplification, 10 μl aliquots of PCR products were subjected to restriction digestion using BtgI (New England BioLabs) at 37 °C for 2 h, which cleaved the *BFP* sequence into two fragments (201 and 123 bp), while leaving the restored *GFP* sequence intact. For gel electrophoresis, identical volumes (2 μl) of digested PCR products were loaded in wells made using a 14-well comb, and an LAS-3000 imager was used to capture images.

Presence of the full-length 324 bp sequence indicated conversion of *BFP* to *GFP* (Fig. 2B). Band intensity was measured using ImageJ software (NIH, https://imagej.net/Citing) and editing efficiency was calculated from band intensity values using the following equation:

$$\frac{\text{Uncut T at 324 bp}}{\text{Uncut T at 324 bp} + \text{Cut C at 201 bp} + \text{Cut C at 123 bp}} \times 100\%$$

2.7 Sanger sequencing

Following PCR using 100 ng cDNA per reaction (volume, 20 μl), PCR products were resolved on 1 % agarose gels and bands were excised and frozen. DNA samples were purified using a QIA quick Gel Extraction kit and their concentrations were quantified with an ND-1000 spectrophotometer. Purified DNA was sequenced using the *BFP* forward primer (GCTTATCGAAATTAATACG) and a Big Dye Terminator v3.1 Cycle Sequencing Kit (Thermo Fisher, Waltham, MA, USA). Sequence Scanner software (version 2, Applied Biosystems) was used to evaluate raw sequencing data. A dual peak (C [unedited] and T [edited]) was detected at the target location when edited and unedited products were both present (Fig. 2C).

Editing efficiency was calculated by measuring peak area and peak height using ImageJ software (NIH, https://imagej.net/Citing), as reported previously (Bhakta et al., 2018) (Eggington, Greene, & Bass, 2011; Rinkevich, Schweitzer, & Scott, 2015) (Fig. 2D).

2.8 Editing efficiency (sense)

$$\text{Considering peak area: } \frac{\text{Area of T}}{\text{Area of T} + \text{Area of C}} \times 100\%$$

$$\text{Considering peak height: } \frac{\text{Peak height of T}}{\text{peak height of T} + \text{Peak height of C}} \times 100$$

Chromatogram peak height was also measured, and editing rates were analyzed using Edit-R (v1.0.10) or MultiEditR (v1.1.0) software, as described by Kluesner et al. (2018, 2021).

2.9 Total RNA-sequencing (RNA-seq)

RNA-seq analysis was conducted by Filgen (Nagoya, Japan). Reads were trimmed using CLC Genomics Server 11 (Qiagen), to remove low-quality reads from the analysis. Reads with the following characteristics were eliminated: including those with bases with Phred scores < 13, containing > 3 "N" nucleotides, and lengths < 100 bp. Function mapping of RNA-seq data was conducted.

RNA-seq mapping parameters were as follows:

Reference type = Genome annotated with genes and transcripts.

Reference sequence = *Homo sapiens* (hg38) sequence + BFP1 sequence.

Gene track = *Homo sapiens* (hg38) (Gene) + BFP1 gene.

mRNA track = *Homo sapiens* (hg38) (mRNA) + BFP1 mRNA.

Mismatch cost = 2.

Insertion cost = 3.

Deletion cost = 3.

Length fraction = 0.8.

Similarity fraction = 0.8.

Global alignment = No.

Auto-detect paired distances = No.
Strand specific = Both.
Maximum number of hits for a read = 10.

For edited *GFP* samples obtained following application of two factors (APOBEC1 + guide RNA) in cells stably expressing BFP (HEK_293T3F) and targeted mRNA from HEK 293 cells stably expressing BFP (BFP_1), the average coding area coverage values based on mapping results were 82.6 % and 7.5 %, respectively, with 85 % of sequence reads mapped onto exons.

Basic Variant Detection, implemented in CLC Genomics Server 11, was used for variant detection, as follows: Minimum coverage (lower limit for the number of reads mapped to the mutation location) = 10; Minimum count (lower limit for the number of reads actually calling the mutation) = 5; Minimum frequency (lower limit for the percentage of reads that mapped with mutations, calculated as counts/coverage) (%) = 1.0. Mutations that met all three conditions were detected.

RNA-seq data were visualized in several ways, including the percentages of expressed genes with at least one edited cytosine (C-to-U or G-to-A) among total single nucleotide variants; box plots comparing the rate of cytosine editing by APOBEC1(DD)-MS2 to that in the editing-negative control, allowing for detection of off-target events; and jitter plots showing the transcriptome wide efficiency values of C-to-U or G-to-A edits (y-axis) found in HEK 293 cells modified by APOBEC1(DD)-MS2 by RNA-seq, using the procedure described by others (Kluesner et al., 2018; Vogel & Stafforst, 2014) (Fig. 3).

3. Discussion

In this chapter, we present an artificial system that includes a deaminase (APOBEC1) to provide a C-to-U RNA-editing enzyme compatible with the MS2 system (Fig. 1). The results of the experiments indicate that addition of a synthetic gRNA can render the APOBEC1 deaminase domain, which has no double-stranded RNA-binding domain, capable of RNA editing. Since the APOBEC1 deaminase domain lacks both a nuclear export signal and a nuclear localization signal, the modified enzyme system is expected to localize to the cytoplasm, where it is most likely to come into contact with the target mRNA (Fig. 1); however, whether minute quantities of the MS2-fused deaminase chimera can localize to the nucleus is currently unknown.

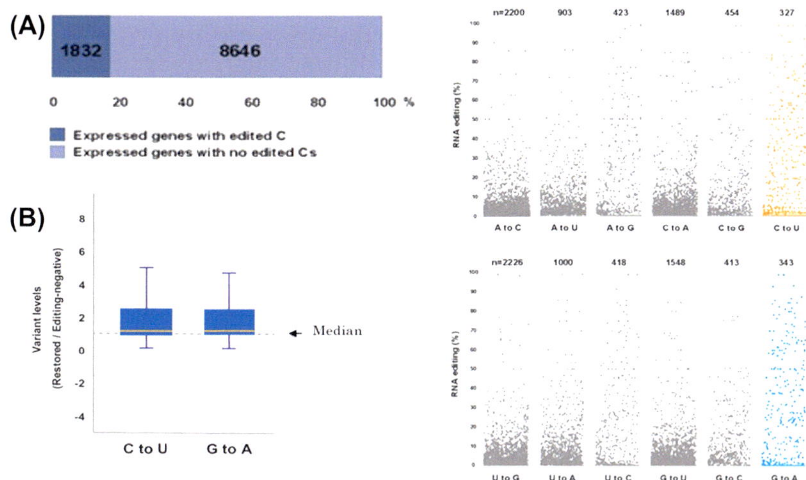

Fig. 3 ARPOBEC1(DD)-MS2 system induces some off-target C-to-U RNA editing in HEK293 cells. (A) Percentages of expressed genes with at least one edited cytosine (C-to-U or G-to-A) in total SNVs. (B) Box plots showing rate of cytosines edited by APOBEC1(DD)-MS2 compared to control (mutated BFP target in HEK 293 cells), yellow line is median. (C) Jitter plots showing transcriptome-wide efficiencies of C-to-U or G-to-A edits (y-axis) identified from RNA-seq experiments in HEK293 cells modified by APOBEC1 (DD)-MS2 vs editing-negative control (BFP target, stably transformed in HEK 293 cells). n: total number of modified cytosines identified.

Regarding restoration of the genetic code using the artificial deaminase enzyme system created in this study, we were only able to detect conversion of (mutant) BFP to GFP when BFP-expressing cells were transfected with both the catalytic domain and the gRNA. Future research should focus on further optimization of the system, including the gRNA sequence, flexibility in using the MS2 moiety with enzymes, and relative amounts of the two components. Minimal auto-editing can be prevented by using appropriate amounts of the enzyme and reporter substrate (Fukuda et al., 2017; Schneider, Wettengel, Hoffmann, & Stafforst, 2014).

This study demonstrated that, in contrast to single transfection with either APOBEC1 deaminase or gRNA, which resulted in their cleavage, transfection with both the APOBEC1 deaminase and gRNA resulted in no cleavage of the recovered GFP sequence by BtgI. This result fully validates the findings of Luyen et al. (2015), and Luyen and Toshifumi (2017), and is consistent with the sequence preference of the BtgI restriction enzyme.

The 324 bp band visually appeared slightly stronger than those at 201 and 123 bp; however, densitometry analysis of band intensity demonstrated a restoration percentage of 20.4 % (Fig. 2B).

Sanger sequencing provided additional proof of the function of the established system. Dual peaks were detected only in cases where editing resulted genetic code restoration. We computed the editing efficiency for both sense and antisense primers based on peak height, showing that our deaminase method altered 21.5 % of BFP. We also analyzed sequence outcomes using Edit-R software (version 10), following the method described by the Moriarty group (Kluesner et al., 2018), and determined the editing percentage of the specific nucleotide by measuring chromatogram peaks. This analysis revealed that an average of 21 % of target C nucleotides were restored to U nucleotides, consistent with the proportions determined based on peak height and peak area (Fig. 2C).

To investigate off-target effects, we sequenced each position of the amplified RNA coding sequences, and detected only a single point mutation upstream of the target C. The mutation was silent with no effect on the amino acid sequence. Hence, the off-target effects were minimal in this system employing RNA editing with the MS2 system, together with gRNA and the APOBEC 1 deaminase domain. The developers of the Cas9-AID system likewise reported a very low percentage of off-target editing (0.14–0.38 %), indicating that off-target effects are relatively rare when this type of approach is used, although they described a genome editing system, while we report a method for RNA editing (Nishida et al., 2016; Shimatani et al., 2017). Further, a group using CRISPR, and adenosine base editors reported an indel rate of only 0.1 % (Gaudelli et al., 2017); however, their system, which also developed for genome editing and used CRISPR with adenine base editors, effectively converted targeted A•T base pairs to G•C in approximately 50 % of cases in a human cell system. Subsequently, Zuo et al. performed in vivo cytosine base editing, and discovered that only one to four indels were present in the resulting embryos, none of which overlapped with anticipated off-target sites (Zuo et al., 2019); in their study, genome-wide off-target analysis was performed using the two-cell embryo injection technique, in conjunction with base editors/cas9, to identify off-target events in vivo.

The research described here presents a method that can effectively correct point mutations (U-to-C) in RNA, and thus would have the potential to reduce symptoms of patients with various diseases caused by such mutations; for example, Menkes kinky disease/Copper deficiency syndrome. Other hereditary disorders resulting from T-to-C mutations, such as insulin resistance and thalassemia α, among others, could also be treated by editing using gRNA and MS2-APOBEC1 to correct altered RNA sequences.

4. Conclusion

This chapter describes how to use the MS2-APOBEC1 system to effectively correct U-to-C point mutations at the RNA level. The system has the potential to alleviate the symptoms of patients with various diseases caused by such mutations through C-to-U RNA editing. Future research should focus on maximizing the editing efficiency of this technique while minimizing the transcriptome-wide off-target effect. This could include adjustment of the MS2 stem-loop and employing gRNA molecules of varying length. The value of this method should also be assessed using animal models.

Acknowledgments

The authors gratefully acknowledge scholarships from the Ministry of Education, Culture, Sports, Science, and Technology (MEXT), Japan.

References

Azad, M. T. A., Bhakta, S., & Tsukahara, T. (2017). Site-directed RNA editing by adenosine deaminase acting on RNA (ADAR1) for correction of the genetic code in gene therapy. *Gene Therapy, 24*(12), 779–786.

Bahn, J. H., Lee, J. H., Li, G., Greer, C., Peng, G., & Xiao, X. (2012). Accurate identification of A-to-I RNA editing in human by transcriptome sequencing. *Genome Research, 22*, 142–150.

Bhakta, S., Azad, M. T. A., & Tsukahara, T. (2018). Genetic code restoration by artificial RNA editing of Ochre stop codon with ADAR1deaminase. *Protein Engineering, Design and Selection, 31*(12), 1–8.

Bhakta, S., & Tsukahara, T. (2022). C-to-U RNA editing: A site directed RNA editing tool for restoration of genetic code. *Genes, 13*(9), 1636.

Blanc, V., & Davidson, N. O. (2010). APOBEC-1-mediated RNA editing. *Wiley Interdisciplinary Reviews: Systems Biology and Medicine, 2*, 594–602.

Brennicke, A., Marchfelder, A., & Binder, S. (1999). RNA editing. *FEMS Microbiology Reviews, 23*, 297–316.

Chen, S. H., Habib, G., Yang, C. Y., Gu, Z. W., Lee, B. R., Weng, S. A., ... Rosseneu, M. (1987). Apolipoprotein B-48 is the product of a messenger RNA with an organ-specific inframe stop codon. *Science (New York, N. Y.), 238*, 363–366.

Eggington, J. M., Greene, T., & Bass, B. L. (2011). Predicting sites of ADAR editing in double-stranded RNA. *Nature Communications, 2*(319), 1–9.

Freyer, R., Kiefer-Meyer, M. C., & Kössel, H. (1997). Occurrence of plastid RNA editing in all major lineages of land plants. *Proceedings of the National Academy of Sciences, 94*, 6285–6290.

Fukuda, M., Umeno, H., Nose, K., Nishitarumizu, A., Noguchi, R., & Nakagawa, H. (2017). Construction of a guide-RNA for site-directed RNA mutagenesis utilizing intracellular A-to-I RNA editing. *Scientific Reports,* 7(41478), 1–13.

Gaudelli, N. M., Komor, A. C., Rees, H. A., Packer, M. S., Badran, A. H., Bryson, D. I., & Liu, D. R. (2017). Programmable base editing of A or T to G or C in genomic DNA without DNA cleavage. *Nature, 551,* 464–471.

Gott, J. M., & Emeson, R. B. (2000). Functions and mechanisms of RNA editing. *Annual Review of Genetics, 34,* 499–531.

John, D., Weirick, T., Dimmeler, S., & Uchida, S. (2017). RNA editor: Easy detection of RNA editing events and the introduction of editing islands. *Briefings in Bioinformatics, 18,* 993–1001.

Katrekar, D., Chen, G., Meluzzi, D., Ganesh, A., Worlikar, A., Shih, Y. R., ... Mali, P. (2019). In vivo RNA editing of point mutations via RNA-guided adenosine deaminases. *Nature Methods, 16*(3), 239–242.

Kluesner, M. G., Nedveck, D. A., Lahr, W. S., Garbe, J. R., Abrahante, J. E., Webber, B. R., & Moriarity, B. S. (2018). EditR: A method to quantify base editing via Sanger sequencing. *CRISPR Journal, 1*(3), 239–250.

Kluesner, M. G., Tasakis, R. N., Lerner, T., Arnold, A., Wüst, S., Binder, M., ... Pecori, R. (2021). MultiEditR: The first tool for the detection and quantification of RNA editing from Sanger sequencing demonstrates comparable fidelity to RNA-seq. *Molecular Therapy Nucleic Acids, 25,* 515–523.

Luyen, T. V., Nguyen, T. T. K., Alam, S., Sakamoto, T., Fujimoto, K., Suzuki, H., & Tsukahara, T. (2015). Changing blue fluorescent protein to green fluorescent protein using chemical RNA editing as a novel strategy in genetic restoration. *Chemical Biology & Drug Design, 86,* 1242–1252.

Luyen, V. T., Nguyen, T. T. K., Azad, M. T. A., Suzuki, H., & Tsukahara, T. (2012). Chemical RNA editing as a possibility novel therapy for genetic disorders. *International Journal of Advanced Computer Science, 2*(6), 237–241.

Luyen, T. V., & Toshifumi, T. (2017). C-to-U editing and site-directed RNA editing for the correction of genetic mutations. *BioScience Trends Adv. Publ. 11*(3), 243–253.

Maas, S., Rich, A., & Nishikura, K. (2003). A-to-I RNA editing: Recent news and residual mysteries. *The Journal of Biological Chemistry, 278,* 1391–1394.

Montiel-Gonzalez, M. F., Quiroz, J. F. D., & Rosenthal, J. J. C. (2019). Current strategies for site-directed RNA editing using ADARs. *Science Direct Methods (San Diego, Calif.), 156,* 16–24.

Montiel-González, M. F., Vallecillo-Viejo, I. C., & Rosenthal, J. J. (2016). An efficient system for selectively altering genetic information within mRNAs. *Nucleic Acids Research, 44*(21), 157–168.

Moris, A., Murray, S., & Cardinaud, S. (2014). AID and APOBECs span the gap between innate and adaptive immunity. *Frontiers in Microbiology, 5,* 534.

Navaratnam, N., Bhattacharya, S., Fujino, T., Patel, D., Jarmuz, A. L., & Scott, J. (1995). Evolutionary origins of apoB mRNA editing:Catalysis by a cytidine deaminase that has acquired a novel RNA-binding motif at its active site. *Cell, 81,* 187–195.

Nishida, K., Arazoe, T., Yachie, N., Banno, S., Kakimoto, M., Tabata, M., ... Kondo, A. (2016). Targeted nucleotide editing using hybrid prokaryotic and vertebrate adaptive immune systems. *Nature Communications, 353*(6305), 8729 (1-7).

Nishikura, K. (2010). Functions and regulation of RNA editing by ADAR deaminases. *Annual Review of Biochemistry, 79,* 321–349.

Patnaik, S. K., & Kannisto, E. (2016). APOBEC3B is a new RNA editing enzyme. In *Proceedings of the RNA, Annual Meeting of RNA Society*, Kyoto, Japan, 28 June–2 July 2016.

Powell, L. M., Wallis, S. C., Pease, R. J., Edwards, Y. H., Knott, T. J., & Scott, J. (1987). A novel form of tissue-specific RNA processing produces a polipoprotein-B48 in intestine. *Cell, 50*, 831–840.

Rinkevich, F. D., Schweitzer, P. A., & Scott, J. G. (2015). Antisense sequencing improves the accuracy and precision of A-to-I editingmeasurements using the peak height ratio method. *BMC Research Notes, 5*(63), 1–6.

Schneider, M. F., Wettengel, J., Hoffmann, P. C., & Stafforst, T. (2014). Optimal guideRNAs for re-directing deaminase activity of hADAR1 and hADAR2 in trans. *Nucleic Acids Research, 42*(10), e87.

Sharma, S., Patnaik, S. K., Taggart, R. T., Kannisto, E. D., Enriquez, S. M., Gollnick, P., & Baysal, B. E. (2015). APOBEC3A cytidine deaminase induces RNA editing in monocytes and macrophages. *Nature Communications, 6*, 6881.

Sharma, S., Patnaik, S. K., Taggart, R. T., & Baysal, B. F. (2016). The double-domain cytidine deaminase APOBEC3G is a cellular site-specific RNA editing enzyme. *Scientific Reports, 6*, 39100.

Shimatani, Z., Kashojiya, S., Takayama, M., Terada, R., Arazoe, T., Ishii, H., ... Kondo, A. (2017). Targeted base editing in rice and tomato using a CRISPR-Cas9 cytidine deaminase fusion. *Nature Biotechnology, 35*, 441–443.

Smith, H. C. (2016). RNA binding to APOBEC deaminases; Not simply a substrate for C-to-U editing. *RNA Biology, 14*(9), 1153–1165.

Stafforst, T., & Schneider, M. F. (2012). An RNA Deaminase conjugates electively repairs point mutations. *Angewandte Chemie International Edition, 51*(44), 11166–11169.

Vogel, P., Hanswollemenke, A., & Stafforst, T. (2017). Switching protein localization by site directed RNA editing under control of Light. *ACS Synthetic Biology, 6*, 1642–1649.

Vogel, P., & Stafforst, T. (2014). Site-directed RNA editing with antagomir deaminases-A tool to study protein and RNA function. *ChemMedChem, 9*(9), 2021–2025.

Wolf, J., Gerber, A. P., & Keller, W. (2002). tadA, an essential tRNA-specific adenosine deaminase from Escherichia coli. *The EMBO Journal, 21*, 3841–3851.

Zuo, E., Sun, Y., Wei, W., Yuan, T., Ying, W., Sun, H., ... Yang, H. (2019). Cytosine base editor generates substantial off-target single-nucleotide variants in mouse embryos. *Science (New York, N. Y.), 364*(6437), 289–292.

Further reading

Hiroki, S., Nagata, K., Sekiguchi, M., Fujioka, R., Matsuba, Y., Hashimoto, S., ... Saido, T. C. (2018). Introduction of pathogenic mutations into the mouse Psen1 gene by Base Editor and Target-AID. *Nature Communications, 9*, 2892.

CHAPTER TWELVE

Identification of RBP binding sites using RNA deaminases

Tao Yu[a,b,c,d], Qishan Liang[a,b,c,d], Shuhao Xu[a,b,c], and Gene W. Yeo[a,b,c,d,e,*]

[a]Department of Cellular and Molecular Medicine, University of California San Diego, La Jolla, CA, United States
[b]Sanford Stem Cell Institute and Stem Cell Program, University of California San Diego, La Jolla, CA, United States
[c]Institute for Genomic Medicine, University of California San Diego, La Jolla, CA, United States
[d]Center for RNA Technologies and Therapeutics, University of California San Diego, La Jolla, CA, United States
[e]Sanford Laboratories for Innovative Medicines, La Jolla, CA, United States
*Corresponding author. e-mail address: geneyeo@ucsd.edu

Contents

Abstract

RNA-binding proteins (RBPs) are critical regulators of gene expression and RNA processing. Identification of their binding sites has important implications for their physiological and disease-related functions. Crosslinking and immunoprecipitation, followed by sequencing (CLIP-seq) and its derivatives, are the most commonly used methods to identify RBP binding sites, but are laborious and require a large amount of starting material. Recent advancements harnessing RNA deaminases in fusion to any RBP of interest, allow for the profiling of RBP binding sites from low-input samples in simpler procedures. Among these efforts, we developed STAMP (Surveying Targets by

Methods in Enzymology, Volume 713
ISSN 0076-6879, https://doi.org/10.1016/bs.mie.2024.11.043
Copyright © 2025 Elsevier Inc. All rights are reserved, including those for text and data mining, AI training, and similar technologies.

APOBEC-Mediated Profiling), which efficiently detects RBP–RNA interactions. This chapter describes the detailed protocol for the STAMP method, including plasmid construction, delivery and sorting, library preparation and bioinformatic data analysis.

1. Introduction

RNA-binding proteins (RBPs) regulate the fate of mRNA through their lifecycle, including transcription, splicing, transport, translation, and decay (Das, Vera, Gandin, Singer, & Tutucci, 2021; Gerstberger, Hafner, & Tuschl, 2014; Ule & Blencowe, 2019). The critical function of RBPs and their interactions with RNA have been implicated in many biological processes and diseases, such as cancer and neurodegenerative disorders (Nussbacher, Batra, Lagier-Tourenne, & Yeo, 2015; Pereira, Billaud, & Almeida, 2017). Therefore, identifying the binding sites of RBPs is crucial for understanding their function and inspiring therapeutic innovations.

Crosslinking and immunoprecipitation-based methods (CLIP) are traditional ways to profile RBP binding sites, including various versions, iCLIP (Ule et al., 2003), PAR-CLIP (Hafner et al., 2010) and eCLIP (Van Nostrand et al., 2016). These methods crosslink RBP-RNA interactions by ultraviolet (UV) light, trimmed RNA to binding site resolution by RNase digestion and use immunoprecipitation to enrich the RBP-bound RNA. These methods allow the capturing of RBP binding sites from static states and are extensively optimized both experimentally and computationally. However, they require a large amount of input material, are laborious and do not have a resolution to identify isoform-specific binding sites.

Recently, we and others developed a new class of methods to profile RBP binding sites using RNA deaminases, including TRIBE (McMahon et al., 2016) and STAMP (Brannan et al., 2021). For the STAMP method we developed, an RBP of interest is fused to the rat APOBEC1 enzyme, and the cells are engineered to express this fusion protein. While the RBP binds to its RNA targets, the fused APOBEC1 generates C-to-U edits near the RBP binding sites. Through RNA sequencing, the RBP binding sites can be bioinformatically identified based on the C-to-U editing profile. This method has been shown to have single cell resolution and be compatible with long-read sequencing for detecting isoform-specific binding sites (Brannan et al., 2021).

We recently also benchmarked other RNA deaminases including variants of ADAR, APOBEC1, APOBEC3A, APOBEC3G and TadA to evaluate their performance to profile RBP binding sites (Medina-Munoz et al., 2024). We observed that different RNA deaminases have different sequence context preferences, while in terms of signal-to-noise ratio, APOBEC1 and TadA7.10 outperformed the other RNA deaminase, with APOBEC1 having higher editing efficiency (Medina-Munoz et al., 2024). With a single enzyme, APOBEC1 might still be the best enzyme available for mapping RBP binding sites. Future studies could work on protein engineering to generate better enzymes or explore the feasibility of combining different enzymes as cocktails.

In this chapter, we describe a detailed protocol of the STAMP experiment. This protocol uses IGF2BP2 and U2AF2 in fusion with APOBEC1 and Illumina RNA sequencing as examples. The protocol can be easily adapted to a different RBP or sequencing platform.

2. Materials and equipment

2.1 Plasmid construction

1. Gene specific primers (IDT):
   GGGG ACA AGT TTG TAC AAA AAA GCA GGC TTC + Forward primer
   GGGG AC CAC TTT GTA CAA GAA AGC TGG GTC + Reverse primer
2. KAPA HiFi HotStart ReadyMix (Roche, KK2602)
3. Gateway™ BP Clonase™ II Enzyme mix (Invitrogen, 11789020)
4. Gateway™ LR Clonase™ II Enzyme mix (Invitrogen, 11791020)
5. Gateway™ pDONR™221 Vector (Invitrogen, 12536017)
6. pLIX403_Capture1_APOBEC_HA_P2A_mRuby (Addgene, 183901)
7. One Shot™ Stbl3™ Chemically Competent *E. coli* (Invitrogen, C737303)
8. Thermal Cycler (Biorad)

2.2 Delivery and sorting

1. DMEM, high glucose (Gibco, 11965092)
2. Fetal Bovine Serum, Value, heat inactivated (Gibco, A5256801)
3. Penicillin-Streptomycin (5000 U/ml) (Gibco, 15070063)
4. DPBS, no calcium, no magnesium (Gibco, 14190144)
5. Lipofectamine™ 3000 Transfection Reagent (Invitrogen, L3000001)
6. Accutase (Stemcell Technologies, 07920)

7. FACS tube with filter cap (Falcon, 08-771-23)
8. BD FACSAria™ II Workstation (BD, 642510) or equivalent cell sorter

2.3 Library preparation

1. TRIzol™ Reagent (Invitrogen, 15596026)
2. Direct-zol RNA Miniprep Kit (Zymo, R2052)
3. Illumina® Ribo-Zero Plus rRNA Depletion Kit (Illumina, 20040526)
4. TruSeq® Stranded mRNA Library Prep (Illumina, 20020594)
5. RNA ScreenTape (Agilent, 5067–5576)
6. D1000 ScreenTape (Agilent, 5067–5582)
7. TapeStation System (Agilent)
8. Illumina NovaSeq 6000 Sequencing System (Illumina)

2.4 Data analysis

STAR (Dobin et al., 2012): https://github.com/alexdobin/STAR.
Sailor (Deffit et al., 2017): https://github.com/YeoLab/sailor.
FLARE (Kofman, Yee, Medina-Munoz, & Yeo, 2023): https://github.com/YeoLab/FLARE.

3. Methods

3.1 Overview

The protocol starts with the molecular cloning to generate RBP-APOBEC1 fusion construct by gateway cloning (Fig. 1A). The plasmid construct could then be delivered into a desired cell line. The construct also carries mRuby reporter expressing together with the RBP-APOBEC1 cassette upon doxycycline induction, allowing for flow sorting of cells expressing high level of the fusion protein (Fig. 1B). The sorted cells are then subjected to RNA extraction and library preparation, followed by Illumina sequencing (Fig. 1C). The RNA sequencing data are then analyzed through our bioinformatic pipeline to identify binding sites and motifs (Fig. 1D).

3.2 Plasmid construction

1. Design gene specific primers for amplification of full-length ORF sequence with stop codon removed and the addition of AttB sequences for gateway cloning
   GGGG ACA AGT TTG TAC AAA AAA GCA GGC TTC + Forward primer

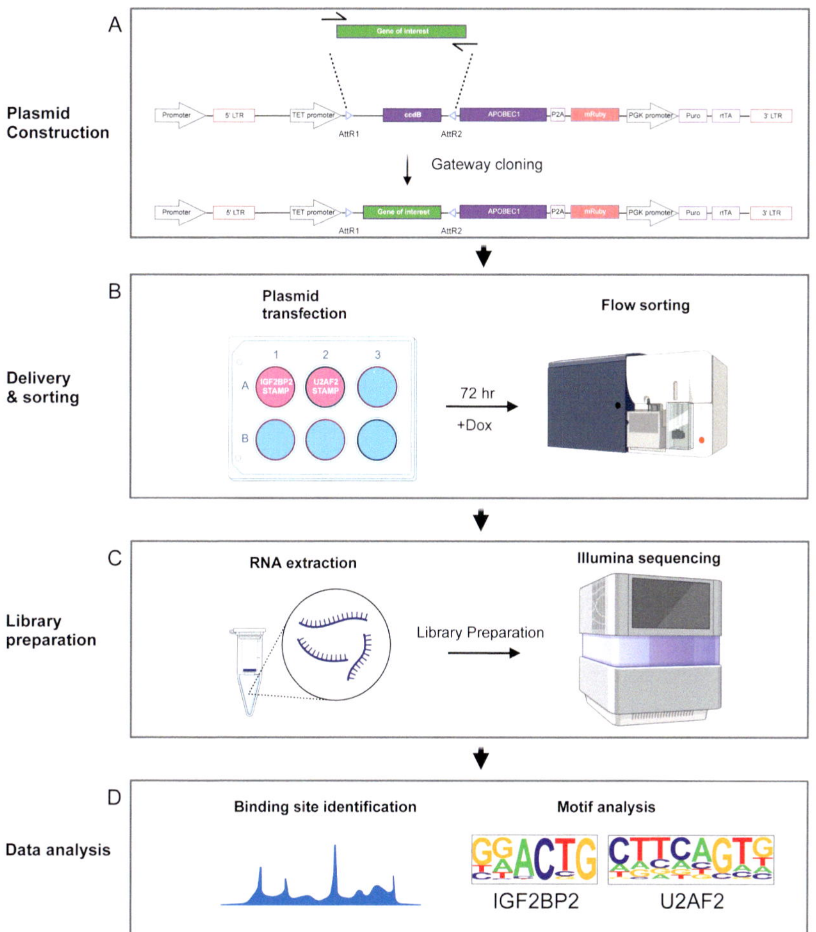

Fig. 1 Overview of the STAMP protocol. (A) A gene of interest will first be amplified and cloned into the STAMP vector by Gateway cloning. (B) Verified plasmids will be transfected to the desired cell type and cultured for 72 h with doxycycline induction of STAMP transgene expression. Fluorescence-activated cell sorting (FACS) will be performed to enrich for cells expressing high levels of the STAMP construct. (C) RNA will be extracted from sorted cells and subjected to library preparation and sequencing. (D) RNA-seq data will be analyzed through the established bioinformatic pipeline to identify binding sites and motifs.

GGGG AC CAC TTT GTA CAA GAA AGC TGG GTC + Reverse primer

2. Amplify the gene of interest via PCR with 2x KAPA Hifi PCR master mix (25 μl reaction volume) from HEK293T cDNA following KAPA Hifi manufacturer's protocol.

Note: for easier amplification, gene fragments can be ordered to serve as the template DNA.

3. Purify PCR product with Qiagen PCR purification kit and perform BP cloning into pDONR 221 plasmid following the manufacturer's protocol (Gateway BP cloning).
4. Pick colonies, perform plasmid extraction with Qiagen plasmid miniprep kit and Sanger sequencing to validate the correct insertion of the gene of interest.
5. Perform LR cloning into the pLIX403_Capture1_APOBEC_HA_P2A_mRuby destination vector following the manufacturer's protocol (Gateway LR cloning).

   Note: The STAMP destination vector is a doxycycline inducible lentiviral vector. For certain cell types, it might not yield sufficient expression of the fusion protein. For example, this STAMP construct expressed weakly in iPSC-derived differentiation models and primary blood cells (unpublished). In those cases, additional subcloning into a different vector for optimal expression may be required.
6. Pick colonies, perform plasmid extraction with Qiagen plasmid miniprep kit and Sanger sequencing to validate the plasmid.

3.3 Delivery and sorting

Methods of delivery may vary between different cell lines, the RBP being studied, the purpose of the research. This protocol will take IGF2BP2 and U2AF2 as examples to study their binding sites in HEK293T cells.

1. Seed HEK293T cells in 6 well plates at 300,000 cells per well density and culture them in DMEM+ 10 % HI FBS media (cell culture media).
2. After 24 h, transfect HEK293T cells with IGF2BP2 or U2AF2 STAMP plasmids (2 μg per well) using Lipofectamine 3000 transfection reagent and add doxycycline to a final concentration of 1 μg/ml. Shake the plate to mix well and incubate for 72 h.
3. Three days post transfection, wash the cells once with 1x DPBS and add 500 μl Accutase to each well to dissociate the cells into single cells suspension.
4. Centrifuge at 200 × g for 3 min, resuspend in 1 ml cell culture media and transfer to FACS tubes through the filtered cap to remove cell clumps.
5. Sort the cells at a flow cytometry facility and gate the mRuby signal to collect the top 20 % of the mRuby positive cells (sorting more than 100,000 cells is recommended, Fig. 2).

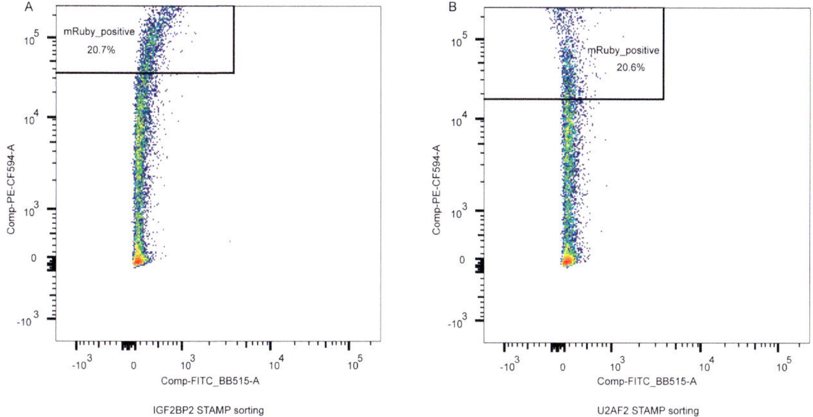

Fig. 2 FACS gating of IGF2BP2 and U2AF2 STAMP experiments. (A, B) Scatter plot showing compensated mRuby signal (PE channel, y axis, log scale) over FITC autofluorescence background (x axis, log scale). Around the top 20% of cells expressing the highest level of mRuby are sorted.

Note: Sorting is not necessary when the transfection is efficient and homogenous. However, it could improve signals in most cases, so it is highly recommended.

6. Centrifuge the cells at 200 × g for 5 min, remove the supernatant, lyse the cells in 200 μl TRIzol and store in −80 °C until RNA extraction and library preparation can be performed.

3.4 Library preparation

1. Extract RNA using Directzol RNA miniprep kit following the manufacturer's instructions.
2. Assess RNA quality using RNA ScreenTape following the manufacturer's protocol. We recommend a RIN > 7 to start the library preparation.
3. Start with 100–300 ng RNA and remove ribosomal RNA using Ribo-Zero Plus rRNA Depletion Kit following the manufacturer's protocol.
4. Elute the RNA with 5 μl nuclease-free water for the final elution step of the rRNA removal.
5. Add 13 μl of FPF buffer and proceed to run the Elution 2 - Frag - Prime program (94 °C × 8 min, Section "Fragment mRNA" Step 14 in "TruSeq Stranded mRNA Reference Guide") and start the Truseq mRNA stranded library preparation from that step on following the manufacturer's protocol.

Note: This protocol focuses on RBP's binding sites on all non-ribosomal RNAs. Total RNA sequencing will allow for the detection of non-polyA RNAs including pre-mRNA and therefore, permit the detection of intronic binding sites with deep sequencing. For RBPs known to bind mature mRNAs, an mRNA library preparation protocol could be followed. For isoform-specific binding site identification, Iso-seq library preparation could be performed for PacBio sequencing.

6. Assess library quality by D1000 ScreenTape following the manufacturer's protocol.
7. Sequencing with Novaseq 6000 System (or other Illumina sequencing platforms).

3.5 Data analysis

1. Download sequencing data (fastq.gz files) into a working directory.
2. Use alignment tools such as Spliced Transcripts Alignment to a Reference (STAR) to align the sequencing reads to the reference genome (e.g., hg38/mm10). Align only single-end reads. If the sequencing data are paired-end reads, use one read from each pair to avoid duplicated counts of editing sites.
3. Generate configuration (.json) files for "workflow_sailor" according to https://github.com/YeoLab/FLARE. Set "edit_type" as "ct". Input the.bam files from the alignment steps. Use the same reference genome as the alignment step in "reference_fasta". Run "workflow_sailor" snakemake pipeline. This yields a directory of sailor output.
4. Generate configuration (.json) files for "workflow_FLARE" according to https://github.com/YeoLab/FLARE. Input the files from sailor output subfolder "8_bw_and_bam", where the ".bed", ".bw" and ".bam" files contain the individual C-to-U edit sites. Set "edit_type" as "CT".
5. FLARE outputs ".tsv" files and ".bed" files with C-to-U edit cluster information. The ".tsv" files include information of chromosome, start coordinate, end coordinate, fraction of target bases edited, strand, number of target bases, number of edited bases, number of edited reads overlapping region, number of reads overlapping region, etc.
6. Perform downstream analysis as needed (Examples of motif analysis and IGV visualization can be found in Fig. 3).

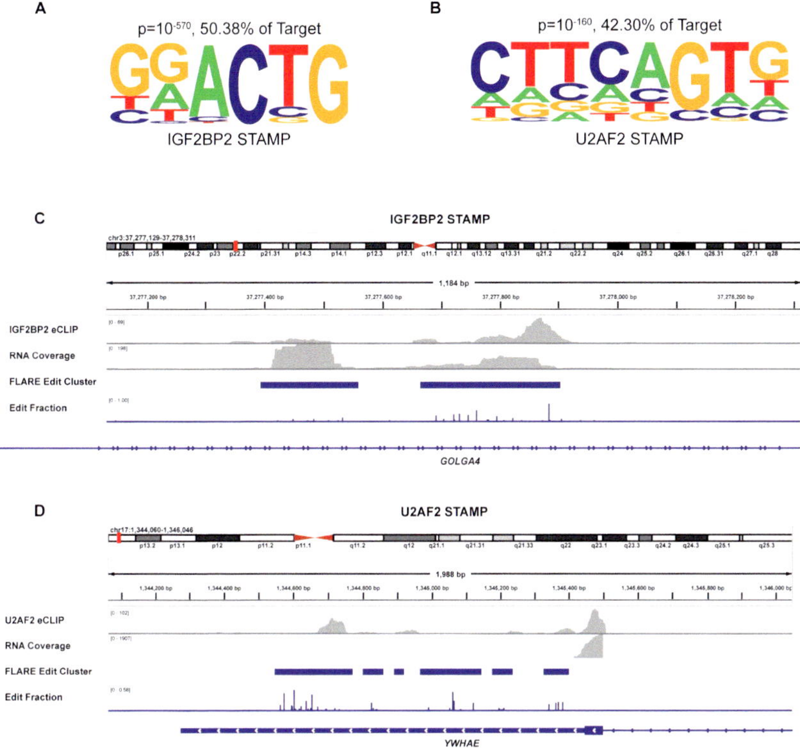

Fig. 3 Example results of IGF2BP2 and U2AF2 STAMP. (A, B) Motif analysis of IGF2BP2 and U2AF2 STAMP confident edit clusters showing significant enrichment of "DRACH" motif for IGF2BP2 and 3′ splice site motif for U2AF2. (C, D) IGV (Integrative Genomics Viewer) tracks showing examples of IGF2BP2 and U2AF2 binding sites detected by both eCLIP (from ENCODE (Van Nostrand et al., 2020)) and STAMP (FLARE edit cluster and edit fraction).

4. Conclusions

Traditional CLIP-based methods are labor-intensive and require substantial amounts of input material, posing challenges for profiling RBP binding sites in limited or precious samples and in high-throughput applications. RNA deaminase-based approaches offer a streamlined alternative for mapping RBP binding sites with greater simplicity. Here, we present a detailed protocol for conducting a successful STAMP experiment in HEK293T cells, using IGF2BP2 and U2AF2 as examples. Sequencing results revealed expected binding motifs and example binding sites, as illustrated in Fig. 3. This method is highly flexible with different sequencing strategies and could be easily adapted to single cell and long-read sequencing. We expect that STAMP serves both as a complementary approach to validate CLIP findings and as a powerful tool for exploring RBP binding sites in novel contexts, including rare cell types.

Funding

The authors acknowledge support from the National Institutes of Health (RF1 MH126719, R01 HG011864, R01 HG004659 to G.W.Y.). This publication includes data generated at the UC San Diego IGM Genomics Center utilizing an Illumina NovaSeq 6000 that was purchased with funding from a National Institutes of Health SIG grant (#S10 OD026929). Computational resources were provided by the Department of Defense High Performance Computing Modernization Program (HPCMP) and the Triton Shared Computing Cluster (TSCC) at the San Diego Supercomputer Center (SDSC) (San Diego Supercomputer Center (2022): Triton Shared Computing Cluster. University of California, San Diego. Service., 2022).

Declaration of interests

G.W.Y. is an SAB member of Jumpcode Genomics and a co-founder, member of the Board of Directors, on the SAB, equity holder, and paid consultant for Eclipse BioInnovations. G.W.Y. is a distinguished visiting professor at the National University of Singapore. G.W.Y.'s interests have been reviewed and approved by the University of California, San Diego in accordance with its conflict-of-interest policies. The authors declare no other competing financial interests.

References

Brannan, K. W., Chaim, I. A., Marina, R. J., Yee, B. A., Kofman, E. R., Lorenz, D. A., et al. (2021). Robust single-cell discovery of RNA targets of RNA-binding proteins and ribosomes. *Nature Methods, 18*(5), 507–519. https://doi.org/10.1038/s41592-021-01128-0.

Das, S., Vera, M., Gandin, V., Singer, R. H., & Tutucci, E. (2021). Intracellular mRNA transport and localized translation. *Nature Reviews. Molecular Cell Biology, 22*(7), 483–504. https://doi.org/10.1038/s41580-021-00356-8.

Deffit, S. N., Yee, B. A., Manning, A. C., Rajendren, S., Vadlamani, P., Wheeler, E. C., et al. (2017). The C. elegans neural editome reveals an ADAR target mRNA required for proper chemotaxis. *Elife, 6*. https://doi.org/10.7554/ELIFE.28625.

Dobin, A., Davis, C. A., Schlesinger, F., Drenkow, J., Zaleski, C., Jha, S., et al. (2012). STAR: Ultrafast universal RNA-seq aligner. *Bioinformatics (Oxford, England), 29*, 15. https://doi.org/10.1093/BIOINFORMATICS/BTS635.

Gerstberger, S., Hafner, M., & Tuschl, T. (2014). A census of human RNA-binding proteins. *Nature Reviews. Genetics, 15*(12), 829–845. https://doi.org/10.1038/nrg3813.

Hafner, M., Landthaler, M., Burger, L., Khorshid, M., Hausser, J., Berninger, P., et al. (2010). Transcriptome-wide identification of RNA-binding protein and microRNA target sites by PAR-CLIP. *Cell, 141*, 129–141. https://doi.org/10.1016/j.cell.2010.03.009.

Kofman, E., Yee, B., Medina-Munoz, H. C., & Yeo, G. W. (2023). FLARE: A fast and flexible workflow for identifying RNA editing foci. *BMC Bioinformatics, 24*, 1–15. https://doi.org/10.1186/S12859-023-05452-4/FIGURES/5.

McMahon, A. C., Rahman, R., Jin, H., Shen, J. L., Fieldsend, A., Luo, W., et al. (2016). TRIBE: Hijacking an RNA-editing enzyme to identify cell-specific targets of RNA-binding proteins. *Cell, 165*, 742–753. https://doi.org/10.1016/J.CELL.2016.03.007.

Medina-Munoz, H. C., Kofman, E., Jagannatha, P., Boyle, E. A., Yu, T., Jones, K. L., et al. (2024). Expanded palette of RNA base editors for comprehensive RBP-RNA interactome studies. *Nature Communications, 15*(1), 1–17. https://doi.org/10.1038/s41467-024-45009-4.

Nussbacher, J. K., Batra, R., Lagier-Tourenne, C., & Yeo, G. W. (2015). RNA-binding proteins in neurodegeneration: Seq and you shall receive. *Trends in Neurosciences, 38*, 226–236. https://doi.org/10.1016/J.TINS.2015.02.003/ASSET/1174461B-C912-48C0-A2A0-7245A129C44A/MAIN.ASSETS/GR3.JPG.

Pereira, B., Billaud, M., & Almeida, R. (2017). RNA-binding proteins in cancer: Old players and new actors. *Trends in Cancer, 3*, 506–528. https://doi.org/10.1016/J.TRECAN.2017.05.003/ASSET/A393FEED-A03A-4A93-A2A9-FB2A0AF54E2C/MAIN.ASSETS/GR1.JPG.

San Diego Supercomputer Center. (2022). Triton Shared Computing Cluster. University of California, San Diego. Service. https://doi.org/10.57873/T34W2R.

Ule, J., & Blencowe, B. J. (2019). Alternative splicing regulatory networks: functions, mechanisms, and evolution. *Molecular Cell, 76*, 329–345. https://doi.org/10.1016/J.MOLCEL.2019.09.017.

Ule, J., Jensen, K. B., Ruggiu, M., Mele, A., Ule, A., & Darnell, R. B. (2003). CLIP identifies Nova-regulated RNA networks in the brain. *Science (New York, N. Y.), 302*, 1212–1215. https://doi.org/10.1126/SCIENCE.1090095.

Van Nostrand, E. L., Freese, P., Pratt, G. A., Wang, X., Wei, X., Xiao, R., et al. (2020). A large-scale binding and functional map of human RNA-binding proteins. 5837818 *Nature, 583*, 711–719. https://doi.org/10.1038/s41586-020-2077-3.

Van Nostrand, E. L., Pratt, G. A., Shishkin, A. A., Gelboin-Burkhart, C., Fang, M. Y., Sundararaman, B., et al. (2016). Robust transcriptome-wide discovery of RNA-binding protein binding sites with enhanced CLIP (eCLIP). *Nature Methods, 13*(6), 508–514. https://doi.org/10.1038/nmeth.3810.

CHAPTER THIRTEEN

Programmable C-to-U editing to track endogenous proteins

Min Hao and Tao Liu*
State Key Laboratory of Natural and Biomimetic Drugs, Chemical Biology Center, Department of Molecular and Cellular Pharmacology, Pharmaceutical Sciences, Peking University, Beijing, P.R. China
*Corresponding author. e-mail address: taoliupku@pku.edu.cn

Contents

Abstract

Protein labeling techniques provide robust tools for studying protein localization, structure, and function. Nonetheless, the challenge persists in labeling endogenous proteins within live cells under their native conditions. Here, we present a universal approach by combining programmable cytidine to uridine (C-to-U) RNA editing with non-canonical amino acids (ncAAs) to achieve site-specific labeling of a diverse array of endogenous proteins using minimal amino acid side-chain tags in living cells. This innovative system, termed RNA Editing mediated ncAAs Protein Tagging (RENAPT), allows for real-time tracking of endogenous proteins by integrating ncAAs into target proteins at specific sites. RENAPT thus emerges as a promising platform with broad applicability for tagging endogenous proteins in live cells, facilitating investigations into their localization and functions.

Methods in Enzymology, Volume 713
ISSN 0076-6879, https://doi.org/10.1016/bs.mie.2024.11.039
Copyright © 2025 Elsevier Inc. All rights are reserved, including those for text and data mining, AI training, and similar technologies.

1. Introduction

Labeling and visualizing a protein of interest in living cells is crucial for studying cellular functions, allowing real-time monitoring of protein dynamics and interactions (Kim, Do Heo, & Cells, 2018; Lee, Kang, Park, & cells, 2019; Peng & Hang, 2016). Fluorescence-based techniques, such as genetic fusion with fluorescent proteins or self-labeling enzymes like Halo-tag (Los et al., 2008), SNAP/CLIP (Gautier et al., 2008; Keppler et al., 2003), DHFR (Calloway et al., 2007; Gallagher, Sable, Sheetz, & Cornish, 2009), FlAsH (Adams et al., 2002), and ReAsH (Martin, Giepmans, Adams, & Tsien, 2005), have advanced protein localization and function studies (Cheng, Kuru, Sachdeva, & Vendrell, 2020). However, these methods typically require ectopic expression of the protein of interest coding region and can disrupt native expression, folding, and function due to fusion size and site specificity (Chatterjee, Guo, Lee, & Schultz, 2013). Moreover, fluorescent proteins are prone to photobleaching under prolonged excitation (Zhang, Zheng, Liu, & Chen, 2015). Endogenous protein labeling under native conditions remains challenging due to the difficulty in developing specific probes and the impermeability of antibodies for live-cell applications.

Genetic code expansion involves using orthogonal aminoacyl-tRNA synthetase/tRNA pairs to incorporate designer non-canonical amino acids (ncAAs) into proteins. This approach allows site-specific amino acid mutagenesis with minimal disruption to protein structure and function, making it the leading method for precise protein labeling (Uttamapinant et al., 2015). Previous studies have developed ncAA-based minimal tags as excellent tools for live-cell protein labeling in mammalian cells (Alamudi et al., 2016; Arsić, Hagemann, Stajković, Schubert, & Nikić-Spiegel, 2022; Cheng et al., 2020; Nikić et al., 2014; Uttamapinant et al., 2015; Werther et al., 2021). However, genetic code expansion typically requires ectopic expression of proteins with nonsense codons or tagged proteins, which may not be ideal for endogenous protein labeling. CRISPR-mediated base editing could be used to enable in situ generation of nonsense codons in target genes, but suffers from specific limitations. Notably, the requirement for a PAM sequence nearby the potential mutation sites, reduces the applicability of single-base editing for generating nonsense codons. Finally, introducing a nonsense codon into essential genes could disrupt cellular function, potentially hindering studies of critical proteins.

In recent years, RNA editing technology has rapidly advanced, offering a solution to the aforementioned limitations by introducing designed nonsense codons into mRNA without altering the original gene sequence of the protein of interest. Precise RNA editing technologies for C-to-U conversions would significantly expand the scope of endogenous protein modifications without permanently altering the cellular genome. Here, we developed the RNA Editing mediated ncAAs Protein Tagging (RENAPT) method for precise labeling of endogenous proteins. This approach involves RNA editing to introduce a nonsense codon, followed by incorporating a non-canonical amino acid (ncAA) using an orthogonal tRNA synthase/tRNA pair. This enables real-time labeling of native proteins with fluorescent amino acids or small dyes via bio-orthogonal reactions, facilitating live-cell imaging (Hao et al., 2024) (Fig. 1). The RENAPT strategy provides a simple, convenient, and powerful tool for studying the function and localization of endogenous proteins without changing their cognate genomic DNA sequences.

2. Materials

2.1 Plasmids

1. RESCUEs: This plasmid, based on the pcDNA3.1 vector, enables efficient expression of dRanCas13b and ADAR2dd under the CMV promoter (Addgene #130662).
2. MbPylRS/tRNA$_{CUA}$: This plasmid, also based on the pcDNA3.1 vector, facilitates the efficient expression of the MbPylRS (Y271A,

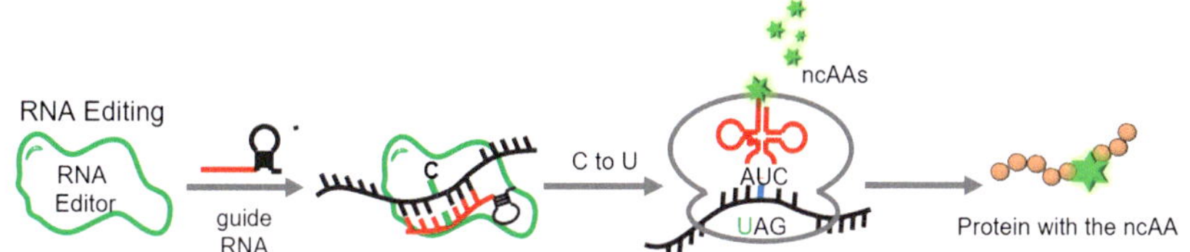

Fig. 1 Schematic representation of RNA Editing mediated ncAAs Protein Tagging (RENAPT) method. The RNA base editor changes CAG/CGA/CAA to UAG/UGA/UAA nonsense codons in the mRNA encoding the protein of interest, which are then recognized by introduced genetic code expansion machinery (orthogonal RS-tRNA pair) to co-translationally incorporate ncAAs into the protein of interest. The fluorescent dye can be either encoded as a fluorescent ncAA or attached to the ncAA through a bio-orthogonal reaction on the protein of interest.

L274M, and C313A) under the CMV promoter and PyltRNA under the U6 promoter (see **Note 1** for DNA sequences of MbPylRS and PyltRNA).
3. Guide RNA: This plasmid contains a guide RNA specific to the corresponding protein under the U6 promoter (Abudayyeh et al., 2019) (See **Note 2**). Cell line and Hippocampal neurons

2.2 Cell line and hippocampal neurons

1. U2OS cells (ATCC Cat# HTB-96).
2. Hippocampal neurons were isolated from postnatal day 0 male C57 mice.

2.3 Cell culture and transfection

1. McCoy's 5a (Gibco cat. no.16600-082) medium supplemented with 10 % v/v FBS.
2. Neurobasal medium containing B27 supplement (ThermoFisher, Waltham, MA)
3. L-glutamate (ThermoFisher)
4. 0.25 % trypsin.
5. OPTI-MEM.
6. B27 (ThermoFisher, Waltham, MA).
7. L-glutamate (ThermoFisher).
8. 1 mg/mL polyethyleneimine (PEI) transfection solution (see **Note 3**).
9. Lipofectamine 2000 (Invitrogen, cat. no. 11668027).
10. 12 mm glass coverslips coated with poly-D-lysine hydrobromide.
11. 24 well tissue culture plate.
12. Confocal dishes (SORFA, cat. no. 201100).
13. 10 cm cell culture plate.
14. 0.2 μm media filter unit.

2.4 Counting and plating

1. Trypan blue

2.5 ncAAs

1. N-(4*E*)-TCO-L-lysine (Confluro, cat. no. 1380349-88-1)
2. SiR-Me-tetrazine (Confluro, cat. no. BSR-7).

2.6 Tracker

1. Mito-Tracker Red CMXRos (Invitrogen, cat. no. M7512).
2. Lyso-Tracker Red (Invitrogen, cat. no. L7528).

3. Lyso-Tracker Green (Beyotime, cat. no. C1047S).
4. ER-Tracker Red (Beyotime, cat. no. C1041).
5. Cell Plasma Membrane Staining Kit with DiI (Beyotime, cat. no. C1991S).
6. Golgi-Tracker Red (Beyotime, cat. no. C1043).
7. Hoechst 33342 (Beyotime, cat. no. C1022).

2.7 Antibodies

1. Rabbit anti-$Na_v1.6$ (SCN8A) antibody (Alomone labs, cat. no. #ASC-009).
2. Rabbit anti-68kDa neurofilament (Abcam, cat. no. ab223343).
3. Rabbit anti-$Ca_v2.1$ (CACNA1A) antibody (Bioss, cat. no. bs-3930R).
4. Mouse anti-ankyrin G antibody (Santa Cruz Biotechnology, cat. no. sc-12719).
5. Mouse anti-MAP2 antibody (proteintech, cat. no. 67015-1-Ig).
6. Donkey anti-rabbit AF488 (Jackson ImmunoResearch Laboratories, cat. no. 711-547-003).
7. Goat anti-mouse Cy3 (proteintech, cat. no. SA00009-1).
8. Goat anti-rabbit Cy3 (proteintech, cat. no. SA00009-2).

2.8 Equipment

1. Biological safety cabinet.
2. Confocal microscope (Zeiss, Oberkochen, Germany).
3. SIM super-resolution microscopy P-204W (HIS-SIM P-204W)

3. Methods

3.1 Cell culture

3.1.1 U2OS cells

1. To revive U2OS cells, thaw 1 mL of frozen cells in a 37 °C water bath for 1 min
2. Mix thawed cells with 9 mL of pre-warmed 10% FBS McCoy's 5a medium in a 10 cm cell culture plate. Grow cells at 37 °C with 5% CO_2 for 16–24 h.
3. At 16–24 h, examine cells under a microscope. More than 90% of cells should attach to the surface of the plate. Replace cell culture media with 5 mL fresh pre-warmed McCoy's 5a medium supplemented with 10% FBS. Incubate cells at 37 °C and 5% CO_2 for 2–3 days until cells reach ~80% confluency.

4. To maintain good cell fitness, passage cells at around 80% confluency (see **Note 4**). Remove culture media; gently rinse cells once with warm PBS. Detach cells by adding 2 mL of warm 0.05% trypsin followed by incubation at 37 °C for 2 min. Separate cells by pipetting up and down gently 10 times, and then inactivate trypsin by adding 3 mL of warm McCoy's 5a medium supplemented with 10% FBS. Dilute 0.5 mL of resuspended cells to 4.5 mL of warm McCoy's 5a medium supplemented with 10% FBS in a new 10 cm cell culture plate (1:10 dilution). Incubate cells at 37 °C and 5% CO_2.
5. Passage cells every 2–3 days when cells reach ~80% confluency. Examine cells frequently under a microscope (see **Note 5**).

3.1.2 Hippocampal neurons

1. Hippocampal neurons isolated from day 0 male C57 mice.
2. Seed onto 12 mm glass coverslips coated with poly-D-lysine hydrobromide placed in 24 well tissue culture plates (at a density of 1×10^5 cells per well) in neurobasal medium containing B27 supplement and L-glutamate.

3.2 Transfection

3.2.1 U2OS cell line

1. Seed $1 \times 10^{\wedge 5}$ cells into individual wells of confocal dishes. Culture the cells at 37 °C in a 5% CO_2 atmosphere for 24 h prior to transfection.
2. Prepare a total of 2 μg DNA (RESCUEs, MbPylRS/$tRNA_{CUA}$, and on-target gRNAs for each endogenous gene at a 1:1:2 ratio) in 150 μL Opti-MEM.
3. Add 6 μL of 1 mg/mL PEI to 150 μL Opti-MEM.
4. Mix the transfection complexes (plasmids and PEI).
5. Incubate at room temperature for 15–20 min to allow the formation of transfection complexes.
6. Add the DNA/PEI mixture drop-wise onto the medium in each well, gently swirling the plate to homogenize the mixture.
7. Incubate the cells at 37 °C in a 5% CO_2 atmosphere for 6 h.
8. Exchange the medium with 250 mM TCO-L-lysine solution diluted 1:1000 in McCoy's 5a medium, achieving a final concentration of 250 μM (see **Note 6**).
9. Incubate the cells at 37 °C in a 5% CO_2 atmosphere for 48 h.

3.2.2 Hippocampal neurons

1. Seed 10^5 cells per well onto 12 mm glass coverslips coated with poly-D-lysine hydrobromide in 24-well tissue culture plates. Incubate the cells at 37 °C in a 5% CO_2 atmosphere for 8 days (DIV 8) (see **Note 7**).
2. Mix DNA (RESCUEs, MbPylRS/$tRNA_{CUA}$, and on-target gRNAs for each endogenous gene) at a 1:1:2 ratio with Lipofectamine 2000, using a ratio of 1 μg DNA to 2 μL Lipofectamine 2000 in neural basal medium.
3. Prior to adding to cells, mix the transfection solution with 100 μL warm neural basal medium and incubate for 5 min at 37 °C with 5% CO_2.
4. After 6 h of incubation, add an equal volume of fresh neural basal medium to the cells.
5. Dilute a 250 mM TCO-L-lysine stock solution (in 0.2 M NaOH containing 15% DMSO) 1:1000 in neural basal medium and apply it to neurons to achieve a final concentration of 250 μM.
6. Incubate primary mouse hippocampal neurons at 37 °C with 5% CO_2 for 4 days prior to click labeling.

3.3 Confocal imaging

1. Cells were washed three times with fresh medium for 1 h each to remove excess TCO-L-lysine (see **Note 8**).
2. The cells were then incubated with tetrazine dye (SiR-Tz) diluted in fresh medium (5 μM) for 10 min at 37 °C with 5% CO_2.
3. After the labeling period, cells were washed twice with fresh medium and further incubated for 2–3 h at 37 °C with 5% CO_2.
4. Cells were stained with the appropriate tracker according to the manufacturer's instructions, such as MitoTracker Red.
5. MitoTracker Red was added to fresh medium at a ratio of 1:5000 after removing the culture media.
6. Incubate the cells at 37 °C with 5% CO_2 for 20 min
7. Remove the MitoTracker Red staining solution and wash cells three times with fresh medium, each wash lasting 5 min
8. Confocal imaging was conducted using a 100 × oil immersion objective (NA 1.46) (Fig. 2).

3.4 Immunocytochemistry staining

1. Neurons were fixed with 0.5% electron microscopy grade PFA for 10 min (see **Note 9**).
2. The fixed neurons were washed three times with PBS for 5 min each.

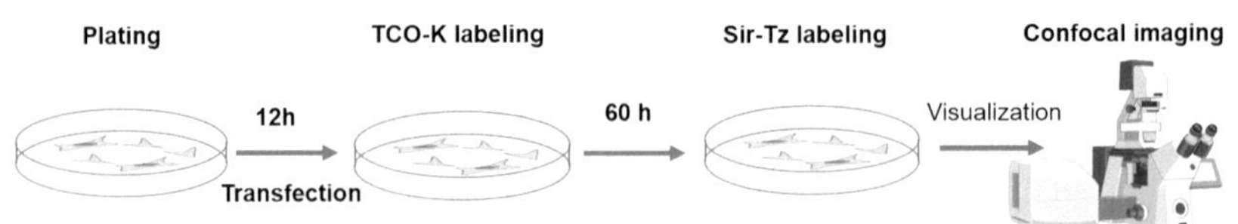

Fig. 2 A schematic overview of the experimental workflow for U2OS cells. Following 12 h seeding into cell-culture plates, cells were transfected with plasmids encoding RESCUEs, gRNA, and MbPylRS /$tRNA_{CUA}$. After 48 h of incubation with TCO-L-lysine, the cells were labeled with the SiR-Me-tetrazine dye through click chemistry. After the labeling step, the cells were stained with corresponding tracker and imaged by confocal laser scanning microscopy.

3. Cells were permeabilized at room temperature for 10 min using immunostaining permeabilization buffer with saponin (Beyotime, cat. no. P0095).
4. After permeabilization, neurons were washed three times with PBS for 5 min each.
5. Washed and fixed neurons were incubated at room temperature for 2 h in blocking buffer containing 5% bovine serum albumin (Yeasen, cat. no. 36101ES60).
6. Primary antibodies (rabbit anti-Na_v1.6, rabbit anti-Ca_v2.1, mouse anti-MAP2 at 1:200 dilution; goat anti-ankG and rabbit anti-68kDa neurofilament at 1:100 dilution) were diluted in blocking buffer.
7. Hippocampal neurons were then incubated overnight at 4 °C with the primary antibodies.
8. Then, neurons were washed three times with PBS for 5 min each.
9. After washing, neurons were incubated for 2 h at room temperature with secondary antibodies diluted in blocking buffer (all secondary antibodies at 1:200 dilution).
10. The neurons were washed three times with PBS for 5 min each and imaged immediately (Fig. 3).

4. Notes

1. The DNA sequence for these aaRSs and tRNAs is given below:

MbPylRS (Y271A, L274M, and C313A) ATGGATAAAAAACCA TTAGATGTTTTAATATCTGCGACCGGGCTCTGGATGTCCA GGACTGGCACGCTCCACAAAATCAAGCACCATGAGGTCTC

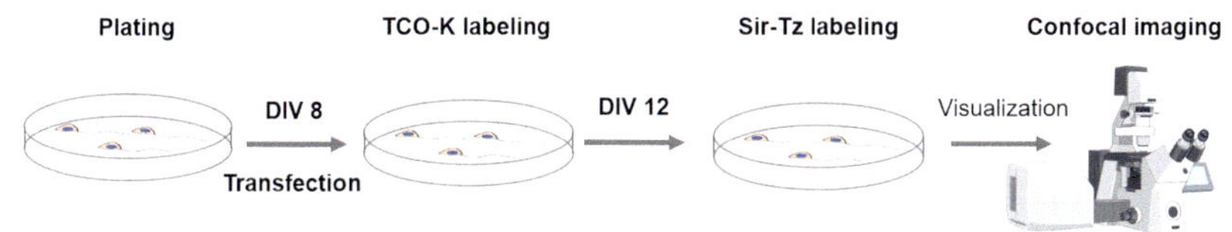

Fig. 3 A schematic overview of the experimental workflow for mouse hippocampal neurons. Following 8 days seeding into cell-culture plates, neurons were transfected with plasmids encoding RESCUEs, gRNA, and MbPylRS/$tRNA_{CUA}$. After 4 days of incubation with TCO-L-lysine, the neurons were labeled with the SiR-Tz dye through click chemistry. After the labeling step, the neurons were stained with anti-Nav1.6/Cav2.1/NFL and anti-ankG/MAP2 antibodies, followed by Alexa Fluor 488 and CY3 conjugated secondary antibodies, and imaged by confocal laser scanning microscopy.

AAGAAGTAAAATATACATTGAAATGGCGTGTGGAGACCA
TCTTGTTGTGAATAATTCCAGGAGTTGTAGAACAGCCAGA
GCATTCAAACATCATAAGTACAGAAAAACCTGCAAACGAT
GTAGGGTTTCGGACGAGGATATCAATAATTTTCTCACAAG
ATCAACCGAAAGCAAAAACAGTGTGAAAGTTAGGGTAGTTt
CTGCTCCAAAGGTCAAAAAAGCTATGCCGAAATCAGTTTC
AAGGGCTCCGAAGCCTCTGGAAAATTCTGTTTCTGCAAAG
GCATCAACGAACACATCCAGATCTGTACCTTCGCCTGCAA
AATCAACTCCAAATTCGTCTGTTCCCGCATCGGCTCCtGC
TCCTTCACTTACAAGAAGCCAGCTTGATAGGGTTGAGGCT
CTCTTAAGTCCAGAGGATAAAATTTCTCTAAATATGGCAA
AGCCTTTCAGGGAACTTGAGCCTGAACTTGTGACAAGAAG
AAAAAACGATTTTCAGCGGCTCTATACCAATGATAGAGAA
GACTACCTCGGTAAACTCGAACGTGATATTACGAAATTTT
TCGTAGACCGGGGTTTTCTGGAGATAAAGTCTCCTATCCT
TATTCCGGCGGAATACGTGGAGAGAATGGGTATTAATAA
TGATACTGAACTTTCAAAACAGATCTTCCGGGTGGATAAA
AATCTCTGCTTGAGGCCAATGCTTGCCCCGACTCTTGCGA
ACTATATGCGAAAACTCGATAGGATTTTACCAGGCCCAAT
AAAAATTTTCGAAGTCGGACCTTGTTACCGGAAAGAGTCT
GACGGCAAAGAGCACCTGGAAGAATTTACTATGGTGAACT
TCGCGCAGATGGGTTCGGGATGTACTCGGGAAAATCTTGA
AGCTCTCATCAAAGAGTTTCTGGACTATCTGGAAATCGAC
TTCGAAATCGTAGGAGATTCCTGTATGGTCTTTGGGGATA
CTCTTGATATAATGCACGGGGACCTGGAGCTTTCTTCGGC
AGTCGTCGGGCCAGTTTCTCTTGATAGAGAATGGGGTATT
GACAAACCATGGATAGGTGCAGGTTTTGGTCTTGAACGCT

TGCTCAAGGTTATGCACGGCTTTAAAAACATTAAGAGGGC ATCAAGGTCCGAATCTTACTATAATGGGATTTCAACCAAT CTATAA

PyltRNA

GGAAACCTGATCATGTAGATCGAATGGACTCTAAATCC GTTCAGCCGGGTTAGATTCCCGGGGTTTCCGCCA

2. gRNAs are designed to recruit dRanCas13b and ADAR2dd and site-specifically edit the mRNA coding the protein of interest.
3. For PEI transfection, it is important to use the specific product from Polysciences (polyethylenimine 40000, cat. no. 24765-1). The DNA/PEI ratio should be optimized for every batch. A 1:3 DNA/PEI ratio usually yields good results.
4. To maintain good cell fitness, subculture cells when they become ~80% confluent. Overgrown cell culture contains unhealthy or dead cells, which leads to low transfection efficiency and negatively affects cell-based assays.
5. After cell revival, do not subculture U2OS cells for more than 2 months or 25 generations. Discard old cell culture and revive new cell culture.
6. We utilize commercially available TCO-L-lysine and prepare a 250 mM stock solution using a solution of 0.2 M NaOH with 15% DMSO. This stock solution is prepared fresh as needed.
7. Every 2 days, replace 500 μL of medium with fresh medium to maintain a total volume of 1 mL.
8. Remove and wash the cells gently to minimize disruption and maintain their normal state.
9. Using 0.5% electron microscopy grade PFA to fix neuronal cells effectively preserves their structural integrity, thereby enhancing the reliability and precision of subsequent electron microscopy analysis.

5. Conclusions

Here we present a detailed protocol for RENAPT, a robust method for site-specific labeling of endogenous proteins, utilizing a minimal synthetic amino acid tag. This technique has been successfully employed in living mammalian cells as well as in primary neurons, showcasing its versatility and potential for broad biological applications.

Acknowledgments

This work was financially supported by the National Key Research and Development Program of China (2022YFA0912400 and 2021YFA0909900), the National Natural Science Foundation of China (22325701, U22A20332, 92156025 and 92253301).

References

Abudayyeh, O. O., Gootenberg, J. S., Franklin, B., Koob, J., Kellner, M. J., Ladha, A., ... Zhang, F. (2019). A cytosine deaminase for programmable single-base RNA editing. *Science (New York, N. Y.), 365*(6451), 382–386. https://doi.org/10.1126/science.aax7063.

Adams, S. R., Campbell, R. E., Gross, L. A., Martin, B. R., Walkup, G. K., Yao, Y., ... Tsien, R. Y. (2002). New biarsenical ligands and tetracysteine motifs for protein labeling in vitro and in vivo: Synthesis and biological applications. *Journal of the American Chemical Society, 124*(21), 6063–6076.

Alamudi, S. H., Satapathy, R., Kim, J., Su, D., Ren, H., Das, R., ... Hoppmann, C. J. N. C. (2016). Development of background-free tame fluorescent probes for intracellular live cell imaging. *Nature Communications, 7*(1), 1–9.

Arsić, A., Hagemann, C., Stajković, N., Schubert, T., & Nikić-Spiegel, I. J. N. C. (2022). Minimal genetically encoded tags for fluorescent protein labeling in living neurons. *Nature Communications, 13*(1), 1–18.

Calloway, N. T., Choob, M., Sanz, A., Sheetz, M. P., Miller, L. W., & Cornish, V. W. J. C. (2007). Optimized fluorescent trimethoprim derivatives for in vivo protein labeling. *Chembiochem: A European Journal of Chemical Biology, 8*(7), 767–774.

Chatterjee, A., Guo, J., Lee, H. S., & Schultz, P. G. (2013). A genetically encoded fluorescent probe in mammalian cells. *Journal of the American Chemical Society, 135*(34), 12540–12543.

Cheng, Z., Kuru, E., Sachdeva, A., & Vendrell, M. (2020). Fluorescent amino acids as versatile building blocks for chemical biology. *Nature Reviews Chemistry, 4*(6), 275–290.

Gallagher, S. S., Sable, J. E., Sheetz, M. P., & Cornish, V. W. (2009). An in vivo covalent TMP-tag based on proximity-induced reactivity. *ACS Chemical Biology, 4*(7), 547–556.

Gautier, A., Juillerat, A., Heinis, C., Corrêa Jr, I. R., Kindermann, M., Beaufils, F., ... biology (2008). An engineered protein tag for multiprotein labeling in living cells. *Chemistry & Biology, 15*(2), 128–136.

Hao, M., Ling, X., Sun, Y., Wang, X., Li, W., Chang, L., ... Chen, L. (2024). Tracking endogenous proteins based on RNA editing-mediated genetic code expansion. *Nature Chemical Biology,* 1–11.

Keppler, A., Gendreizig, S., Gronemeyer, T., Pick, H., Vogel, H., & Johnsson, K. J. N. B. (2003). A general method for the covalent labeling of fusion proteins with small molecules in vivo. *Nature Biotechnology, 21*(1), 86–89.

Kim, J., Do Heo, W. J. M., & Cells (2018). Synergistic ensemble of optogenetic actuators and dynamic indicators in cell biology. *Molecules and Cells, 41*(9), 809.

Lee, K. J., Kang, D., Park, H.-S. J. M., & cells (2019). Site-specific labeling of proteins using unnatural amino acids. *Molecules and Cells, 42*(5), 386.

Los, G. V., Encell, L. P., McDougall, M. G., Hartzell, D. D., Karassina, N., Zimprich, C., ... Wood, K. V. (2008). HaloTag: a novel protein labeling technology for cell imaging and protein analysis. *ACS Chemical Biology, 3*(6), 373–382.

Martin, B. R., Giepmans, B. N., Adams, S. R., & Tsien, R. Y. (2005). Mammalian cell–based optimization of the biarsenical-binding tetracysteine motif for improved fluorescence and affinity. *Nature Biotechnology, 23*(10), 1308–1314.

Nikić, I., Plass, T., Schraidt, O., Szymański, J., Briggs, J. A., Schultz, C., & Lemke, E. A. (2014). Minimal tags for rapid dual-color live-cell labeling and super-resolution microscopy. *Angewandte Chemie International Edition, 53*(8), 2245–2249.

Peng, T., & Hang, H. C. (2016). Site-specific bioorthogonal labeling for fluorescence imaging of intracellular proteins in living cells. *Journal of the American Chemical Society, 138*(43), 14423–14433.

Uttamapinant, C., Howe, J. D., Lang, K., Beránek, V., Davis, L., Mahesh, M., ... Chin, J. W. (2015). Genetic code expansion enables live-cell and super-resolution imaging of site-specifically labeled cellular proteins. *Journal of the American Chemical Society, 137*(14), 4602–4605.

Werther, P., Yserentant, K., Braun, F., Grußmayer, K., Navikas, V., Yu, M., ... Wombacher, R. (2021). Bio-orthogonal red and far-red fluorogenic probes for wash-free live-cell and super-resolution microscopy. *ACS Central Science, 7*(9), 1561–1571.

Zhang, G., Zheng, S., Liu, H., & Chen, P. R. (2015). Illuminating biological processes through site-specific protein labeling. *Chemical Society Reviews, 44*(11), 3405–3417.

Printed in the United States
by Baker & Taylor Publisher Services